PRIMORDIAL NUCLEI AND THEIR GALACTIC EVOLUTION

Cover figure adapted from Schramm, p. 3:
Big Bang Nucleosynthesis abundance yields versus baryon density (Ω_b) for a homogeneous universe.

Space Sciences Series of ISSI

Volume 4

The International Space Science Institute is organized as a foundation under Swiss law. It is funded through recurrent contributions from the European Space Agency, the Swiss Confederation, the Swiss National Science Foundation, and the Canton of Bern. For more information, see the homepage at http://ubeclu.unibe.ch/issi/index.html.

PRIMORDIAL NUCLEI AND THEIR GALACTIC EVOLUTION

Proceedings of an ISSI Workshop
6–10 May 1997, Bern, Switzerland

Edited by

NIKOS PRANTZOS
Institut d'Astrophysique de Paris, France

MONICA TOSI
Osservatorio Astronomico di Bologna, Italy

RUDOLF VON STEIGER
International Space Science Institute, Bern, Switzerland

Dedicated to the memory of David N. Schramm (1945–1997)

Springer Science+Business Media, B.V.

Space Sciences Series of ISSI

Library of Congress Cataloging-in-Publication Data

ISBN 978-0-7923-5114-6   ISBN 978-94-011-5116-0 (eBook)
DOI 10.1007/978-94-011-5116-0

Printed on acid-free paper

TABLE OF CONTENTS

IV: GALACTIC DISK, GALACTIC EVOLUTION

V: SOLAR NEBULA

VI: LOCAL INTERSTELLAR MEDIUM

EPILOGUE

viii

ISSI Workshop
Primordial Nuclei and Their Galactic Evolution
6–10 May 1997, Bern, Switzerland
Group Photographs

1.	V. Manno	15.	T. Walker	29.	C. Deliyannis
2.	B. Pagel	16.	B. Fields	30.	T. Beers
3.	R. Rebolo	17.	F. Spite	31.	L. Linsky
4.	W. Sargent	18.	T. Donahue	32.	D. Duncan
5.	N. Prantzos	19.	E. Terlevich	33.	F. Palla
6.	A. Vidal-Madjar	20.	E. Skillman	34.	Mrs. Beers
7.	R. Terlevich	21.	M. Tosi	35.	Mrs. Donahue
8.	G. A. Tammann	22.	F. Primas	36.	J. Audouze
9.	T. Wilson	23.	C. Charbonnel	37.	S. Wenger
10.	J. Linsky	24.	S. Vauclair	38.	G. Nusser Jiang
11.	R. Rood	25.	G. Steigman	39.	D. Taylor
12.	R. v. Steiger	26.	G. Gloeckler	40.	R. Cayrel
13.	C. Hogan	27.	J. Geiss	41.	?
14.	S. Levshakov	28.	T. X. Thuan		

Not on these pictures: P. Bochsler, D. Gautier, M. Rees, H. Reeves,
D. Schramm, F. Thielemann, J. Truran, D. Tytler

FOREWORD

The present volume, the fourth one in the "Space Sciences Series of ISSI" (International Space Science Institute), contains the proceedings of a workshop on "Primordial Nuclei and Their Galactic Evolution", which was held at ISSI in Bern on 6–10 May 1997. This topic was chosen following some general enquiries with the scientific community concerning its desirability and timeliness. Five convenors, D. Duncan, C. Hogan, J. Linsky, N. Prantzos, and H. Reeves (chair) subsequently set up the workshop, nominated a list of invitees, structured the workshop into a series of introductory talks and into six topical working groups (early Universe – extragalactic objects – low-Z stars – galactic disk and galactic evolution – solar nebula – local interstellar medium), and described the tasks of the working groups in a list of keywords.

It is the main task of ISSI to bring together space scientists, ground-based observers, and theorists from different fields and to give them the opportunity to discuss and compare their results, thus contributing to the achievement of a deeper understanding, adding value to those results through multi-disciplinary research in an atmosphere of international co-operation. In that spirit the convenors selected participants working in fields ranging from Big Bang theory to observers of today's Solar System, thus spanning the widest possible range both in time and space. The fields were first presented in 18 introductory talks, whereupon the workshop was split into the working groups to which ample time was allocated for individual and joint splinter meetings. In that way, data from at least a dozen space missions (Apollo, ASCA, COBE, Exosat, Galileo, Hipparcos, HST, ISO, ISEE-3, ORFEUS, ROSAT, Ulysses, ...) and from a large number of ground-based telescopes were presented, compared and discussed. Each working group was concluded by a rapporteur presentation in the plenary, summarising the discussions, the achieved progress, the remaining gaps in our understanding, and pointing out directions for future research. The present volume is a collection of the papers resulting from the invited, contributed, and rapporteur presentations, each of which was reviewed by an independent referee. The papers are arranged into six sections, one for each working group, thus closely resembling the structure of the workshop. Finally, the volume is concluded with a workshop summary as an epilogue.

We wish to express our sincere thanks to all those who have made this volume possible. First of all, we should like to thank the authors for writing original articles, for keeping to the various deadlines, and for producing unusually neat camera-ready versions. We also thank the reviewers for their critical and timely reports, which have significantly contributed to the quality of this volume. Finally, it is our pleasure to thank ISSI, in particular its directors J. Geiss and B. Hultqvist, for taking the initiative to host and to support this workshop, the ISSI staff, V. Manno,

G. Nusser Jiang, M. Preen, D. Taylor, and S. Wenger, for the local organization of this workshop, and U. Pfander and X. Schneider for their assistance in the preparation of this volume.

At the time when this volume was going to press, we (as the rest of the scientific community) were saddened by the death of David Schramm. It was a great loss, not only to his family and friends, but also to physics in general and the field of primordial nucleosynthesis in particular.

Indeed, perhaps more than anybody else in the past quarter century, Dave contributed to improving the model of the Hot Early Universe to the status of a theory, bringing together the disciplines of particle physics, nuclear physics and cosmology. His most fundamental contribution to physics is probably the calculation of the number of neutrino families, using arguments from primordial nucleosynthesis. At a time when two families of elementary particles were known and most physicists assumed that many more particle families would be found, Dave and his colleagues boldly predicted in the seventies that only one more family should be expected. The prediction was spectacularly confirmed in the late eighties, by experiments at CERN and in Stanford, revealing the predictive power of the Hot Early Universe theory.

Dave's work on primordial nucleosynthesis was crucial to the establishment of this "third pillar" of modern Big Bang cosmology (after Hubble's cosmic expansion and the cosmic microwave background). Furthermore, his calculation of the amount of "ordinary" (baryonic) matter in the universe showed that it accounts for only a small fraction of the total, thus requiring the existence of some form of "exotic" matter. Despite his unshakeable faith on the validity of the Big Bang theory (making him its most prominent defender), Dave always tried to put it on the most firm observational basis, seeking for new observational data. It is in that spirit that he participated in the ISSI conference in May 1997, delivering once more the "message" with his characteristic brilliant and pedagogical style. Dedicating this volume to his memory is then the least tribute we can offer to Dave Schramm, one of the towering figures of modern astrophysics.

February 1998
N. Prantzos, M. Tosi, R. von Steiger

I: EARLY UNIVERSE

BIG BANG NUCLEOSYNTHESIS AND THE DENSITY OF BARYONS IN THE UNIVERSE

D.N. SCHRAMM

The University of Chicago
5640 S. Ellis Avenue
Chicago, IL 60637, USA

Abstract. Now that extragalactic deuterium observations are being made, Big Bang Nucleosynthesis (BBN) is on the verge of undergoing a transformation. Previously, the emphasis was on demonstrating the concordance of the Big Bang Nucleosynthesis model with the abundances of the light Isotopes extrapolated back to their primordial values using stellar and Galactic evolution theories. Once the primordial deuterium abundance is converged upon, the nature of the field will shift to using the much more precise primordial D/H to constrain the more flexible stellar and Galactic evolution models (although the question of potential systematic error in ^{4}He abundance determinations remains open). The remarkable success of the theory to date in establishing the concordance has led to the very robust conclusion of BBN regarding the baryon density. The BBN constraints on the cosmological baryon density are reviewed and demonstrate that the bulk of the baryons are dark and also that the bulk of the matter in the universe is non-baryonic. Comparison of baryonic density arguments from Lyman-α clouds, x-ray gas in clusters, and the microwave anisotropy are made and shown to be consistent with the BBN value.

Key words: Cosmology, Nucleosynthesis, Light Elements

Abbreviations: HST – Hubble Space Telescope; ISM – Interstellar Medium; LEP – Large Electron Positron Collider; SLC – Stanford Linear Collider; ROSAT – German X-ray satellite; ASCA – Japanese X-ray satellite; ESA – European Space Agency; CDM – cold dark matter

1. Introduction

The study of Big Bang Nucleosynthesis and the light element abundances is undergoing a major transformation. The bottom line remains: primordial nucleosynthesis has joined the Hubble expansion and the microwave background radiation as one of the three pillars of Big Bang cosmology. Of the three, Big Bang Nucleosynthesis (BBN) probes the universe to far earlier times ($\sim$ 1 sec) than the other two and led to the interplay of cosmology with nuclear and particle physics. Furthermore, since the Hubble expansion is also part of alternative cosmologies such as the steady state, it is BBN and the microwave background that really drive us to the conclusion that the early universe was hot and dense. The new extragalactic deuterium observations not only cement this picture and give added convergence on a value of the baryon density, Ω_b, they also enable BBN to become a constraint on stellar and Galactic evolution scenarios. It is this latter point that is the core of the transformation. Furthermore, new alternative methods of estimating the cosmic baryon density are now coming into use and are independently confirming the BBN prediction.

Space Science Reviews **84:** 3–14, 1998.
© 1998 *Kluwer Academic Publishers*

The current review will draw heavily on the recent review by Schramm (1997) in the *Proceedings of the National Academy of Sciences*.

2. Overview

Although the extragalactic D/H observations have naturally attracted the most attention, it should not be forgotten that there are also recent heroic observations of ^{6}Li, Be and B, as well as ^{3}He and new ^{4}He determinations. Let us now briefly review the history, with special emphasis on the remarkable agreement of the observed light element abundances with the calculations. This agreement works only if the baryon density is well below the cosmological critical value. We will also note how a convergence on extragalactic D/H will enable powerful new constraints on stellar and Galactic evolution.

It should be noted that there is a symbiotic connection between BBN and the 3K background dating back to Gamow and his associates, Alpher and Herman. The initial BBN calculations of Gamow's group (Alpher *et al.*, 1948) assumed pure neutrons as an initial condition and thus were not particularly accurate, but their inaccuracies had little effect on the group's predictions for a background radiation.

Once Hayashi (1950) recognized the role of neutron-proton equilibration, the framework for BBN calculations themselves has not varied significantly. The work of Alpher, Follin and Herman (1953) and Tayler and Hoyle (1964), preceeding the discovery of the 3K background, and of Peebles (1966) and Wagoner, Fowler and Hoyle (1967), immediately following the discovery, and the more recent work of our group of collaborators (Copi *et al.*, 1997; Copi *et al.*, 1994; Walker *et al.*, 1991; Olive *et al.*, 1990; Schramm and Wagoner, 1977; Olive *et al.*, 1981; Yang *et al.*, 1984; Kawano *et al.*, 1988) all do essentially the same basic calculation, the results of which are shown in Figure 1.

As far as the calculation itself goes, solving the reaction network is relatively simple by the standards of explosive nucleosynthesis calculations in supernovae, with the changes over the last 25 years being mainly in terms of more recent nuclear reaction rates as input, not as any great calculational insight, although the current Kawano code (Kawano *et al.*, 1988) is somewhat streamlined relative to the earlier Wagoner code. In fact, the earlier Wagoner code (Wagoner *et al.*, 1967) is, in some sense, a special adaptation of the larger nuclear network calculation developed by Truran (1965; Truran *et al.*, 1966) for work on explosive nucleosynthesis in supernovae. With the exception of Li yields and non-yields of Be and B (Steigman *et al.*, 1993), the reaction rate changes over the past 25 years have not had any major effect [see Yang *et al.* (1984) and Krauss and his collaborators (Krauss and Romanelli, 1990; Kernan and Krauss, 1994), or Copi, Schramm, and Turner (1994) for a discussion of uncertainties]. The one key improved input is a better neutron lifetime determination (Mampe *et al.*, 1989; Mampe *et al.*, 1993). There has been much improvement in the t(α, γ)^{7}Li reaction rate, but as the width of the curves

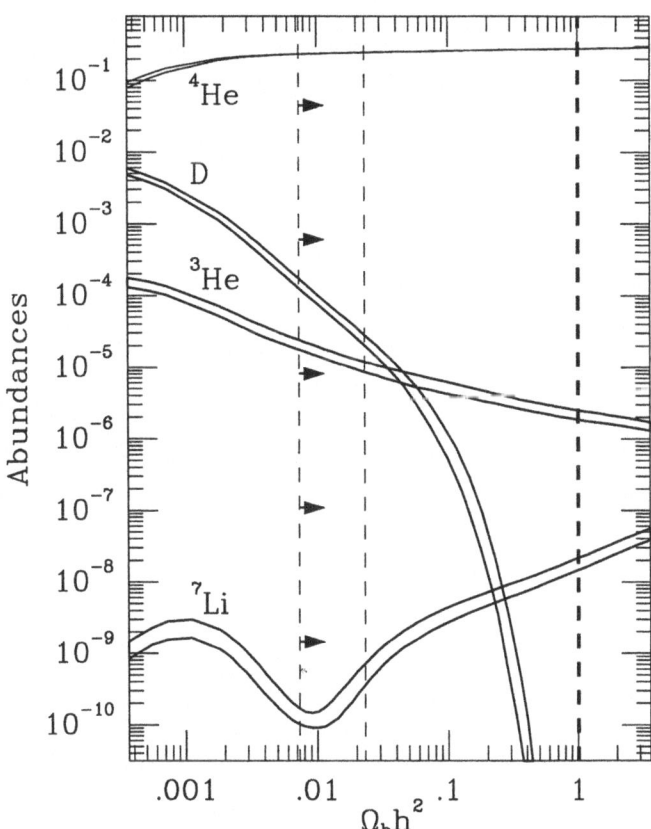

Figure 1. Big Bang Nucleosynthesis abundance yields versus baryon density (Ω_b) and $\eta \equiv n_b/n_\gamma$ for a homogeneous universe. ($h \equiv H_0/100$ km/sec/Mpc; thus, the concordant region of $\Omega_b h^2 \sim 0.015$ corresponds to $\Omega_b \sim 0.06$ for $H_0 = 50$ km/sec/Mpc.) The figure is from Copi, Schramm and Turner (1994). Note that the concordance region is slightly larger than in Walker *et al.* (1991) due primarily to inclusion of possible systematic errors on Li/H. The width of the curves represents the uncertainty due to input nuclear physics in the calculation. Recent measurements by Tytler, Fan and Burles (1996) narrow the vertical concordance region towards the high Ω_b side.

in Figure 1 shows, the ^{7}Li yields are still the poorest determined, both because of this reaction and even more because of the poorly measured $^3\text{He}(\alpha, \gamma)^7\text{Be}$.

With the exception of the effects of elementary particle assumptions, to which we will return, the real excitement for BBN over the last 25 years has not really been in redoing the basic calculation. Instead, the true action is focused on understanding the evolution of the light element abundances and using that information to make powerful conclusions. In the 1960's, the main focus was on ^{4}He which is very insensitive to the baryon density. The agreement between BBN predictions and observations helped support the basic Big Bang model but gave no significant information, at that time, with regard to density. In fact, in the mid-1960's, the other light isotopes (which are, in principle, capable of giving density information) were generally assumed to have been made during the T-Tauri phase of stellar evolution (Fowler *et al.*, 1962), and so, were not then taken to have cosmological significance. It was during the 1970's that BBN fully developed as a tool for probing the universe. This possibility was in part stimulated by Ryter *et al.* (1970) who showed that the T-Tauri mechanism for light element synthesis failed. Furthermore, D abundance determinations improved significantly with solar wind measurements (Geiss and Reeves, 1971; Black, 1971) and the interstellar work from the Copernicus satellite (Rogerson and York, 1973). (Recent HST observations reported by Linsky *et al.* (1993) and Linsky (1998) have compressed the local ISM D error bars considerably.) Reeves, Audouze, Fowler and Schramm (1973) argued for cosmological D and were able to place a constraint on the baryon density excluding a universe closed with baryons. Subsequently, the D arguments were cemented when Epstein, Lattimer and Schramm (1976) proved that no realistic astrophysical process other than the Big Bang could produce significant D. This baryon density was compared with dynamical determinations of density by Gott, Gunn, Schramm and Tinsley (1974). See Figure 2 for an updated $H_0 - \Omega$ diagram.

In the late 1970's, it appeared that a complementary argument to D could be developed using ^{3}He. In particular, it was argued (Rood *et al.*, 1976) that, unlike D, ^{3}He was made in stars; thus, its abundance would increase with time. Unfortunately, recent data on ^{3}He in the interstallar medium (Gloeckler and Geiss, 1996) has shown that ^{3}He has been constant for the last 5 Gyr. Thus, low mass stars are not making a significant addition, contrary to these previous theoretical ideas. Furthermore, Rood, Bania and Wilson (1992) have shown that interstellar ^{3}He is quite variable in the Galaxy, contrary to expectations for a nucleus produced mainly by low mass stars. However, the work on planetary nebulae shows that at least some low mass stars do produce ^{3}He. Nonetheless, the current observational situation clearly shows that arguments based on theoretical ideas about ^{3}He evolution should be avoided [c.f. Hata *et al.* (1995), where their "crisis" is really about ^{3}He problems (and excessively small assumed uncertainties in ^{4}He), not BBN]. Since ^{3}He now seems not to have a well understood history, simple ^{3}He or ^{3}He+D inventory arguments are misleading at best. However, one is not free to go to arbitrary low baryon densities and high primordial D and ^{3}He, since processing of D and ^{3}He in massive stars also produces metals which are constrained (Copi *et al.*, 1995; Scully *et al.*, 1996) by the total metal content observed in the hot intra-cluster gas, if not the Galaxy. In the near future, this problem with ^{3}He evolution will be severely constrained by the extragalactic D/H. In particular, the Tytler and Burles D/H $= 3.2 \pm 0.6 \times 10^{-5}$ (discussed at the 1997 Trento Workshop) is only slightly higher than the pre-solar D/H $= 2.1 \pm 0.5 \times 10^{-5}$ (Geiss and Gloeckler, 1998) and less than a factor of 2 above the current interstellar D/H $= 1.5 \pm 0.1 \times 10^{-5}$ (Linsky, 1998). This tells us that the production of the current metal content of the Galaxy did not destroy much D. This implies either a very different initial mass function in the early history of the Galaxy to make the metals, or much primordial infall throughout the history of the Galaxy to replenish the deuterium abundance.

Figure 2. An updated version of $H_0 - \Omega$ diagram of Gott, Gunn, Schramm and Tinsley (1974) showing that Ω_b does not intersect $\Omega_{VISIBLE}$ for any value of H_0 and that $\Omega_{TOTAL} > 0.1$, so non-baryonic dark matter is also needed (Shi *et al.*, 1995).

It was interesting that the abundances of the other light elements led to the requirement that ^7Li be near its minimum of $^7Li/H \sim 10^{-10}$, which was verified by the Pop II Li measurements of Spite and Spite (1982; Rebolo *et al.*, 1988; Hobbs and Pilachowski, 1988), hence yielding the situation emphasized by Yang *et al.* (1984) that the light element abundances are consistent over nine orders of magnitude with BBN, but only if the cosmological baryon density, Ω_b, is constrained to be around 6% of the critical value (for $H_0 \simeq 50$ km/sec/Mpc). The Li plateau argument was further strengthened with the observation of ^6Li in a Pop II star by Smith, Lambert and Nissen (1993). Since ^6Li is much more fragile than ^7Li, and yet it survived, no significant nuclear depletion of ^7Li is possible (Steigman *et al.*, 1993; Olive and Schramm, 1992; Lemoine *et al.*, 1997). This observation of ^6Li was verified by Hobbs and Thorburn (1994). Lithium depletion mechanisms are also severely constrained by the recent work of Spite *et al.* (1996) showing that the lithium plateau is also found in Pop II tidally locked binaries. Thus, meridional mixing is not causing significant lithium depletion. Recently Nollett *et al.* (1997) have discussed how ^{6}Li itself might eventually become another direct probe of BBN depending on the eventual low energy measurement of the $D(\alpha, \gamma)^6$Li cross section and on spectroscopy improvements for extreme metal-poor dwarfs. With the new extragalactic D/H, one should be able to turn the ^{7}Li argument around and argue how much depletion and/or what model atmosphere is necessary. It is again clear from this argument that large amounts of depletion did not occur, contrary to the earlier models of Delyannis (1995).

Another development back in the 70's for BBN was the explicit calculation of Steigman, Schramm and Gunn (1977) showing that the number of neutrino generations, N_ν, had to be small to avoid overproduction of ^{4}He. [Earlier work (Tayler and Hoyle, 1964; Schvartzman, 1969; Peebles, 1971) had commented about a dependence on the energy density of exotic particles but had not done an explicit calculation probing N_ν.] To put this in perspective, one should remember that the mid-1970's also saw the discovery of charm, bottom and tau, so that it almost seemed as if each new detector produced new fundamental particle discoveries, and yet, cosmology was arguing against this "conventional" wisdom. Over the years, the limit on N_ν improved with ^{4}He abundance measurements, neutron lifetime measurements, and with limits on the lower bound to the baryon density, hovering at $N_\nu \lesssim 4$ for most of the 1980's and dropping to slightly lower than 4 just before LEP and SLC turned on (Walker *et al.*, 1991; Olive *et al.*, 1990; Schramm and Kawano, 1989; Pagels, 1991). This was verified by the LEP Collaboration results (1992) (see also the CERN preprint CERN-PPE/96-183, 1996) where now the overall average is $N_\nu = 2.987 \pm 0.02$. A recent examination of the cosmological neutrino limit by Copi *et al.* (1997) in the light of the recent ^{3}He and D/H work shows that the BBN limit remains between 3 and 4 for all reasonable assumption options. It should be noted that this limit remains robust despite the uncertainties on ^{4}He systematics, since those uncertainties are still relatively small compared to a ΔN_ν of unity, although they are not small compared to significant shifts in ^{4}He implications for Ω_b.

The recent apparent convergence of the extra-galactic D/H measurements towards the lower values (Tytler, 1998; Hogan, 1997) D/H $\sim 3 \times 10^{-5}$ is beginning to collapse the Ω_b band in Figure 1 to a relatively narrow strip on the high Ω_b side (see arrows). However, such a full collapse at present is probably a bit premature. In any case, it is clear that deuteronomy (the study of deuterium in the cosmos) is a success since: 1) deuterium is clearly cosmological as it is seen in low metalicity and high redshift Lyman-α clouds; 2) the primordial D/H is higher than the present ISM D/H, as predicted by theory; and 3) the range of values for primordial D/H, regardless of whether or not the high or low ones win out, is consistent with the range of expectations based on the other light nuclei.

One potential problem that the "low D/H, high Ω_b" solution raises is the fact that the central primordial ^{4}He mass fraction is ~ 0.23, rather than ~ 0.245, which the Tytler and Burles (Tytler *et al.*, 1996; Tytler, 1998) D/H value would prefer for concordance. However, as Copi *et al.* (1997) emphasize, systematic uncertainties in Y_p cannot rule out such an excursion. But clearly we have to look carefully at ^{4}He. The recent work of Izotov's group (Thuan *et al.*, 1996) on $Y_p \sim 0.24$ shows how uncertain the present situation is, but the resolution remains to be found. How high Y_p can be and still be consistent with the He observations in extragalactic H-II regions is still quite debatable, although most agree that Y_p up to 0.25 is not impossible. Schram and Turner (1998) show that a Bayesian analysis of the plausible upper limit on Y_p centers on 0.25.

The power of homogeneous BBN comes from the fact that essentially all of the physics input is well determined in the terrestrial laboratory. The appropriate temperature regimes, 0.1 to 1 MeV, are well explored in nuclear physics laboratories. Thus, what nuclei do under such conditions is not a matter of guesswork, but is precisely known. In fact, it is known for these temperatures far better than it is for the centers of stars like our Sun. The center of the Sun is only a little over 1 keV, thus, below the energy where nuclear reaction rates yield significant results in laboratory experiments, and only the long times and higher densities available in stars enable anything to take place.

3. Density of Baryons

The bottom line that emerges from the above discussion is that (Copi *et al.*, 1997)

$$0.01 \lesssim \Omega_b h_0^2 \lesssim 0.025$$

where $h_0 \equiv H_0(\text{km/sec/Mpc})/100$. If the Tytler arguments on D/H do indeed hold up, then this will compress towards the high side, say $\Omega_b h^2 \sim 0.02 \pm 0.005$. Let us now compare with other ways of estimating Ω_b.

Attempts to circumvent the conclusion of homogeneous BBN by invoking a first-order quark-hadron phase transition (Applegate *et al.*, 1988; Alcock *et al.*, 1987) have merely illustrated the robustness of the conclusions. Figure 3 illustrates this fact, showing that for an optimized first-order quark-hadron phase transition, the abundances are only fit for the same range in $\Omega_b h^2$ as in the homogeneous case. Only if the lithium constraint is completely ignored can higher $\Omega_b h^2$ values work, and even then, only a factor of 2 is possible.

3.1. LYMAN-α CLOUDS

Recent work by Bi and Davidsen (1997), by Quashnock and Vanden Berk (1997), and by Weinberg *et al.* (1997) also argues that the density of gas in the form of Lyman-α clouds at high redshift is consistent with the high end of the Big Bang Nucleosynthesis range on Ω_b. This would appear to resolve the long time problem of where are the "dark baryons." It is well known that $\Omega_{\text{VISIBLE}} \lesssim 0.01$, which, when compared to Ω_{BBN}, implies that the bulk of the baryons are not associated

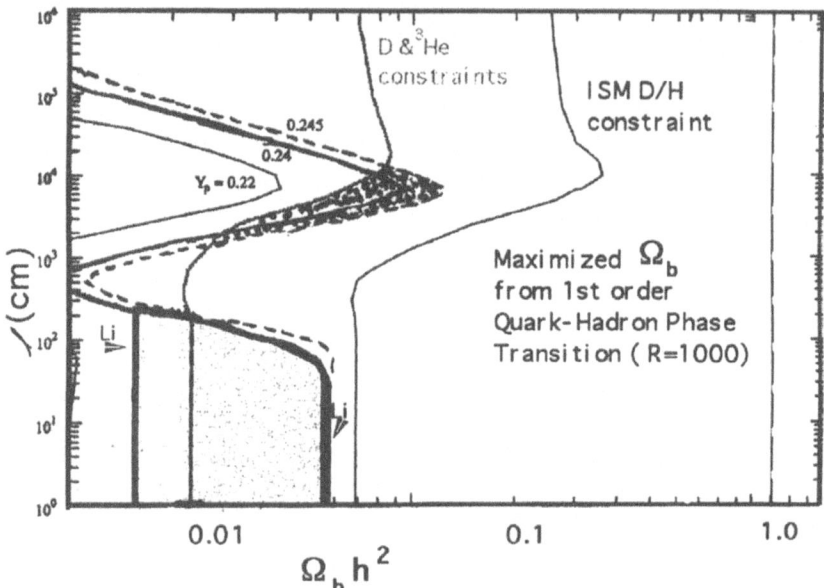

Figure 3. An assumed first order quark-hadron phase transition still yields the same allowed range on $\Omega_b h^2$ when the same abundance constraints are used. The parameter ℓ is the separation of nucleation sites (the most sensitive additional parameter in the calculation). The dotted region is an additional allowed region if lithium is completely ignored. Even then, the constraint on Ω_b is within a factor of 2 of the standard value.

with stellar material. At least at high redshift this unseen material appears to have been found in these Lyman-α clouds. In conjunction with the Lyman-α clouds, it should also be noted that singly ionized helium is seen in the intergalactic gas, thus supporting the BBN fact that helium is primordial, and also supporting the point that significant numbers of baryons were between galaxies at high redshift (Jakobsen *et al.*, 1994; Bi and Davidsen, 1997).

3.2. Hot Gas in Clusters

Hot gas has been found in clusters of galaxies by ROSAT and ASCA. The temperature of the gas can be used to estimate the gravitational potential of the clusters if it is assumed that the gas is virialized and purely supported by thermal pressure. Similarly, the intensity of the emission can be used to estimate the density of the gas. White *et al.* (1993) have shown that the typical values for x-ray clusters yield a hot gas to total mass ratio M_{HOT}/M_{TOT} of about 0.2.

Cluster masses can be estimated either from the temperature of the hot gas or from dynamics or from gravitational lensing. All yield the cluster implied density, $\Omega_{CLUSTER}$, of $\sim 0.25 \pm 0.10$. Thus, the implied baryon density from x-ray gas in clusters, $\Omega_{b,CLUSTER} \simeq \frac{M_{HOT}}{M_{TOT}} \times \Omega_{CLUSTER} \sim 0.05$, in good agreement with the BBN value for $H_0 = 50$ km/s/Mpc and for the Tytler D/H.

3.3. MICROWAVE ANISOTROPIES

The method with the most potential for checking Ω_b is the measurement of the acoustic peaks in the microwave background anisotropy at angular scales near 1° or less (Jungman *et al.*, 1996a; 1996b). The height of the first Doppler (acoustic) peak for gaussian fluctuation models is directly related to $\Omega_b h^2$, thus a direct check on BBN. Current experiments at the South Pole, at Saskatoon, and using balloons seem to favor values near the high side of the BBN range, but the present uncertainties are too large to make any strong statements. However, the next generation of satellites, NASA's MAP and ESA's PLANCK (formerly COBRAS/SAMBA) should be able to fix $\Omega_b h^2$ to better than 10% (if the sky is gaussian), which should provide a dramatic test of Big Bang Nucleosynthesis.

4. Dark Matter

The robustness of the basic BBN arguments and the new D/H measurements have given renewed confidence to the limits on the baryon density constraints. Let us convert this density regime into units of the critical cosmological density for the allowed range of Hubble expansion rates. This is shown in Figure 2. Figure 2 also shows the lower bound on the age of the universe of 10 Gyr from both nucleochronology and from globular cluster dating (Shi *et al.*, 1995) and a lower bound on H_0 of 38 km/s/Mpc from extreme type Ia supernova models with pure $1.4 M_\odot$ carbon white dwarfs being converted to ^{56}Fe. The constraint on Ω_b means that the universe *cannot be closed with baryonic matter*. This point was made over twenty years ago (Reeves *et al.*, 1973) and has proven to be remarkably strong. If the universe is truly at its critical density, then nonbaryonic matter is required. This argument has led to one of the major areas of research at the particle-cosmology interface, namely, the search for non-baryonic dark matter. In fact, from the lower bound on Ω_{TOTAL} from cluster dynamics of $\Omega_{TOTAL} > 0.15$, it is clear that non-baryonic dark matter is required unless $H_0 < 40$ km/s/Mpc, regardless of whether $\Omega_{TOTAL} = 1$. The need for non-baryonic matter is strengthened on even larger scales (Davis *et al.* 1997). Figure 2 also shows the range of $\Omega_{VISIBLE}$ and shows that there is no overlap between Ω_b and $\Omega_{VISIBLE}$. Hence, the bulk of the baryons are dark, that is, not in the form of stars.

Another interesting conclusion (Gott *et al.*, 1974) regarding the allowed range in baryon density is that it is in agreement with the density implied from the dynamics of single galaxies, *including their dark halos*. The recent MACHO (Alcock *et al.*, 1993; Aubourg *et al.*, 1993) reports of halo microlensing may well indicate that at least some of the dark baryons are in the form of baryonic objects in the halo. However, Gates, Gyuk and Turner (1995), and Alcock *et al.* (1997, 1993) show that the observed distribution of MACHOs favors less than 50% of the halo being in the form of MACHOS, but a 100% MACHO halo cannot be completely excluded, yet.

D.N. SCHRAMM

For dynamical estimates of Ω, one estimates the mass from $M \sim \frac{v^2 r}{G}$ where v is the relative velocity of the objects being studied, r is their separation distance, and G is Newton's constant. The proportionality constant depends on orientation, relative mass, etc. For large systems such as clusters, one uses averaged quantities. For single galaxies v would represent the rotational velocity and r, the radius of the star or gas cloud. It is this technique which yields the cluster bound on Ω shown in Figure 2. As mentioned above, the value of $\Omega_{CLUSTER} \sim 0.25$ is also obtained in those few cases where alignment produces giant gravitational-lens arcs (Fahlman *et al.*, 1994) and also from the gravitational potential implied by the temperature of the observed x-ray gas in the clusters. As Davis (1997) showed, if the large scale velocity flows measured from the IRAS survey are due to gravity, then $\Omega_{IRAS} \gtrsim 0.2$. For $H_0 > 40$ km/s/Mpc, $\Omega_{CLUSTER}$ already requires $\Omega_{TOTAL} > \Omega_{BARYON}$ and hence the need for non-baryonic dark matter.

An Ω of unity is, of course, preferred on theoretical grounds since that is the only long-lived natural value for Ω, and inflation (Guth, 1981; Linde, 1982) or something like it provided the early universe with the mechanism to achieve that value and thereby solve the flatness and smoothness problems. Note that our need for exotica is not dependent on the existence of dark galactic halos and that high values of H_0 increase the need for non-baryonic dark matter.

It is also interesting to note that the convergence of Ω on cluster scales at $\sim 0.25 \pm 0.1$ has important implications. If Ω_{TOTAL} is really unity, it would necessitate clusters not being a fair sample of the universe. Since standard CDM implies cluster scales as fair samples, this would imply a more complex structure formation picture. Options include biasing on cluster scales, a very hot dark matter component, or even a smooth background component such as a Λ_0 term, or a vacuum energy from a late-time phase transition (Hill and Schramm, 1985; Frieman *et al.*, 1996).

Acknowledgements

I would like to thank my collaborators, Craig Copi, Ken Nollett, Martin Lemoine, David Dearborn, Brian Fields, Dave Thomas, Gary Steigman, Brad Meyer, Keith Olive, Angela Olinto, Bob Rosner, Michael Turner, George Fuller, Sean Scully, Karsten Jedamzik, Rocky Kolb, Grant Mathews, Bob Rood, Jim Truran and Terry Walker for many useful discussions. I would further like to thank Art Davidsen, Poul Nissen, Jeff Linsky, David Tytler, Len Cowie, Scott Burles, Craig Hogan, Julie Thorburn, Doug Duncan, Lew Hobbs, Evan Skillman, Bernard Pagel and Don York for valuable discussion regarding the astronomical observations.

This work is supported by the NASA and the DoE (nuclear) at the University of Chicago, and by the DoE and by NASA grant NAG5-2788 at Fermilab.

References

Alcock, C. *et al.*: 1993, *Nature* **365**, 621–623.

Alcock, C. *et al.*: 1997, *Astrophys. J.* **486**, 697–726.

Alcock, C.R., Fuller, G. and Mathews, G.: 1987, *Astrophys. J.* **320**, 439–447.

Alpher, R.A., Bethe, H. and Gamow, G.: 1948, *Phys. Rev.*, **73**, 803–804.

Alpher, R.A., Follin, J.W. and Herman, R.C.: 1953, *Phys. Rev.*, **92**, 1347–1361.

Applegate, J.H., Hogan, C.J. and Scherrer, R.J.: 1988, *Astrophys. J.* **329**, 572–579.

Aubourg, E. *et al.*: 1993, *Nature* **365**, 623–625.

Bi, H.G. and Davidsen, A.F.: 1997, *Astrophys. J.* **479**, 523–542.

Black, D.: 1971, *Nature* **234**, 148–149.

Copi, C.J., Schramm, D.N. and Turner, M.S.: 1994, *Science* **267**, 192–199.

Copi, C.J., Schramm, D.N. and Turner, M.S.: 1995, *Astrophys. J.* **455**, L95–L98.

Copi, C.J., Schramm, D.N. and Turner, M.S.: 1997, *Phys. Rev. D* **55**, 3389–3393.

Davis, M.: (1997) in A. Olinto (ed.). *Proc. of the 18th Texas Symposium on Relativisitic Astrophysics*, in press (World Scientific, Singapore).

Delyannis, C.P.: 1995, in *The Light Element Abundances* (Proc. of an ESO/EIPC Workshop held in Marciana Marina, Isola d'Elba, May 1994), ed. P. Crane, ESO Astrophysics Symposia (Berlin: Springer-Verlag), 395–409.

Epstein, R., Lattimer, J. and Schramm, D.N.: 1976, *Nature* **263**, 198–202.

Fahlman, G. Kaiser, N., Squires, G. and Woods, D.: 1994, *Astrophys. J.* **437**, 56–62.

Fowler, W.A., Greenstein, J. and Hoyle, F.: 1962 *Geophys. J.R.A.S.* **6**, 148–220.

Frieman, J.A., Hill, C.T., Stebbins, A. and Waga, I.: 1996, *Phys. Rev. Lett.* **75**, 2077–2080.

Gates, E., Gyuk, G. and Turner, M.: 1995, *Phys. Rev. Lett.* **74**, 3724–3727.

Geiss, J. and Gloeckler, G.: 1998, in *Primordial Nuclei and Their Galactic Evolution*, Eds. N. Prantzos, M. Tosi and R. von Steiger (Dordrecht: Kluwer Academic), in press.

Geiss, J. and Reeves, H.: 1971, *Astron. and Astrophys.* **18**, 126–132.

Gloeckler, G. and Geiss, J.: 1996, *Nature* **381**, 210–212.

Gott, J.R., III, Gunn, J., Schramm, D.N. and Tinsley, B.M.: 1974, *Astrophys. J.* **194**, 543–553.

Guth, A.: 1981, *Phys. Rev. D* **23**, 347–356.

Hata, N., Scherrer, R.J., Steigman, G., Thomas, D., Walker, T.P., Bludman, S. and Langacker, P.: 1995, *Phys. Rev. Lett.* **75**, 3977–3980.

Hayashi, C.: 1950, *Prog. Theor. Phys.* **5**, 224–235.

Hill, C.T. and Schramm, D.N.: 1985, *Phys. Rev. D* **31**, 564–580.

Hobbs, L. and Pilachowski, C.: 1988, *Astrophys. J.* **326**, L23–L26.

Hobbs, L. and Thorburn, J.: 1994, *Astrophys. J. Lett.* **428**, L25–L28.

Hogan, C.: 1997, in A. Olinto (ed.). *Proc. of the 18th Texas Symposium on Relativisitic Astrophysics*, in press (World Scientific, Singapore).

Jakobsen, P., Boksenberg, A., Deharveng, J.M., Greenfield, P., Jedrzewski, R. and Paresce, F.: 1994, *Nature* **370**, 35–39.

Jungman, G., Kosowsky, A., Kamionkowski, M., Spergel, D.N.: 1996a, *Phys. Rev. Lett.* **76**, 1007–1010.

Jungman, G., Kosowsky, A., Kamionkowski, M., Spergel, D.N.: 1996b, *Phys. Rev. D* **54**, 1332–1344.

Kawano, L., Schramm, D.N. and Steigman, G.: 1988, *Astrophys. J.* **327**, 750–754.

Kernan, P. and Krauss, L.: 1994, *Phys. Rev. Lett.* **72**, 3309–3312.

Krauss, L.M. and Romanelli, P.: 1990, *Astrophys. J.* **358**, 47–59.

Kronberg, P.P.: 1994, *Rep. Prog. Phys.* **57**, 325–382.

Lemoine, M, Schramm, D.N., Truran, J.W. and Copi, C.J.: 1997, *Astrophys. J.* **478**, 554–562.

The LEP Collaboration: ALEPH, DELPHI, L3, and OPAL: 1992, *Phys. Lett. B* **276**, 247–253.

Linde, A.: 1990, *Particle Physics and Inflationary Cosmology* (Harwood Academic Publishers, N.Y.).

Linsky, J.: 1998, in *Primordial Nuclei and Their Galactic Evolution*, Eds. N. Prantzos, M. Tosi and R. von Steiger (Dordrecht: Kluwer Academic), in press.

Linsky, J., Brown, A., Gayley, K., Diplas, A., Savage, B., Ayres, T., Landsman, W., Shore, S. and Heap, S.: 1993, *Astrophys. J.* **402**, 694–709.

Mampe, W. *et al.*: 1993, *JETP Lett.* **57**, 82–87.

Mampe, W., Ageron, P., Bates, C., Pendlebury, J.M. and Steyerl, A.: 1989, *Phys. Rev. Lett.* **63A**, 593–596.

Mushotsky, R.: 1993, in C.W. Akerlof and M.A. Srednicki (eds.), *Relativistic Astrophysics and Particle Cosmology: Texas PASCOS 92, Annals of the N.Y Academy of Sciences* **688**, 184–194.

Nollett, K., Lemoine, M. and Schramm, D.N.: 1997, *Phys. Rev. C.* in press, Aug. 1, 1997.

Olive, K. and Schramm, D.N.: 1992, *Nature* **360**, 439–442.

Olive, K., Schramm, D.N., Steigman, G., Turner, M. and Yang, J.: 1981, *Astrophys. J.* **246**, 557–568.

Olive, K., Schramm, D.N., Steigman, G. and Walker, T.: 1990, *Phys. Lett. B* **236**, 454–460.

Pagel, B.: 1991, *Physica Scripta*, **T36**, 7–15.

Peebles, P.J.E.: 1966, *Phys. Rev. Lett.* **16**, 410–413.

Peebles, P.J.E.: 1971, *Physical Cosmology* (Princeton University Press).

Quashnock, J.M. and Vanden Berk, D.E.: 1997, *Astrophyhs. J.*, submitted.

Rebolo, R., Molaro, P. and Beckman, J.: 1988, *Astron. and Astrophys.* **192**, 192–205.

Reeves, H., Audouze, J., Fowler, W.A. and Schramm, D.N.: 1973, *Astrophys. J.* **179**, 909–930.

Reeves, H., Fowler, W.A. and Hoyle, F.: 1970, *Nature* **226**, 727–729.

Rogerson, J. and York, D.: 1973, *Astrophys. J.* **186**, L95–L98.

Rood, R.T., Bania, T., and Wilson, J.: 1992, *Nature* **355**, 618–620.

Rood, R.T., Steigman, G. and Tinsley, B.M.: 1976, *Astrophys. J.* **207**, L57.

Ryter, C., Reeves, H., Gradstajn, E. and Audouze, J.: 1970, *Astron. and Astrophys* **8**, 389–397.

Schramm, D.N.: 1997, 'Primordial Nucleosynthesis.' *PNAS*, **94**, in press.

Schramm, D.N. and Kawano, L.: 1989, *Nucl. Inst. and Methods A* **284**, 84–88.

Schramm, D.N. and Wagoner, R.V.: 1977, *Ann. Rev. of Nuc. Sci.* **27**, 37–74.

Schvartzman, V.F.: 1969, *JETP Letters* **9**, 184–186.

Scott, D. and White, M.: 1995, *Gen. Rel. and Grav.* **27**, 1023–1030.

Scully, S.T., Cassé, M., Olive, K.A., Schramm, D.N., Truran, J. and Vangioni-Flam, E.: 1996, *Astrophys. J.* **462**, 960–968.

Shi, X., Schramm, D.N., Dearborn, D. and Truran, J.W.: 1995, *Comments on Astrophys.* **17**, 343–360.

Smith, V.V., Lambert, D.L. and Nissen, P.E.: 1993, *Astrophys. J.* **408**, 262–276.

Spite, M., Nissen, P.E. and Spite, F.: 1996, *Astron. and Astrophys.* **307**, 172–183.

Spite, J. and Spite, M.: 1982, *Astron. and Astrophys.* **115**, 357–366.

Steigman, G., Fields, B.D., Schramm, D.N., Olive, K. and Walker T.: 1993, *Astrophys. J.* **415**, L35–L38.

Steigman, G., Schramm, D.N. and Gunn, J.: 1977, *Phys. Lett. B* **66**, 202–204.

Strickland, R. and Schramm: 1997, *Astrophys. J.*, in press.

Tayler, R. and Hoyle, F.: 1964, *Nature* **203**, 1108–1110.

Thuan, T.X., Izotov, Y.I. and Lipovetsky, V.A.: 1996, *Astrophys. J.* **463**, 120–133.

Truran, J.W.: 1965, Doctoral Thesis, Yale University.

Truran, J. W., Cameron, A.G.W. and Gilbert, A.: 1966, *Can. Jour. of Phys.* **44**, 563–592.

Tytler, D.: 1998, in *Primordial Nuclei and their Galactic Evolution*, Eds. N. Prantzos, M. Tosi and R. von Steiger (Kluwer, Dordrecht), in press.

Tytler, D., Fan, X.-M. and Burles, S.: 1996, *Nature*, **381**, 207.

Wagoner, R., Fowler, W.A. and Hoyle, F.: 1967, *Astrophys. J.* **148**, 3–49.

Walker, T., Steigman, G., Schramm, D.N., Olive, K. and Kang, H.S.: 1991, *Astrophys. J.* **376**, 51–69.

Weinberg, D.H., Miralda–Escué, J., Hernquist, L. and Katz, N.: 1997, *Astrophys. J.*, **490**, 564–570.

White, S.D.M., Navarro, J.F., Evrard, A.E. and Frenck, C.S.: 1993, *Nature* **366**, 429–433.

Yang, J., Turner, M., Steigman, G., Schramm, D.N. and Olive, K.: 1984, *Astrophys. J.* **281**, 493–511.

Address for correspondence: University of Chicago, 5640 South Ellis Avenue - AAC 140, Chicago, IL 60637, USA

OBSERVED DENSITIES IN THE UNIVERSE

G. A. TAMMANN
Astronomisches Institut der Universität Basel,
Venusstr. 7, CH-4102 Binningen, Switzerland

Abstract. Baryons observed in Ly α absorbers contribute to the density parameter Ω_0 by $\Omega_{bar} \gtrsim$ 0.06 in close agreement with the value of 0.06 from primordial nucleosynthesis (H_0=55 km s^{-1} Mpc^{-1}, $\Lambda = 0$ assumed throughout). A number of methods are known to measure Ω_0 from density fluctuations; bound structures tend to yield lower values ($\Omega_m \approx$ 0.2-0.4), field galaxies over large scales higher, but still undercritical values ($\Omega_m \approx 0.6\pm0.2$). The best compromise value is $\Omega_0 \approx 0.5$, but the present methods are blind to diffusely distributed, exotic matter which still could make $\Omega_0 = 1$. A satisfactory solution of Ω_0 (and Λ) will only come from a fundamental cosmological test (e. g. the Hubble diagram of [evolution-corrected] supernovae type Ia) in combination with the CMB fluctuation spectrum.

1. Introduction

The mean density of the Universe is either expressed by the deceleration parameter q_0, which is defined as

$$q_0 \equiv \frac{\ddot{R}_0}{R_0^2} = \frac{4\pi G \rho_0}{3H_0^2} - \frac{\Lambda c^2}{3H_0^2} \tag{1}$$

(the subscript 0 denotes present values; R is the scale factor of the Universe; ρ is the mean total matter/energy density and Λ is Einstein's cosmological constant), or by the density parameter Ω_0, which is defined as

$$\Omega_0 \equiv \frac{\rho_0}{\rho_{0,crit}} = 2q_0 + \frac{2\Lambda c^2}{3H_0^2} \ . \tag{2}$$

The present "critical density" $\rho_{0,crit}$ is the density which corresponds to a zero-curvature Universe. It is given, if $\Lambda = 0$, by

$$\rho_{0,crit} = 5.7 \times 10^{-30} \left(\frac{H_0}{55}\right)^2 \text{g cm}^{-3}. \tag{3}$$

Here – as throughout this paper – a Hubble constant of $H_0 = 55$ km s^{-1} Mpc^{-1} is adopted (Tammann and Federspiel, 1997; Sandage and Tammann, 1997; Saha *et al.*, 1997; Federspiel *et al.*, 1997). The results are given such that the reader can scale them to his preferred value of H_0. – Note that q_0 and Ω_0 were extremely close to $1/2$ and 1, respectively, in the very early Universe.

While q_0 and Ω_0 are closely related, the observer distinguishes between the two parameters. The value of q_0 is determined from the so-called "cosmological tests", i.e. from the dynamics or from the curvature of the Universe. It has the advantage

Space Science Reviews **84**: 15–30, 1998.
© 1998 Kluwer Academic Publishers.

of yielding, at least in principle, the total density. Ω_0 is derived from the structures in the Universe and – in most cases – from their influence on the local dynamics. The determination of Ω_0 may be easier, but it is also weaker because the Universe may contain highly evasive matter as smoothly distributed, dark, weakly interacting particles. The true value of Ω_0 may therefore always be larger than observed.

It follows from equation(2) that Ω_0 can be split into two terms

$$\Omega_0 = \Omega_{total} = \Omega_m + \Omega_\Lambda, \tag{4}$$

where Ω_m stands for the matter/energy density and Ω_Λ is the effective energy density contributed by a cosmological constant Λ. In a matter-dominated Universe any additional energy density is negligible. Ω_m is then the sum of

$$\Omega_m = \Omega_{bar} + \Omega_{ex}, \tag{5}$$

where Ω_{bar} reflects the baryonic matter and Ω_{ex} any "exotic" (non-baryonic) matter. The latter is often referred as "dark matter", but since baryons may be dark, the term is ambiguous.

The cosmological constant Λ has never been detected and its present value is in any case small. Present observational limits from gravitational lensing place an upper limit of $\Omega_\Lambda = 0.65$ at 95% confidence (Carroll et al., 1992; Maoz and Rix, 1993; Ostriker and Steinhardt, 1995; Kochanek, 1996). A recent determination from the Hubble diagram of distant supernova of type Ia gives even $\Omega_\Lambda = 0.06^{+0.28}_{-0.34}$ (Perlmutter, 1997). Λ has had during the last decade somewhat the role of a fudge factor to satisfy the needs of the inflationists, allowing always a zero-curvature Universe by choosing Ω_Λ such as to make

$$\Omega_m + \Omega_\Lambda = 1. \tag{6}$$

The critical case was almost a dictate of inflation theory over the last decade. Now hybrid inflationary models are discussed which allow $\Omega_{tot} < 1$. Even from a theoretical standpoint undercritical Universes are therefore "in", which destroys one of the arguments for $\Omega_\Lambda > 0$. (The terms "undercritical" and "overcritical" are used instead of the standard terms "open" and "closed", because the latter expressions depend on the topology. Even an undercritical Universe can be closed).

Because the baryon density Ω_{bar} is of particular interest for the topic of this Conference it is dealt with in Section 2. The status of the cosmological tests leading to q_0 is discussed in Section 3. Various routes towards Ω_0 are discussed in Sections 4 and 5. The conclusions are in Section 6.

2. The Baryon Density Ω_{bar}

2.1. Ω_{bar} FROM PRIMORDIAL NUCLEOSYNTHESIS

Much of this Conference is devoted to the question what primordial nucleosynthesis tells us about Ω_{bar}. It is not attempted here to enter this discussion. The arguments about non-homogeneous distribution of D and other isotopes are still raging, but a value of

$$\Omega_{bar} = (0.075 \pm 0.015) \left(\frac{55}{H_0}\right)^2 \tag{7}$$

(Geiss and Gloeckler, 1998) is secure to better than a factor of 2. With this accuracy it is one of the strongest – if not the strongest – pillars of the Ω determinations.

2.2. Ω_{bar} IN FRONT OF QUASARS

It is natural to assume that the intervening hydrogen clouds, which are observed in absorption in quasar spectra out to redshifts of $z = 4$ (the so called damped Ly α systems), are the progenitors of the present discs of spiral galaxies. This is supported by the baryonic surface brightness being similar for both kinds of objects. One question is raised by the fact that the former have about a ten times higher co-moving space density than the latter. One of several possible solutions is the current view that many (all?) elliptical galaxies have formed from the merging of spiral galaxies or their progenitors. In any case the damped Ly α systems cannot overrepresent the baryon density.

Observations of damped Ly α systems in seven quasars with redshifts $2.5 \leqslant z \leqslant 4.55$ give, in combination with detailed calculations of a standard cold-dark-matter model

$$\Omega_{bar} = 0.058 \geqslant \left(\frac{H_0}{55}\right)^{1.5} \tag{8}$$

(Sargent, 1997; Rauch et al., 1997). This is a *lower* limit because only the hydrogen-ionizing flux of known quasars is considered; any increase of the UV flux by a factor f would increase Ω_{bar} by a factor $f^{1/2}$.

There may be in addition *diffuse* intergalactic matter. An upper limit for neutral hydrogen of $\Omega_{bar} < 0.03$ is set by the absence of the Lyman break in quasar spectra (Coles and Ellis, 1997). An even tighter upper limit of $\Omega_{bar} < 0.02$ is available from an ionized-He observation in a high-redshift quasar; the corresponding lower limit is as low as $\Omega_{bar} < 0.002$ ($H_0 = 55$) at $z = 2.9$ (Reimers et al., 1997).

The intergalactic diffuse material could contribute as much as $\Omega_{bar} \leqslant 0.1$ if the hydrogen is ionized and cool enough not to distort the microwave background radiation spectrum (Fabian and Barcons, 1991).

The conclusion of this paragraph is that the lower limit of $\Omega_{bar} = 0.06$ is much better determined than its upper limit.

2.3. Ω_{bar} FROM Ω_m

If the density of the matter Ω_m is assumed to be known as well as its fraction in baryons, Ω_{bar} follows directly. In Section 4.5 it is concluded that the total baryonic mass fraction in clusters is between $\geqslant 15\%$ and 50%. If this ratio holds universally the extreme values of $0.2 < \Omega_m \leqslant 1$ correspond to $0.03 < \Omega_{bar} \leqslant 0.5$. Again the lower limit on Ω_{bar} is stronger than the upper one since the baryonic fraction in clusters may be higher than assumed, and it would be difficult to argue that the fraction of baryons inside clusters is lower than outside.

2.4. Ω_{bar} FROM CONSISTENCY ARGUMENTS

Combining the allowed ranges of various cosmological observables Steigman *et al.* (1998) concluded that consistency is best obtained with $H_0 = 57$, $\Omega_m = 0.61$ and $\Omega_{bar} = 0.10$.

The most interesting inference of Sections 2.2–2.4 is that the *lower* bound of Ω_{bar} agrees quite well with the preferred value from primordial nucleosynthesis. The high abundances of D found by some observers and the corresponding, low value of Ω_{bar} is therefore unlikely to be representative of the early *"ylem"*. The most stringent *upper* bound of Ω_{bar} is set by primordial nucleosynthesis.

3. Cosmological Tests Leading to q_0

The different cosmological tests leading to q_0 were described by Sandage (1961) in a paper which has become the basis of modern observational cosmology. The status of five such tests is briefly reviewed in the following.

3.1. q_0 FROM THE HUBBLE DIAGRAM

If extragalactic objects with sufficiently small luminosity dispersion σ_M – so-called *standard candles* – are plotted in a (log redshift) versus (apparent magnitude) diagram they define, in a linearly expanding Universe, a line with slope 0.2. At large redshifts Friedmann models predict a small deviation from this line even in an empty Universe. Larger deviations, particularly positive redshift excesses at given apparent magnitude, indicate that the Universe expanded faster at early epochs, i. e. that the expansion is decelerated. This allows to determine the deceleration parameter q_0.

In a series of papers Sandage (1973 and references therein) has shown that first-ranked (brightest) cluster galaxies have small luminosity scatter ($\sigma_M \sim 0.3$ mag) at any given redshift. Their observations, which were pushed to the technical limits of the time, provided a formal value of $q_0 = 1 \pm 0.5$ (Sandage *et al.*, 1976). However it was realized already then that this determination is not to be trusted because of stellar-population evolution and merging effects among cluster galaxies, which destroy the notion of standard candles being independent of age. Attempts to correct for luminosity evolution *and* cannibalism have remained unconvincing.

One of the original goals of the Hubble Space Telescope (HST) was to use distant supernovae of type Ia (SNe Ia) for the determination of q_0 (Tammann, 1979). Blue

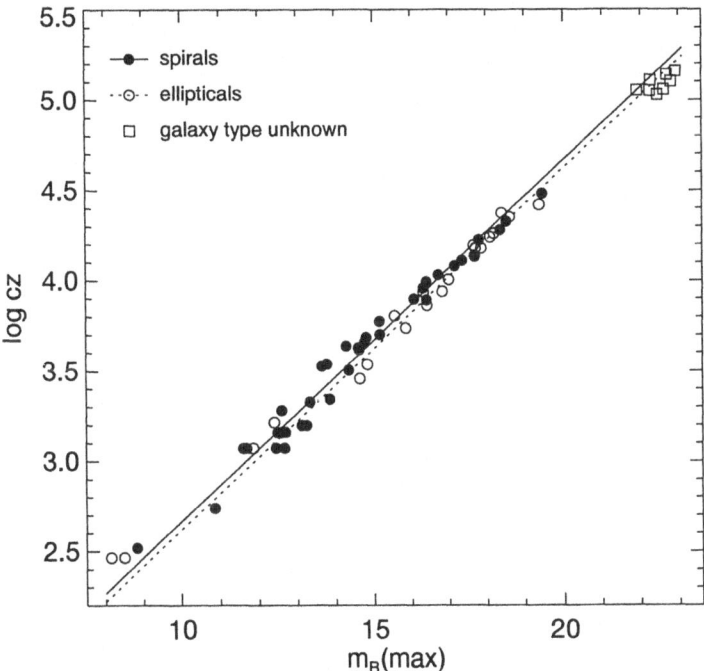

Figure 1. The Hubble diagram of SNe Ia in spiral and elliptical galaxies. The position of the first SNe Ia with $z \sim 0.5$ is shown. A line of slope 0.2 has been fitted to the nearer SNe Ia to guide the eye.

SNe Ia are indeed the best standard candles known with $\sigma_M \sim 0.2$ mag (Sandage and Tammann, 1993). There is no obvious reason why they should have a secular luminosity evolution, unless their luminosity is metal-dependent (Höflich *et al.*, 1997). First results from HST are now coming in. Garnavich *et al.* (1997) find from three SNe Ia $q_0 = -0.1 \pm 0.5$. From 20 SNe Ia out to $z = 0.7$ and allowing for a light curve "stretch factor" as some control of secular variations Perlmutter (1997) obtains in case of a Friedmann Universe $q_0 = 0.35 \pm 0.15$ (or for a Einstein-de Sitter Universe $\Omega_0 = 0.94, \Omega_\Lambda = 0.06 \pm 0.30$).

The continued search for distant *and* nearby SNe Ia offers presently the most realistic hope to determine the curvature of the Hubble diagram and hence directly the value of q_0 and eventually even the value of Λ.

3.2. q_0 FROM GALAXY COUNTS

The number $N(m)$ of galaxies brighter than apparent magnitude m increases more rapidly in an undercritical than in an overcritical Universe. Counts provide therefore q_0 in principle (Sandage, 1961), if the counted galaxies had a time-independent and therefore z-independent luminosity distribution and if the dimming of the redshift effect (K-correction) could be allowed for analytically. Both conditions not being met, one must proceed by constructing model counts of galaxies for different cosmologies allowing for individual K-corrections and the luminosity evolution of different stellar populations, and by comparing these predictions with actual counts. The conclusion from modern counts to very faint levels is that $q_0 \sim 0.05$; the case of $q_0 = \frac{1}{2}$ is only acceptable in case of additional "number evolution", i.e. that there was a distant class of galaxies which has since disappeared through mergers or other processes (Djorgovski *et al.*, 1995; Pozzetti *et al.*, 1996; Metcalfe *et al.*, 1996).

3.3. q_0 FROM DIAMETERS

If standard rods are brought to increasing distances, their angular sizes θ decrease, but in overcritical Universes they begin to increase again beyond a certain distance. The exact behavior of θ is a function of q_0 (Sandage, 1961). The difficulty is, of course, to define standard rods. Early attempts to use the large-scale structure of radio galaxies (Miley, 1971), or galaxy clusters (e. g. Wagner and Perrenod, 1981), or compact radio sources, giving $q_0 \simeq 0.5$ (Kellermann, 1993), are probably dominated by size evolution of the sources. The apparent sizes of active galactic nuclei gave a poorly confined value of $q_0 = 0.16 \pm 0.71$ (Gurvits, 1994). The metric Petrosian diameters of galaxies have not yet provided a definite answer (Djorgovski and Spinrad, 1981; Sandage and Perelmuter, 1991). Bender *et al.* (1997) explored a new technique to correct luminous E galaxies for evolution and obtained then from their angular sizes a still wide margin of $0 < q_0 < 0.70$ with 90% confidence.

3.4. q_0 FROM $N(z)$

In principle the most sensitive cosmological test, not covered by Sandage (1961), is the number $N(z)$ of all objects with redshifts smaller than z in function of z. In first approximation one can write

$$\frac{dN}{dz} \propto (1+z)^{\gamma}, \tag{9}$$

where $\gamma = 1$ for $q_0 = 0$ and $\gamma = 0.5$ for $q_0 = 0.5$. The crux here is to define a *complete* sample out to redshift z. First results of galaxy counts (Loh and Spillar, 1986) have therefore remained inconclusive.

Absorption lines in the spectra of quasars from intervening condensations seem like a gift of heaven, because their number is complete out to z_{quasar}. Yet first results have yielded unacceptable values of γ (Lu *et al.*, 1991). Mg II lines gave inconclusively $0 \leqslant q_0 \leqslant 0.5$ (Steidel and Sargent, 1992). Counts of Lyman α lines have now made clear that the result depends entirely on the evolution of the intervening matter. Stengler-Larrea *et al.* (1995) obtained $q_0 = 0$ on the assumption of no evolution and $q_0 = 0.5$ with significant evolution.

3.5. q_0 FROM THE AGE TEST

The last cosmological test to be included here is the age test. The expansion age of the Universe, which is in Friedmann models a function of H_0 and q_0 only, must accommodate the age of the oldest objects. The oldest known and best dated objects are the globular clusters in our Galaxy. A review of recent determinations of their and other ages gives 13 ± 2 Gyr (Tammann, 1997). With a estimated gestation time of 0.5 Gyr this requires an age of the Universe of $T = 13.5 \pm 2$ Gyr. With $H_0 = 55 \pm 7$ km s^{-1} Mpc^{-1} (i.e. $1/H_0 = 18.1 \pm 2.3$ Gyr) one obtains $TH_0^{-1} = 0.75 \pm 0.15$. This requires (Sandage, 1961, eq. 65)

$$q_0 = 0.25^{+0.55}_{-0.20} . \tag{10}$$

Contrary to opposite claims, the age test does not exclude overcritical models. Only if the (arbitrary) assumption is made that $H_0 > 55$ km s^{-1} Mpc^{-1} and/or $T > 12$ Gyr the critical case becomes marginally excluded.

The combined evidence of the cosmological tests favors an undercritical Universe, but they do not exclude $q_0 = 1/2$ with any certainty.

4. Ω_m from Structures

Structures induce peculiar (virial and other) motions and bind the X-ray gas, they evolve in function of the mean density, and they bend the light. These effects give a handle on their gravitating mass and hence on the mean density Ω_m, if moreover the assumption is made that there is not additional dark, more smoothly distributed matter.

Several routes arrive at the mean mass-to-light ratio $\mathfrak{M}/L$ of the structures. (The mass-to-[blue] light ratio is expressed in solar units, i.e. the Sun has $\mathfrak{M}_\odot/L_\odot \equiv 1$). If $\mathfrak{M}/L$ is the same in clusters and outside it can be multiplied by the mean universal luminosity density $\mathcal{L}$ of the Universe. $\mathcal{L}$ is found by integrating over the luminosity function of galaxies, which gives (Yahil et al., 1980; Tammann, 1982)

$\mathcal{L}$ in elliptical (E) galaxies	$2.2 \times 10^7 \left(\dfrac{55}{H_0}\right) L_{B\odot}$ Mpc^{-3}
$\mathcal{L}$ in spiral galaxies (corr. for internal absorption)	$8.8 \times 10^7 \left(\dfrac{55}{H_0}\right) L_{B\odot}$ Mpc^{-3} (11)
Sum	$1.1 \times 10^8 \left(\dfrac{55}{H_0}\right) L_{B\odot}$ Mpc^{-3}

These values agree to better than factors of 1.5 between different authors, except that some neglect the dust absorption in spirals.

With equation (11) it takes $\mathfrak{M}/L = 630 \left(\frac{H_0}{55}\right)$ to obtain the critical density of equation (3). Note that $\Omega_m = \mathcal{L} \times \mathfrak{M}/L$ is independent of H_0!

4.1. Ω_m IN LUMINOUS STARS

The combined mass and light of all stars in galaxies give within the optical (Holmberg) diameter $\mathfrak{M}/L < 5$. The *luminous* matter contributes therefore very little to Ω_m ($\Omega_{lum} < 0.01$). This implies immediately that most baryons are dark. The nature of the unseen baryons is discussed by Carr (1994).

4.2. Ω_m IN GALAXIES

Within its optical Holmberg radius of 15 kpc the Galaxy has $\mathfrak{M}/L \sim 12$ (Honma and Sofue, 1996). This value seems quite typical for spirals and E galaxies as well. Yet the Galactic mass, as judged by its rotation curve, out to a radial distance of 50 kpc requires $\mathfrak{M}/L = 35$ (Carr, 1994). Indeed, mass tracers beyond the Holmberg radius like H I discs, globular clusters, X-ray halos, satellites (if bound!), and gravitational lensing indicate a general increase of $\mathfrak{M}/L$ with radial distance, i. e. the presence of dark halos. Allowing for halos the total $\mathfrak{M}/L$ of spirals becomes ~ 100 out to 100 kpc (based mainly on the Galaxy, M 31, and M 101 and hence independent of H_0); the corresponding value of ellipticals is $\mathfrak{M}/L = 180\left(\frac{H_0}{55}\right)$ out to radii of $300\left(\frac{55}{H_0}\right)$ (Bahcall, 1997). Inserting these values pairwise into equitation (11) gives an overall $\mathfrak{M}/L = 120\left(\frac{H_0}{55}\right)$ (because of the high weight of spirals this is relatively insensitive to H_0).

The corresponding value of $\Omega_m = 0.19\left(\frac{55}{H_0}\right)$ is about three times higher than Ω_{bar} from nucleosynthesis (equation 7). Galaxies seem therefore to contain a dominant fraction of non-baryonic matter. Contrary to this the MACHO* experiment is interpreted to indicate that 50% of the dark Galactic halo is baryonic (Alcock *et al.*, 1996; 1997), presumably in form of cooled White dwarfs. However, their derived mass depends on their adopted transverse velocity; if this is overestimated the objects could also be Brown Dwarfs and their total (baryonic) mass would then be reduced by a factor of ~ 5 (Silk, 1997). Whatever the contribution is of stars to the dark matter in the halos of spirals, it is not clear yet to which population these stars belong (Rudy *et al.*, 1997).

The value of Ω_m from galaxies depends critically on the adopted radius to which the halo is integrated. Errors of factors 2 are therefore possible.

4.3. Ω_m IN THE LOCAL GROUP

The dynamics of the Local Group with radius ~ 1 Mpc requires a mass of $4 \times 10^{12}\,\mathfrak{M}_\odot$ (Peebles, 1993). This mass does not stand against all criticisms (Dekel *et al.*, 1997), but taking it at face value and dividing it by the integrated blue light of the Local Group of $10^{11}\,L_\odot$ gives $\mathfrak{M}/L = 40$ (independent of H_0), which corresponds to a very low $\Omega_m = 0.06\left(\frac{55}{H_0}\right)$ and with equation (7) to $\Omega_m/\Omega_{bar} \approx 1$.

---

* MACHOs are compact (baryonic) objects whose existence is inferred from the very specific variability of background stars due to gravitational lensing.

The Local Group contains essentially only spirals. The value $\mathfrak{M}/L = 40$ reflects therefore only on galaxies of this type. If one assumes, as in paragraph 4.2, that $\mathfrak{M}/L$ is ~ 2 times higher in E galaxies than in spirals and if one weights $(\mathfrak{M}/L)_{\text{spiral}}$ with the luminosity densities in equation (11), one obtains a still low overall mean mass-to-light ratio of $\mathfrak{M}/L = 48$ $(\Omega_m = 0.08)$.

4.4. Ω_m FROM THE VIRGOCENTRIC INFALL VELOCITY

The excess mass of the nearby Virgo cluster has caused a deceleration of the recession velocity of the Local Group as seen from the cluster. This effect has misleadingly been dubbed as the "Virgocentric infall velocity" v_{vc}. It has been determined by several authors, but the value of $v_{vc} = 220 \pm 50$ km s^{-1} still stands (Tammann and Sandage, 1985).

If in addition the density excess δ and the observed mean cluster velocity v_{Virgo} are known, the gravitating excess mass within the Solar circle can be determined. The density ratio $\delta = \frac{\delta N}{N}$ is obtained by counting galaxies inside and outside the Solar circle. One obtains $\delta = 3$ from Shapley-Ames galaxies (Yahil et al., 1980) and from other galaxy catalogs (Davis et al., 1980); the value of $\delta = 1.4$ from IRAS galaxies (Strauss et al., 1992) may be an underestimate because it relies heavily on spirals only. With the approximate relation

$$\Omega_m = \left[\frac{3v_{vc}}{(v_{\text{Virgo}} + v_{vc})\delta} \right]^{1.5} \tag{12}$$

and $v_{\text{Virgo}} = 950 \pm 50$ km s^{-1} (Binggeli et al., 1987) one obtains then $\Omega_m \sim 0.1$ (cf. also Yahil, 1981).

The lower value of Ω_m has been questioned by arguing that the observed value of v_{vc} does not only reflect the attraction of the Virgo cluster but also that from other more distant clusters. This argument is not convincing because the vector v_{vc} points towards the Virgo cluster to within $16°$ (Richter et al., 1987); this would then have to be an accident.

4.5. Ω_m FROM CLUSTERS

The mass in aggregates of galaxies like groups, clusters and superclusters can be determined from the virial theorem and the requirement – assuming hydrostatic equilibrium – that the observed X-ray gas be bound. A review of the multitude of such determinations shows that $\mathfrak{M}/L$ levels off beyond scales of $\geqslant 0.4 \left(\frac{55}{H_0} \right)$ Mpc at a median value of $\mathfrak{M}/L \sim 150 \left(\frac{H_0}{55} \right)$ (David et al., 1995; Carlberg et al., 1996; Bahcall, 1997).

More recently cluster masses have become available from gravitational lensing. An excellent review is given by Mellier et al. (1997). The most reliable results come so far from gravitationally lensed arcs. At cluster scales of $\sim 1 \left(\frac{55}{H_0} \right)$ Mpc they give $\mathfrak{M}/L \sim 130 \left(\frac{H_0}{55}\right)$ confirming the plateau of the cluster mass-to-light ratios beyond scales of $0.4 \left(\frac{55}{H_0}\right)$ Mpc. One may therefore wonder if values of $\mathfrak{M}/L \sim 240 \left(\frac{H_0}{55}\right)$ over scales of $\sim 1.5 \left(\frac{55}{H_0}\right)$ Mpc are reliable which are obtained from image distortions due to weak gravitational lensing. This effect is indeed excessively difficult to measure.

If $\mathfrak{M}/L = 150$ is taken at face value it seems to follow that $\Omega_m = 0.24$. Yet the value is seemingly in contraction with the Local Group which suggests a 2-3 times lower value. But the difference can be reduced if $\mathfrak{M}/L$ of ellipticals is higher by this factor than that of spirals, because the clusters are dominated by ellipticals, as said before, and the Local Group by spirals. If one assumes $(\mathfrak{M}/L)_{\mathrm{spirals}} = 0.5 \, (\mathfrak{M}/L)_E = 75 \left(\frac{H_0}{55}\right)$ and weights spirals and ellipticals with the respective luminosity densities in equation (7) one obtains an overall mean mass-to-light ratio of $\mathfrak{M}/L = 90 \left(\frac{H_0}{55}\right)$ and hence $\Omega_m = 0.14$ (cf. also Tammann 1982; Carlberg et al. 1997).

It is noteworthy, as pointed out by Bahcall et al. (1995), that Ω_m from clusters is apparently not larger than Ω_m from individual galaxies. This suggests that clusters do not contain any additional dark matter beyond that which galaxies possess in their halos.

4.6. Ω_m FROM Ω_{bar}

Inversely to paragraph 2.3 it is possible to infer the value of Ω_m from Ω_{bar} if $f_{\mathrm{bar}} = \Omega_m / \Omega_{\mathrm{bar}}$ is known. In clusters one can determine the mass of luminous matter $\mathfrak{M}_{\mathrm{lum}}$ (stars and X-ray gas) as well as the total binding mass $\mathfrak{M}_{\mathrm{tot}}$. The difference is made up of dark matter $\mathfrak{M}_{\mathrm{dark}}$, which still may contain a baryonic part in form of dark condensed objects (e. g. MACHOs) or in cool gas in the cluster center. One can therefore conclude only

$$f_{\mathrm{bar}} \geqslant \frac{\mathfrak{M}_{\mathrm{lum}}}{\mathfrak{M}_{\mathrm{tot}}} . \tag{13}$$

White et al. (1993) have found for the Coma cluster $f_{\mathrm{bar}} = 0.009 + 0.129 \left(\frac{55}{H_0}\right)^{3/2}$ where the first term on the right side stands for stars and the second term for the X-ray gas. The corresponding value for the Perseus cluster is somewhat higher at $f_{\mathrm{bar}} = 0.25$ for $H_0 = 55$ (Cruddace et al., 1997). Since the X-ray gas mass is higher than that in stars one can set $\mathfrak{M}_{\mathrm{lum}} \approx \mathfrak{M}_{\mathrm{gas}}$. With this simplification White and Fabian (1995) found for 13 clusters a mean value of $f_{\mathrm{gas}} = 0.15 \left(\frac{55}{H_0}\right)^{3/2}$. For the cluster Abell 2218 the ratio is marginally less (Squires et al., 1996). The X-ray gas mass inferred from the Sunyaev-Zeldovich effect yields $f_{\mathrm{gas}} = 0.11 \left(\frac{55}{H_0}\right)$ (Myers et al., 1997). Two poor clusters indicate a mean of $\sim 0.22 \left(\frac{55}{H_0}\right)^{3/2}$ but with significant variation (Loewenstein and Mushotzky, 1997). Several poor groups of galaxies give a median value of $f_{\mathrm{gas}} = 0.13 \left(\frac{55}{H_0}\right)^{3/2}$ (Mulchaey et al., 1996). Yet Pedersen et al. (1997) find in the sparse but compact, gas-poor group of NGC 3258 only $f_{bar} = 0.06^{+0.4}_{-0.2}$. On the other hand it has been argued that $f_{bar} = 0.5$ in our Galaxy (Alcock *et al.*, 1996; 1997) and also in other galaxies and clusters (David, 1997).

The lower limit of the baryon fraction from all this is $f_{bar} \geqslant f_{lum} \approx 0.15$. This value is derived from E-rich aggregates, while the upper bound of $f_{bar} = 0.5$ hinges directly on the MACHOs in our spiral Galaxy. Since the Universe is dominated by spirals the lower limit may be too low. But accepting the limits as given here yields $\Omega_m = \Omega_{bar}/f_{bar} = 0.12 - 0.40$, excluding $\Omega_m = 1$. This exclusion has been called the "baryon catastrophe", but it is a catastrophe only for the preconception that $\Omega_m = 1$.

4.7. Ω_m FROM GRAVITATIONALLY LENSED QUASARS

The number–redshift statistics of quasars with more than one image due to gravitational lensing is a strong function of the cosmological constant Λ and a weaker function of Ω_m (Turner, 1990). So far the results are limited by small-number statistics, the only firm conclusion being $\Omega_m \geqslant 0.15$ if $\Lambda = 0$ (Kochanek, 1996). Additional results may be expected from lensed flat-spectrum radio sources (Kochanek, 1997).

4.8. Ω_m FROM THE EVOLUTION OF CLUSTERS

The evolution of clusters of galaxies in look-back-time is much faster in dense Universes than in nearly empty ones. Therefore their number density in function of redshift z provides a powerful constraint on Ω_m. The observed, only mild evolution out to $z \lesssim 1$ requires an undercritical Universe with $\Omega_m = 0.3 \pm 0.1$ (Bahcall *et al.*, 1997). From the temperature evolution of X-ray clusters one obtains $\Omega_m = 0.3$ (Weihsueh *et al.*, 1997) and $\Omega_m = 0.50 \pm 0.14$ (Henry, 1997). For other attempts to derive Ω_m from the fluctuation growth rate the reader is referred to Dekel *et al.* (1997).

4.9. Ω_m FROM THE COSMIC MICROWAVE BACKGROUND FLUCTUATION SPECTRUM

Observations of the fluctuation spectrum of the Cosmic Microwave Background (CMB) are constantly augmented. They reflect several cosmological parameters like H_0, Ω_m, Ω_{bar}, and Λ. Present extensive model calculations allowing also for external constraints, give a consistent solution (for $\Lambda = 0$) of $H_0 = 58 \pm 11$ and $\Omega_m = 0.66 \pm 0.16$ (Lineweaver, 1997).

The methods to derive Ω_m from the mass-to-light ratios of bound structures in Sections 4.2 to 4.5 give consistently $\Omega_m \lesssim 0.2$. Models of cluster evolution suggest twice this value (Section 4.8) and the CMB fluctuations an even higher, but still significantly undercritical value. The mass-to-light ratios are derived over scales of $\lesssim 20$ Mpc and one could postulate that the bound aggregates have dark massive halos extending over larger scales. Yet since only $\sim 25\%$ of the mass contributing to Ω_m was attributed to clusters the excess mass could not only reside in clusters, but had to be a widespread phenomenon. Therefore arbitrary amounts of *diffuse*, dark, non-baryonic matter of unknown nature could be added to the above determinations of Ω_m. It must be stressed however that the estimates of Ω_m in Section 4 do not support frequent claims of Ω_m increasing with scale size; the 20 Mpc scale of the Virgocentric flow model gives in fact the lowest value of Ω_m.

5. Ω_m from Field Galaxies

5.1. Ω_m FROM THE COSMIC VIRIAL THEOREM

The virial theorem of bound structures can be extended in a statistical sense to field galaxies by combining the two (or three)-point correlation function of these galaxies with their pairwise line-of-sight velocity dispersion. In this way one obtains $\Omega_m = 0.15$ (Peebles, 1993) or $\Omega_m = 0.25$ (Fisher *et al.*, 1994; Davis, 1997).

5.2. Ω_m DENSITY FLUCTUATIONS AND PECULIAR VELOCITIES

Much work has gone into deriving Ω_m from the observed density fluctuations of luminous field galaxies and their gravitationally induced streaming velocities. The method has the advantage to cover large scales of up to 200 Mpc, but the disadvantage that the determination of peculiar velocities requires reliable distances which must be corrected for severe selection bias occurring over large scales. Present results include $\Omega_0 = 0.39 \pm 0.05$ (Willick *et al.*, 1997), $\Omega_0 = 0.60 \pm 0.10$ (da Costa *et al.*, 1997), and $\Omega_0 = 0.89 \pm 0.12$ (Sigad *et al.*, 1997).

5.3. Ω_m FROM REDSHIFT DISTORTIONS

The value of Ω_m can be inferred, independent of relative distances, from the fact that peculiar motions affect the redshifts, but not the angular positions of galaxies. The correlation functions and the power spectra of galaxies must therefore show characteristic distortions when they are viewed in redshift space rather than in real space. This allows to determine Ω_m from the first and second moments of the velocity distribution function of pairs of galaxies in combination with model simulations (Kaiser, 1987). This method has led to $\Omega_m = 0.35 \pm 0.28$ (Fisher *et al.*, 1994) and $\Omega_m = 0.74 \pm 0.45$ (Bromley *et al.*, 1997).

6. Conclusions

The matter density in gravitationally bound structures is, from arguments presented in Sections 4.2–4.5, $\Omega_m = 0.2 \pm 0.1$. With $\Lambda = 0$ adopted throughout this is also the value of the density parameter Ω_0. The value is insensitive to H_0 ($H_0 =$

$55 \ \mathrm{km \, s^{-1} \, Mpc^{-1}}$ adopted throughout) and agrees with Ω_0 preferred by Coles and Ellis (1997). The baryonic contribution of $\Omega_{\mathrm{bar}} = 0.06 \left(\frac{55}{H_0}\right)^2$ from primordial nucleosynthesis is independently confirmed as a lower limit, but could be somewhat higher (Section 2). From this the existence of dark, non-baryonic matter is therefore not proven beyond doubt.

However, the evidence of the number and temperature evolution of clusters and particularly of the CMB fluctuations (Sections 4.8 and 4.9) suggest a higher value of $\Omega_0 = 0.5 \pm 0.2$. Also the sophisticated analyses of field galaxies require a minimum value of $\Omega_0 \geqslant 0.4$ (Section 5). In this case $\sim 85\%$ of dark, non-baryonic matter must be invoked.

The low and high values of Ω_0 consider different scales and could be reconciled if the dark matter had clustering scale lengths between 20 and 100 Mpc. As one possibility neutrinos with $\leqslant 3$ eV could form such very extended cluster halos, but would contribute with so low a mass only $\Omega_\nu \leqslant 0.1$.

Also the higher values of Ω_{m} from field galaxies could still be a lower limit because their interpretation again depends on the number density fluctuation of luminous galaxies $\delta_{\mathrm{gal}} = \delta n / \overline{n}$ over a scale of for instance 15 Mpc. If the true density fluctuations $\delta_{\mathrm{tot}} = \delta \rho / \overline{\rho}$ (including dark matter) differs by a "linear biasing factor" b, Ω_0 becomes $b^{1.33} \times \Omega_{\mathrm{m}}$ in first approximation. The rich literature on the bias factor allows values of $0.5 < b < 1.5$, and does therefore not exclude the case $\Omega_0 = 1$.

The different determinations of Ω are summarized in Table I. From Table I it is clear that the situation of Ω_0 is inconclusive. As a compromise it seems reasonable to adopt an undercritical Universe with $\Omega_0 \approx 0.5$ (cf. also Dekel *et al.*, 1997), which implies a large fraction of non-baryonic matter, needed also to explain the evolution of structures. If indeed $\Omega_{\mathrm{m}} < 0.6$ and $\Omega_\Lambda < 0.4$ (Perlmutter, 1997; cf. Section 1) an Einstein - de Sitter Universe with $\Omega_0 = 1$ is excluded. The possibility remains, however, that there is still more *diffuse*, dark, non-baryonic matter which has gone unnoticed so far and which suffices to make $\Omega_0 = 1$.

A satisfactory determination of q_0 (and Λ) and hence of Ω_0 will only come, once the evolutionary effects of standard candles or rods are under control, from measuring spacetime directly, i. e. from one of the cosmological tests. The highest

Table I

Available Evidence for Ω ($\Lambda = 0$)

Nucleosynthesis	$\Omega_{\mathrm{bar}} = 0.06 \left(\frac{55}{H_0}\right)^2$
Damped Ly α absorption lines	$\Omega_{\mathrm{bar}} \geqslant 0.06 \left(\frac{H_0}{55}\right)^{3/2}$
Bound structures ($b = 1$)	$\Omega_{\mathrm{m}} = 0.2 \pm 0.1$
Cluster evolution ($b = 1$)	$\Omega_{\mathrm{m}} = 0.4 \pm 0.1$
Field galaxies ($b = 1$)	$\Omega_{\mathrm{m}} = 0.6 \pm 0.2$
Cosmological tests	$\Omega_0 \approx 0.6$ (incl. 1.0)
CMB fluctuations	$\Omega_0 = 0.66 \pm 0.16$ ($H_0 = 58$)

hopes are presently set on the Hubble diagram of supernovae of type Ia (Section 3.1) in combination with the CMB fluctuation spectrum.

Acknowledgements

The author thanks the Swiss National Science Foundation for its support. He thanks the Organizers of this Conference for a most stimulating week, and Mr. B. Reindl for his efficient help to edit this paper.

References

Alcock, C. *et al.*: 1996, *ApJ* **461**, 84.

Alcock, C. *et al.*: 1997, *ApJ* **486**, 697.

Bahcall, N.: 1997, in: *Critical Dialogues in Cosmology*, ed. N. Turok, Singapore: World Scientific, p. 221.

Bahcall, N. A., Fan, X. and Cen, R.: 1997, *Bull. Am. Astron. Soc.* **29**, 847.

Bahcall, N. A., Lubin, L. M. and Dorman, V.: 1995, *ApJ* **447**, L81.

Bender, R., Saglia, R. P., Ziegler, B., Belloni, P., Greggio, L., Hopp, U. and Bruzual, G.: 1997, preprint, astro-ph/97 08 237.

Binggeli, B., Tammann, G. A. and Sandage, A.: 1987, *AJ* **94**, 251.

Bromley, B. C., Warren, M. S. and Zurek, W. H.: 1997, *ApJ* **475**, 414.

Carlberg, R. G., Yee, H. K. C., Ellingson, E., Abraham, R., Gravel, P., Morris, S. and Pritchet, C. J.: 1996, *ApJ* **462**, 32.

Carlberg, R. G. *et al.*: 1997, *ApJ* **476**, L7.

Carr, B. J.: 1994, *Ann. Rev. Astron. Astrophys.* **32**, 531.

Carroll, S. M., Press, W. and Turner, E. L.: 1992, *Ann. Rev. Astr. & Astrophys.* **30**, 499.

Coles, P. and Ellis, G.: 1997, *Is the Universe Open or Closed?*, Cambridge: Cambridge University Press.

Cruddace, R. G., Kowalski, M. P., Fritz, G. G., Snyder, W. A., Fenimore, E. E. and Ulmer, M. P.: 1997, *ApJ* **476**, 479.

da Costa, M. N., Nusser, A., Freudling, W., Giovanelli, R., Haynes, M. P., Salzer, J. J. and Wegner, G.: 1997, *Monthly Not. Roy. Astron. Soc.*, in press, astro-ph/97 07 299.

David, L. P.: 1997, *ApJ* **484**, L11.

David, L. P., Jones, C. and Forman, W.: 1995, *ApJ* **445**, 578.

Davis, M.: 1997, in: *18th Texas Symposium on Relativistic Astrophysics*, eds. A. Olinto, J. Frieman and D. Schramm, Singapore: World Scientific, in press.

Davis, M., Tonry, J., Huchra, J. P. and Latham, D. W.: 1980, *ApJ* **238**, L113.

Dekel, A., Burstein, D. and White, S. D. M.: 1997, in: *Critical Dialogues in Cosmology*, ed. N. Turok, Singapure: World Scientific, p. 175.

Djorgovski, S. and Spinrad, H.: 1981, *ApJ* **251**, 417.

Djorgovski, S. *et al.*: 1995, *ApJ* **438**, L13.

Fabian, A. C. and Barcons, X.: 1991, *Rep. Prog. Phys.* **54**, 1069.

Federspiel, M., Tammann, G. A. and Sandage, A.: 1997, *ApJ*, in press, astro-ph/97 09 040.

Fisher, K. B., Davis, M., Strauss, M. A., Yahil, A. and Huchra, J. P.: 1994, *Monthly Not. Roy. Astron. Soc.* **267**, 927.

Garnavich, P. M. *et al.*: 1997, preprint, astro-ph/97 10 123.

Geiss, J. and Gloeckler, G.: 1998, *Space Sci. Rev.*, this volume.

Gurvits, L. I.: 1994, *ApJ* **425**, 442.

Henry, J. P.: 1997, *ApJ* **489**, L1.

Höflich, P., Thielemann, F. K. and Wheeler, J. C.: 1997, *ApJ*, in press, astro-ph/97 09 233.

Honma, M. and Sofue, Y.: 1996, *Publ. Astron. Soc. Japan* **48**, L103.

Kaiser, N.: 1987, *Monthly Not. Roy. Astron. Soc.* **227**, 1.
Kellermann, K. I.: 1993, *Nature* **361**, 134.
Kochanek, C. S.: 1996, *ApJ* **466**, 638.
Kochanek, C. S.: 1997, preprint, astro-ph/96 11 231.
Lineweaver, C. H.: 1997. Talk presented at the I. A. U. Symp. No. 183, Kyoto.
Loewenstein, M. and Mushotzky, R. F.: 1997, *ApJ* **471**, L83.
Loh, E. D. and Spillar, E. J.: 1986, *ApJ* **307**, L1.
Lu, L., Wolfe, A. M. and Turnshek, D. A.: 1991, *ApJ* **367**, 19.
Maoz, D. and Rix, W.-H.: 1993, *ApJ* **416**, 425.
Mellier, Y., Van Waerbeke, L., Benardeau, F. and Fort, B.: 1997, in press, astro-ph/96 09 197.
Metcalfe, N., Shanks, T., Campos, A., Fong, R. and Gardner, J. P.: 1996, *Nature* **383**, 236.
Miley, G. K.: 1971, *Monthly Not. Roy. Astron. Soc.* **152**, 477.
Mulchaey, J. S., Davis, D. S., Muskotzky, R. F. and Burstein, D.: 1996, *ApJ* **456**, 80.
Myers, S. T., Baker, J. E., Readhead, A. C. S., Leitch, E. M. and Herbig, T.: 1997, preprint, astro-ph/97 03 123.
Ostriker, J. P. and Steinhardt, P. J.: 1995, *Nature* **377**, 600.
Pedersen, K., Yoshii, Y. and Sommer-Larsen, J.: 1997, *ApJ* **485**, L17.
Peebles, P. J. E.: 1993, *Principles of Physical Cosmology*, Princeton: Princeton University Press.
Perlmutter, S.: 1997, Talk presented at the I. A. U. Symposium No. 183, *Cosmological Parameters and Evolution of the Universe*, Kyoto.
Pozzetti, L., Bruzual, G. and Zamorani, G.: 1996, *Monthly Not. Roy. Astron. Soc.* **281**, 953.
Rauch, M. *et al.*: 1997, *ApJ* **489**, 7.
Reimers, D., Köhler, S., Wisotzki, L., Groote, D., Rodriguez-Pascual, P. and Wamsteker, W.: 1997, *Astron. Astrophys.* **327**, 890.
Richter, O.-G., Tammann, G. A. and Huchtmeier, W. K.: 1987, *Astron. Astrophys.* **171**, 33.
Rudy, R. J., Woodward, C. E., Hodge, T., Fairfield, S. W. and Harker, D. E.: 1997, *Nature* **387**, 159.
Saha, A., Sandage, A., Labhardt, L., Tammann, G. A., Macchetto, F. D. and Panagia, P.: 1997, *ApJ* **486**, 1.
Sandage, A.: 1961, *ApJ* **133**, 355.
Sandage, A.: 1973, *ApJ* **183**, 731.
Sandage, A., Kristian, J. and Westphal, J. A.: 1976, *ApJ* **205**, 688.
Sandage, A. and Perelmuter, J.-M.: 1991, *ApJ* **370**, 455.
Sandage, A. and Tammann, G. A.: 1993, *ApJ* **415**, 1.
Sandage, A. and Tammann, G. A.: 1997, in: *Critical Dialogues on Cosmology Conference*, ed. N. Turok, Princeton: Princeton University Press, p. 130.
Sargent, W. L. W.: 1997, Talk presented at this Conference.
Sigad, Y., Eldar, A., Dekel, A., Strauss, M. A. and Yahil, A.: 1997, preprint, astro-ph/97 08 141.
Silk, J.: 1997, private communication.
Squires, G., Kaiser, N., Babul, A., Fahlman, G., Woods, D., Neumann, D. M. and Böhringer, H.: 1996, *ApJ* **461**, 572.
Steidel, C. C. and Sargent, W. L. W.: 1992, *ApJ Suppl.* **80**, 1.
Steigman, G., Hata, J. and Felten, J. E.: 1998, *Space Sci. Rev.*, this volume.
Stengler-Larrea, E. A. *et al.*: 1995, *ApJ* **444**, 64.
Strauss, M. A., Davis, M., Yahil, A. and Huchra, J. P.: 1992, *ApJ* **385**, 421.
Tammann, G. A.: 1979, in: *Scientific Research with the Space Telescope*, eds. M. S. Longair and J. W. Warner, Washington: U. S. Government Printing Office, p. 263.
Tammann, G. A.: 1982, in: *Landolt-Börnstein*, eds. W. Schaifers and H. H. Voigt, Vol. 2c of *New Series*, p. 359.
Tammann, G. A.: 1997, in: *Big Bang and Alternative Cosmologies – a Critical Appraisal*, eds. T. Padmanabhan and J. Narlikar, in press.
Tammann, G. A. and Federspiel, M.: 1997, in: *The Extragalactic Distance Scale*, eds. M. Livio, M. Donahue and N. Panagia, Cambridge: Cambridge University Press, p. 137.
Tammann, G. A. and Sandage, A.: 1985, *ApJ* **294**, 81.
Turner, E. L.: 1990, *ApJ* **365**, L43.
Wagner, R. M. and Perrenod, S. C.: 1981, *ApJ* **251**, 424.

Weihsueh, A. C., Ostriker, J. P. and Strauss, M. A.: 1997, preprint, astro-ph/97 08 250.

White, S. D. M. and Fabian, A. C.: 1995, *Monthly Not. Roy. Astron. Soc.* **278**, 72.

White, S. D. M., Navarro, J. F., Evrad, A. E. and Frenk, C.: 1993, *Nature* **366**, 429.

Willick, J. A., Strauss, M. A., Dekel, A. and Kolatt, T.: 1997, *ApJ* **486**, 629.

Yahil, A.: 1981, in: *Tenth Texas Symposium on Relativistic Astrophysics*, eds. R. Ramaty and F. C. Jones, New York: New York Academy, p. 169.

Yahil, A., Sandage, A. and Tammann, G. A.: 1980, *ApJ* **242**, 448.

NON-BBN CONSTRAINTS ON THE KEY COSMOLOGICAL PARAMETERS

GARY STEIGMAN
Departments of Physics and Astronomy
The Ohio State University, Columbus, OH 43210, USA

NAOYA HATA
Institute for Advanced Study, Princeton, NJ 08540

JAMES E. FELTEN
Code 685, NASA Goddard Space Flight Center, Greenbelt, MD 20771

Abstract. Since the baryon-to-photon ratio η_{10} is in some doubt at present, we ignore the constraints on η_{10} from big bang nucleosynthesis (BBN) and fit the three key cosmological parameters (h, Ω_M, η_{10}) to four other observational constraints: Hubble parameter (h_o), age of the universe (t_o), cluster gas (baryon) fraction ($f_o \equiv f_G h^{3/2}$), and effective shape parameter (Γ_o). We consider open and flat CDM models and flat ΛCDM models, testing goodness of fit and drawing confidence regions by the $\Delta\chi^2$ method. CDM models with $\Omega_M = 1$ (SCDM models) are accepted only because we allow a large error on h_o, permitting $h < 0.5$. Open CDM models are accepted only for $\Omega_M \gtrsim 0.4$. ΛCDM models give similar results. In all of these models, large $\eta_{10}(\gtrsim 6)$ is favored strongly over small $\eta_{10}(\lesssim 2)$, supporting reports of low deuterium abundances on some QSO lines of sight, and suggesting that observational determinations of primordial ^{4}He may be contaminated by systematic errors. Only if we drop the crucial Γ_o constraint are much lower values of Ω_M and η_{10} permitted.

Key words: Baryon-to-photon ratio, Universal matter density, Hubble parameter

1. Introduction

In the context of the hot big bang cosmology, if the number of light-neutrino species has its standard value $N_\nu = 3$, the predicted primordial abundances of four light nuclides (D, ^{3}He, ^{4}He, and ^{7}Li) depend on only one free parameter, η_{10}, the universal ratio (at present) of nucleons (baryons) to photons (in units 10^{-10}). In principle, η_{10} is overdetermined by the observed or inferred primordial abundances of the four light nuclides. Indeed, Steigman, Schramm, and Gunn (1977) have exploited this fact to use big bang nucleosynthesis (BBN) to constrain N_ν. The status quo ante is that observations, principally of D and ^{4}He, have rendered η_{10} one of the best known of the key cosmological parameters: $\eta_{10} = 3.4 \pm 0.3$ (Walker *et al.*, 1991; the error bars being roughly "1σ").

At present, when the microwave background temperature $T = 2.728$ K (Fixsen *et al.*, 1996), the universal baryonic mass-density parameter Ω_B ($\equiv 8\pi G\rho_B/3H_0^2$) is related to η_{10} by

$$\Omega_B h^2 = 3.667 \times 10^{-3} \, \eta_{10} = 0.0125 \pm 0.0011. \tag{1}$$

However, recently there have emerged reasons to suspect that η_{10} may not be so well determined, and even that the standard theory of BBN may not provide a very good fit to the current data (Hata *et al.*, 1995). There are several options available for resolving this apparent conflict between theory and observation. Although some change in standard physics could offer resolution (e.g., a reduction in the effective value of N_ν during BBN below its standard value 3; cf. Hata *et al.*, 1995), Hata *et al.* (1995) note that large systematic errors may compromise the abundance data (cf. Copi, Schramm, and Turner, 1995).

This controversy has been sharpened by new observations giving the deuterium abundances on various lines of sight to high-redshift QSOs. In principle, these data should yield the primordial D abundance, but current results span an order of magnitude. If the low value (D/H by number $\approx 2 \times 10^{-5}$; Tytler, Fan, and Burles, 1996; Burles and Tytler, 1996) is correct, then $\eta_{10} \approx 7$ in the standard model, but then it seems impossible to reconcile the inferred abundance of ^{4}He [Olive and Steigman, 1995 (OS)] with (standard) BBN for this value of η_{10} unless there are large systematic errors in the ^{4}He data. If, instead, the high figures (D/H $\approx 2 \times 10^{-4}$; Carswell *et al.*, 1994; Songaila *et al.*, 1994; Rugers and Hogan, 1996) are correct, then D and ^{4}He are consistent with $\eta_{10} \approx 2$, but modellers of Galactic chemical evolution have a major puzzle: How has the Galaxy reduced D from its high primordial value to its present (local) low value without producing too much ^{3}He (Steigman and Tosi, 1995), without using up too much interstellar gas (Edmunds, 1994; Prantzos, 1996), and without overproducing heavy elements (cf. Tosi, 1996, and references therein)? It appears that η_{10}, though known to order of magnitude, may now be among the less well-known cosmological parameters. Despite this, large modern simulations which explore other cosmological parameters are often limited to a single value of $\eta_{10} = 3.4$ (e.g., Borgani *et al.*, 1997).

Given this unsettled situation Steigman, Hata, and Felten (1997; hereafter SHF) have proposed that it may be constructive to abandon nucleosynthetic constraints on η_{10} entirely and to put η_{10} onto the same footing as the other cosmological free parameters, applying joint constraints on all these parameters based on other (non-BBN) astronomical observations and on theory and simulation. Armed with η_{10} determined in this manner we may then "predict" the primordial abundances of the light nuclides and compare with the data to test the consistency of standard BBN. In this contribution to the proceedings of the ISSI Workshop on Primordial Nuclei and Their Galactic Evolution (Bern, Switzerland, 6–10 May 1997) we present a brief description of our approach along with a summary of our results for a "standard" (fiducial) choice of the observational constraints. For further details (especially of the many variations on the standard case to be described herein) and references the reader is encouraged to consult SHF.

2. The Method: An Overview

Our approach is to let the three key cosmological parameters (h, Ω_M, η_{10}) range freely, fit the constraints (observables other than nucleosynthetic) to be described below, test goodness of fit by χ^2, and draw formal confidence regions for the parameters by the usual $\Delta\chi^2$ method. Most of the SHF results are not surprising, and related work has been done before (White *et al.*, 1996; Lineweaver *et al.*, 1997; White and Silk, 1996; Bludman, 1997), but not with these three free variables and the full χ^2 formalism. Further, some recent cosmological observations and simulations, particularly those related to the "shape parameter" Γ and the cluster baryon fraction (CBF), seem to pose a challenge to popular models, and there is some doubt whether any simple model presently fits all data well. Our approach, which begins by discarding nucleosynthetic constraints, provides a new way of looking at these problems. For example, the CBF and Γ constraints have not been applied jointly in earlier work which often also adopts a precise value for η_{10} (or Ω_B).

SHF find that, given their conservative (generous) choice of error bar on h, the SCDM model is disfavored but by no means excluded. But even with this generous error bar, large values ($\gtrsim 6$) of η_{10} ($\Omega_B h^2 \gtrsim 0.022$) are favored strongly over low values ($\lesssim 2$; $\Omega_B h^2 \lesssim 0.007$). This suggests that the low D abundances measured by Tytler *et al.* (1996) and by Burles and Tytler (1996) may be correct, and that the observed (extrapolated) primordial helium-4 mass fraction [$Y_P \approx 0.23$; cf. OS and Olive, Skillman, and Steigman, 1997 (OSS)], thought to be well determined, may be systematically too low for unknown reasons.

3. CDM Models: Parameters and Observables

3.1. PARAMETERS

The CDM models we consider are defined by three free parameters: Hubble parameter h; mass-density parameter $\Omega_M = 8\pi G \rho_M / 3 H_0^2$; and baryon-to-photon ratio η_{10}, related to Ω_B by equation (1). Here Ω_M by definition includes all "dynamical mass": mass which behaves dynamically like ordinary matter in the universal expansion; Ω_M is not limited to clustered mass only. Other free parameters having to do with structure formation, such as the tilt parameter n, could be added (White *et al.*, 1996; Kolatt and Dekel, 1997; White and Silk, 1996), but generally we have tried to avoid introducing many free parameters.

3.2. OBSERVABLES

We consider four observables (constraints) with measured values and with errors which are assumed to be Gaussian: (1) Hubble parameter h_0; (2) age of the universe t_0; (3) gas-mass fraction $f_0 \equiv f_G h^{3/2}$ in rich clusters; and (4) "shape parameter" Γ_0 from structure studies. In SHF we considered a fifth constraint: the dynamical mass-density parameter Ω_0 as inferred from cluster measurements or from large-scale flows. Here, we ignore this constraint but we shall comment on its relation to our "standard" results.

3.2.1. *Observed Hubble Parameter* h_o

For the Hubble parameter the observable h_o is simply fit with the parameter h. Measurements of h still show scatter which is large compared with their formal error estimates (Bureau, Mould, and Staveley-Smith, 1996; Tonry et al., 1997; Kundić et al., 1997; Tammann and Federspiel, 1997). This indicates systematic errors. To be conservative (permissive), we adopt $h_o = 0.70 \pm 0.15$. Perhaps a smaller error could be justified; below we will comment on the consequences of shrinking the error bar.

3.2.2. *Observed Age of the Universe* t_o

The age for these $\Lambda = 0$ models is a function of h and Ω_M given by: $t = 9.78\, h^{-1} \times f(\Omega_M; \Lambda{=}0)$ Gyr [Weinberg, 1972, equations (15.3.11) & (15.3.20)]. We take the observed age of the oldest globular clusters as $t_{GC} = 14 \pm 2$ Gyr (Bolte and Hogan, 1995; Jimenez, 1997; D'Antona, Caloi, and Mazzitelli, 1997; Chaboyer et al., 1997; Cowan et al., 1997; cf. Nittler and Cowsik, 1997). The universe is older than the oldest globular clusters by an unknown amount Δt. Although most theorists believe that Δt must be quite small (1 or 2 Gyr at most), we are unaware of any conclusive argument which guarantees this. To keep things simple, SHF introduced asymmetric error bars: $t_o = 14^{+7}_{-2}$ Gyr, allowing enough extra parameter space at large ages to accommodate a conservative range of Δt; extremely large ages will be eliminated by the h_o constraint in any case.

3.2.3. *Observed Gas Mass Fraction in Clusters* f_o

As is suggested by simulations, we assume that rich clusters provide a fairly unbiased sample of **the universal ratio of baryonic to dark matter**. Thus, we use the cluster (hot) gas fraction, f_G, not as a constraint on Ω_M, but as a constraint on the universal baryon fraction, the ratio Ω_B/Ω_M. We emphasize that the following argument assumes that rich clusters provide a fair sample of the universal baryon fraction but does **not** assume that most of the mass in the universe, or any specific fraction of it, is in rich clusters.

The measurement of f_G poses some problems; for discussion and references, see SHF. We have followed the approach of Evrard, Metzler, and Navarro (1996), who used gas-dynamical simulations to model the observations. They find that the largest contribution to the error in f_G arises from the measurement of the cluster's *total* mass, and they suggest that this error can be reduced by using an improved estimator and by restricting the measurement to regions of fairly high overdensity. Evrard (1997) applies these methods to data for real clusters and finds $f_G\, h^{3/2} = 0.060 \pm 0.003$. To be conservative, we double his error bars and adopt for our constraint

$$f_o \equiv f_G\, h^{3/2} = 0.060 \pm 0.006. \tag{2}$$

We are interested in the baryonic mass fraction, Ω_B/Ω_M, but even in rich clusters not all baryons are in the form of gas, and selection factors may operate in bringing baryons and dark matter into clusters. White *et al.* (1993) introduced a "baryon enhancement factor" Υ to describe these effects. Υ may be defined by

$$f_{G0} = \Upsilon \Omega_G / \Omega_M, \tag{3}$$

where Ω_G is the initial contribution of *gas* to Ω_M (note that $\Omega_G \leq \Omega_B$) and f_{G0} is the gas mass fraction in the cluster immediately after formation. Υ is really the *gas* enhancement factor, because the simulations do not distinguish between baryonic condensed objects if any (galaxies, stars, machos) and non-baryonic dark-matter particles. All of these are lumped together in the term $(\Omega_M - \Omega_G)$ and interact only by gravitation.

If all the baryons start out as gas $(\Omega_G = \Omega_D)$, and if gas turns into condensed objects only *after* cluster formation, then equation (3) may be rewritten:

$$f_G + f_{GAL} = \Upsilon \Omega_B / \Omega_M, \tag{4}$$

where f_G is the present cluster gas-mass fraction and f_{GAL} the present cluster mass fraction in baryonic condensed objects of all kinds (galaxies, stars, machos). White *et al.* (1993) took some pains to estimate the ratio f_G / f_{GAL} within the Abell radius of the Coma cluster, counting only galaxies (no stars or machos) in f_{GAL}. They obtained

$$f_G / f_{GAL} = 5.5 \, h^{-3/2}. \tag{5}$$

This is large, so unless systematic errors in this estimate are very large, the baryonic content of this cluster (at least) is dominated by the hot gas. Carrying f_{GAL} along as an indication of the size of the mean correction for all clusters, and solving equations (4) and (5) for $f_G h^{3/2}$, we find

$$f_G h^{3/2} = [1 + (h^{3/2}/5.5)]^{-1} (\Upsilon \Omega_B / \Omega_M) \, h^{3/2}, \tag{6}$$

where Ω_B is given from η_{10} and h by equation (1).

This is the appropriate theoretical function of the free parameters to fit to the observations. We set $\Upsilon = 0.9$ in our "standard" case. Although this is representative of results from simulations, Cen (1997) finds that the determination of f_G from X-ray observations may be biased toward high f_G by large-scale projection effects; i.e., the calculated f_G exceeds the true f_G present in a cluster by a bias factor which can be as large as 1.4. Although Evrard *et al.* (1996) and Evrard (1997) have not observed such a bias in their simulations, SHF explored its effect on our analysis by using for Υ, instead of 0.9, an "effective value" $\Upsilon \approx 0.9 \times 1.4 \approx 1.3$.

3.2.4. *Shape Parameter Γ_o from Large-Scale Structure*

The last observable we use is the "shape parameter" Γ, which describes the transfer function relating the initial perturbation spectrum $P_i(k) \propto k^n$ to the present spectrum $P(k)$ of large-scale power fluctuations, as observed, e.g., in the galaxy correlation function. When the spectral index n of $P_i(k)$ has been chosen, Γ is determined by fitting the observed $P(k)$.

Results of observations may be cast in terms of an "effective shape parameter" Γ (White *et al.*, 1996) which we take as our observable. Studies show that for the usual range of CDM models, with or without Λ, the expression for Γ is

$$\Gamma \approx \Omega_M h \, \exp\left[-\Omega_B - (h/0.5)^{1/2}(\Omega_B/\Omega_M)\right] - 0.32\,(n^{-1} - 1) \tag{7}$$

(Peacock and Dodds, 1994; Sugiyama, 1995; Liddle *et al.*, 1996a,b; White *et al.*, 1996; Liddle and Viana, 1996; Peacock, 1997). For $n \approx 1$, if Ω_B and Ω_B/Ω_M are small, we have $\Gamma \approx \Omega_M h$. The Harrison-Zeldovich (scale-invariant, untilted) case is $n = 1$, which we adopt for our standard case.

In contrast to the "standard" case in SHF, here we adopt the more conservative choice (larger error bars) for the observed value of Γ,

$$\Gamma_o = 0.25 \pm 0.05 \tag{8}$$

(cf. Peacock and Dodds, 1994; Maddox, Efstathiou, and Sutherland, 1996). This is based on the galaxy correlation function, and it assumes that light traces mass. Equations (7) and (8) imply, very roughly, that $\Omega_M h \approx 0.25$.

The shape-parameter constraint is in a sense the least robust of the constraints we have discussed since it is not part of the basic Friedmann model. Rather, it depends on a theory for the primordial fluctuations and how they evolve. If the Friedmann cosmology were threatened by this constraint, we believe that those who model large-scale structure would find a way to discard it. Therefore SHF have also explored the consequences of removing this constraint and replacing it with one on Ω_M derived from the M/L ratio in clusters and the luminosity density of the universe (Carlberg, Yee, and Ellingson, 1997) or one based on studies of large-scale flows around voids (Dekel and Rees, 1994; cf. Dekel, 1997).

4. CDM Models: Results

4.1. STANDARD CONSTRAINTS

For our standard case we have four observational constraints: $h_o = 0.70 \pm 0.15$, $t_o = 14^{+7}_{-2}$ Gyr, $f_o \equiv f_G h^{3/2} = 0.060 \pm 0.006$, and $\Gamma_o = 0.25 \pm 0.05$. For this standard case we assume $n = 1$ and $\Upsilon = 0.9$. The results for our three cosmological parameters are displayed in Figures 1 and 2 where the 68% and 95% confidence regions ("CRs") are shown. Also shown are the projected CRs obtained by computing χ^2 for single observables alone, or for pairs of observables. These are not true CRs but are intended to guide the reader in understanding how the various constraints influence the closed contours which show our quantitative results.

Our best-fit values for the three key cosmological parameters are: $\eta_{10} = 8.2^{+3.2}_{-2.2}$, $\Omega_M = 0.48^{+0.22}_{-0.15}$ and $h = 0.58 \pm 0.22$. Although the condition $\Omega_M h \approx 0.25$ poses

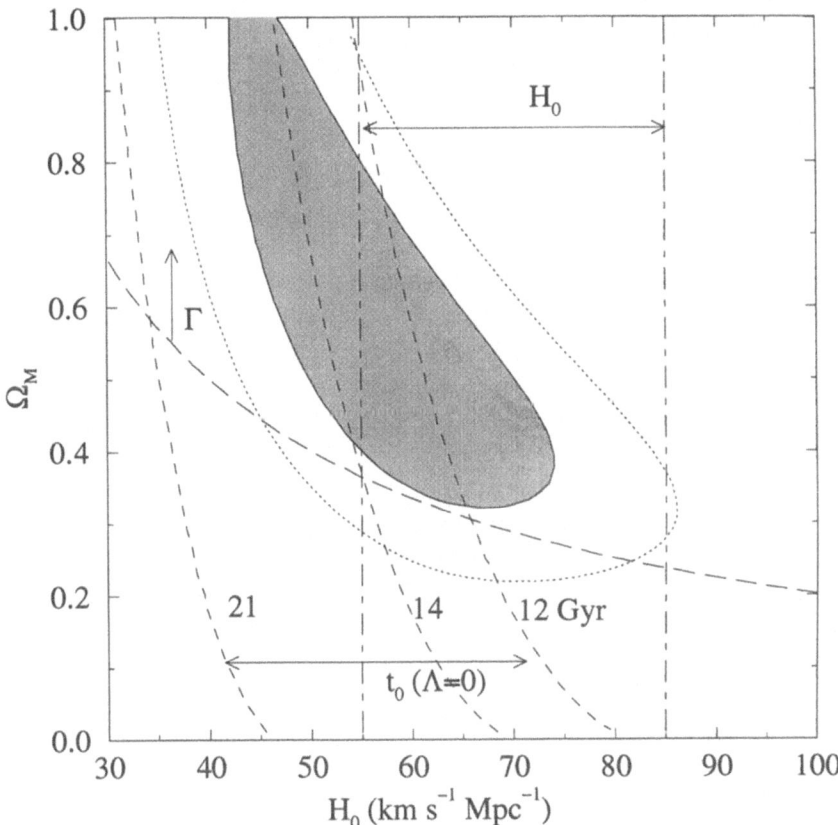

Figure 1. 68% (shaded) and 95% (dotted) confidence regions ("CRs") in the (H_0, Ω_M) plane for CDM models with our four standard constraints. The CRs are closed curves. Individual constraints in this plane are also shown schematically.

some threat to the SCDM ($\Omega_M = 1$) model, Figure 1 shows that this threat is far from acute given our more accurate form of the Γ constraint in equation (7), as long as the error on h_0 is large (0.15) and BBN constraints are discarded. The exponential term in equation (7) becomes significant because the f_G constraint forces Ω_B to increase with Ω_M allowing the product $\Omega_M h$ to exceed 0.25. This has been noted before (White *et al.*, 1996; Lineweaver *et al.*, 1997). The SCDM model with $\Omega_M = 1$ and $h \approx 0.45$ is acceptable but the high value for the baryon-to-photon ratio, $\eta_{10} \approx 13$, is in conflict with the inferred primordial abundances of **all** the light nuclides. Note that, although the uncertainties are large, low values of η_{10} are disfavored (see Figure 2).

If we add the Dekel-Rees estimate of Ω_M ($\Omega_o \gtrsim 0.4$), the five-constraint fit favors somewhat higher values of Ω_M and η_{10} and slightly lower values of h. In contrast, if instead we include the cluster estimate ($\Omega_o = 0.2 \pm 0.1$; Carlberg, 1997; cf. Carlberg, Yee, and Ellingson, 1997), we find a barely acceptable fit ($\chi^2_{min} = 5.0$

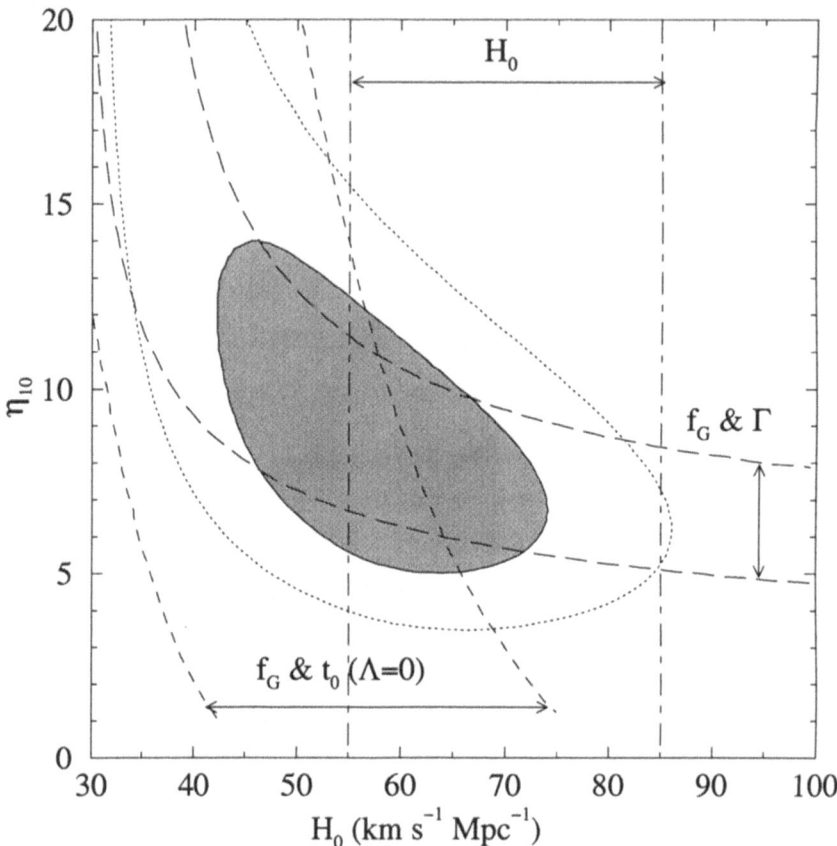

Figure 2. Same as Figure 1, but in the (H_0, η_{10}) plane. Individual and paired constraints are also shown schematically.

for 2 DOF, 92% CL), which favors lower values of Ω_M and η_{10} and slightly higher values of h.

4.2. VARIATIONS

Tilt in the primordial spectrum has been investigated in many papers (Liddle *et al.*, 1996a,b; White *et al.*, 1996; Kolatt and Dekel, 1997; White and Silk, 1996; Liddle and Viana, 1996). We considered the effect of a moderate "red tilt" ($n = 0.8$ instead of $n = 1$). This has the effect (see equation 7) of raising slightly the 68% and 95% contours in Figure 1. With this tilt the Γ constraint favors higher Ω_M, so that the SCDM model is allowed for h up to nearly 0.5. The favored likelihood range for η_{10} is now also higher, though $\eta_{10} \approx 7$ is still allowed. However, the higher allowed range for η does threaten the consistency of BBN. Conversely, a "blue" tilt, $n > 1$ (Hancock *et al.*, 1994), would move the CR downward and allow models with $\Omega_M \le 0.3$ at high h.

Changing to a gas enhancement factor $\Upsilon = 1.3$ (modest positive enhancement of gas in clusters) instead of 0.9 does not change the contours in Figure 1 by much since Γ is only weakly coupled to Ω_B through the exponential term in Γ. Although the effect is to lower the contours in Figure 1 slightly and to move downward the acceptable range for η_{10}, $\eta_{10} \leq 4$ is still excluded, disfavoring the low D abundance inferred from some QSO absorbers and favoring a higher helium abundance than is revealed by the H II -region data.

The possibility that the fraction of cluster mass in baryons in galaxies, isolated stars, and machos (f_{GAL}) might be larger – even much larger – than is implied by equation (5) would affect the CRs in much the same way as a *small* Υ, favoring even higher values of Ω_M and η_{10}.

The Γ constraint is crucial for our standard results favoring high Ω_M and high η. If, for example, we drop the Γ constraint and in its place use the cluster estimate $\Omega_o = 0.2 \pm 0.1$, low Ω_M and low η are now favored (see SHF).

The acceptability – or not – of the SCDM model depends crucially on the choice of Hubble parameter. SHF have experimented with replacing the standard constraint on H_0 with $h_o = 0.70 \pm 0.07$. Now, the SCDM model is strongly excluded.

4.3. ΛCDM MODELS

SHF have also considered models with nonzero Λ, limiting their investigation to the popular flat ($k = 0$) "ΛCDM" models with $\Omega_\Lambda = 1 - \Omega_M$, where $\Omega_\Lambda \equiv \Lambda/(3H_0^2)$. For these models there are still only three free parameters and the four constraints discussed earlier are still in force, except that the product of the age and the Hubble parameter is a different function of $\Omega_M = 1 - \Omega_\Lambda$: $t = 9.78\,h^{-1}f(\Omega_M; k = 0)$ Gyr [Carroll, Press, and Turner, 1992, equation (17)]. For a given $\Omega_M < 1$, the age is longer for the flat ($k = 0$) model than for the $\Lambda = 0$ model. The results differ very little from those in Figures 1 and 2. The longer ages do allow the CRs to slide farther down toward large h and small Ω_M. Because of the longer ages at low Ω_M (high Ω_Λ), Ω_o from clusters can now be accepted as a fifth constraint. In this case (see SHF) large Ω_M and small h are now excluded while $\eta_{10} > 4$ is still favored strongly.

5. Conclusions

If BBN constraints on the baryon density are removed (or relaxed), the interaction among the shape-parameter (Γ) constraint, the cluster baryon fraction (f_G) constraint, and the value of η_{10} assumes critical importance. These constraints still permit a flat CDM model, but only as long as $h < 0.5$ is allowed by observations of h. The f_G constraint means that large Ω_M implies fairly large Ω_B. Therefore the exponential term in Γ becomes important allowing $\Omega_M = 1$ to satisfy the Γ constraint. However, values of $\eta_{10} \approx 8 - 15$ are required (see Figures 1 and 2). The best-fit SCDM model has $h \approx 0.45$ and $\eta_{10} \approx 13$, which is grossly inconsistent with the predictions of BBN and the observed abundances of D, ^{4}He, and ^{7}Li. For $h > 0.5$ a fit to SCDM is no longer possible. The SCDM model is severely challenged.

The Γ and age constraints also challenge low-density CDM models. The Γ constraint permits $\Omega_M < 0.4$ only for high h, while the age constraint forbids high h, so $\Omega_M \gtrsim 0.4$ is required. The bound $\Omega_M \gtrsim 0.4$ conflicts with the added cluster constraint $\Omega_0 = 0.2 \pm 0.1$ at the 98% CL, suggesting strongly that there is additional mass not traced by light.

Although a few plausible variations on the CDM models do not affect the constraints very much, removing the Γ constraint would have a dramatic effect. Both high and low values of Ω_M would then be permitted. The Γ constraint plays a crucial role in our analysis.

At either low or high density, the situation remains about the same for the ΛCDM models. Because the ages are longer, we can tolerate $\Omega_M \approx 0.3$ for $h = 0.85$. The ΛCDM model therefore accepts more easily the added constraint $\Omega_0 = 0.2 \pm 0.1$. Improved future constraints on Ω_Λ will come into play here.

Having bounded the baryon density using data independent of constraints from BBN, we may explore the consequences for the light-element abundances. In general, our fits favor large values of η_{10} ($\gtrsim 6$) over small values ($\lesssim 2$). While such large values of the baryon density are consistent with estimates from the Ly-α forest, they do create some tension for BBN. For deuterium there is no problem, since for $\eta_{10} \gtrsim 6$ the BBN-predicted abundance, $(D/H)_P \lesssim 3 \times 10^{-5}$ (2σ), is entirely consistent with the low abundance inferred for some of the observed QSO absorbers (Tytler et $al.$, 1996; Burles and Tytler, 1996). Similarly, the BBN-predicted lithium abundance, $(Li/H)_P \gtrsim 2.5 \times 10^{-10}$ (2σ), is consistent with the observed surface lithium abundances in the old, metal-poor stars (allowing, perhaps, some minimal destruction or dilution of the prestellar lithium). However, the real challenge comes from ^{4}He where the BBN prediction for $\eta_{10} \gtrsim 6$, $Y_P \gtrsim 0.248$ (2σ), is to be contrasted with the H II region data which suggest $Y_P \lesssim 0.238$ (OS, OSS).

Acknowledgements

The research of G.S. at Ohio State is supported by DOE grant DE-FG02-91ER-40690. N.H. is supported by the National Science Foundation Contract No. NSF PHY-9513835. J.E.F. acknowledges fruitful visits to the Physics Department at Ohio State University and J.E.F. and G.S. acknowledge useful visits to the Aspen Center for Physics.

References

Bludman, S. A.: 1997, *ApJ*, submitted, astro/ph 9706047.

Bolte, M., and Hogan, C. J.: 1995, *Nature* **376**, 399.

Borgani, S., *et al.*: 1997, *New Astr.* **1**, 321.

Bureau, M., Mould, J. R., and Staveley-Smith, L.: 1996, *ApJ* **463**, 60.

Burles, S., and Tytler, D.: 1996, *ApJ* **460**, 584.

Carlberg, R. G.: 1997, private communication.

Carlberg, R. G., Yee, H. K. C., and Ellingson, E.: 1997, *ApJ* **478**, 462.

Carroll, S. M., Press, W. H., and Turner, E. L.: 1992, *ARA&A* **30**, 499.

Carswell, R. F., Rauch, M., Weymann, R. J., Cooke, A. J., and Webb, J. K.: 1994, *MNRAS* **268**, L1.

Cen, R.: 1997, *ApJ* **485**, 39.

Chaboyer, B., Demarque, P., Kernan, P. J., and Krauss, L. M.: 1997, *ApJ*, submitted, astro-ph/9706128.

Copi, C. J., Schramm, D. N., and Turner, M. S.: 1995, *Phys. Rev. Lett.* **75**, 3981.

Cowan, J. J., McWilliam, A., Sneden, C., and Burris, D. L.: 1997, *ApJ* **480**, 246.

D'Antona, F., Caloi, V., and Mazzitelli, I.: 1997, *ApJ* **477**, 519.

Dekel, A.: 1997, in *Galaxy Scaling Relations: Origins, Evolution and Applications*, ed. L. da Costa (Berlin: Springer), in press, astro-ph/9705033.

Dekel, A., and Rees, M. J.: 1994, *ApJ* **422**, L1.

Edmunds, M. G.: 1994, *MNRAS* **270**, L37.

Evrard, A. E.: 1997, *MNRAS* **292**, 289.

Evrard, A. E., Metzler, C. A., and Navarro, J. F.: 1996, *ApJ* **469**, 494.

Fixsen, D. J., Cheng, E. S., Gales, J. M., Mather, J. C., Shafer, R. A., and Wright, E. L.: 1996, *ApJ* **473**, 576.

Hancock, S., Davies, R. D., Lasenby, A. N., Gutierrez de la Crux, C. M., Watson, R. A., Rebolo, R., and Beckman, J. E.: 1994, *Nature* **367**, 333.

Hata, N., Scherrer, R. J., Steigman, G., Thomas, D., Walker, T. P., Bludman, S., and Langacker, P.: 1995, *Phys. Rev. Lett.* **75**, 3977.

Jimenez, R.: 1997, Invited lecture at the Cosmology School in Casablanca 97, astro-ph/9701222.

Kolatt, T., and Dekel, A.: 1997, *ApJ* **479**, 592.

Kundić, T., *et al.*: 1997, *ApJ* **482**, 75.

Liddle, A. R., Lyth, D. H., Roberts, D., and Viana, P. T. P.: 1996a, *MNRAS* **278**, 644.

Liddle, A. R., Lyth, D. H., Viana, P. T. P., and White, M.: 1996b, *MNRAS* **282**, 281.

Liddle, A. R., and Viana, P. T. P.: 1996, in *Aspects of Dark Matter in Astro- and Particle Physics* (Heidelberg, September 1996), ed. H. V. Klapdor-Kleingrothaus and Y. Ramachers (Heidelberg: World Scientific), in press, astro-ph/9610215.

Lineweaver, C. H., Barbosa, D., Blanchard, A., and Bartlett, J. G.: 1997, *A&A* **322**, 365.

Maddox, S. J., Efstathiou, G., and Sutherland, W. J.: 1996, *MNRAS* **283**, 1227.

Nittler, R. L., and Cowsik, R.: 1997, *Phys. Rev. Lett.* **78**, 175.

Olive, K. A., and Steigman, G.: 1995, *ApJS* **97**, 49. (OS)

Olive, K. A., Skillman E. D., and Steigman, G.: 1997, *ApJ* **483**, 788. (OSS)

Peacock, J. A.: 1997, *MNRAS* **284**, 885.

Peacock, J. A., and Dodds, S. J.: 1994, *MNRAS* **267**, 1020.

Prantzos, N.: 1996, *A&A* **310**, 106.

Rugers, M., and Hogan, C. J.: 1996, *ApJ* **459**, L1.

Songaila, A., Cowie, L. L., Hogan, C. J., and Rugers, M.: 1994, *Nature* **368**, 599.

Steigman, G., Schramm, D. N., and Gunn, J.: 1977, *Phys. Lett.* **B66**, 202.

Steigman, G., and Tosi, M.: 1995, *ApJ* **453**, 173.

Steigman, G., Hata, N., and Felten, J. E.: 1997, *ApJ*, submitted, astro-ph/9708016. (SHF)

Sugiyama, N.: 1995, *ApJS* **100**, 281.

Tammann, G. A., and Federspiel, M. 1997, in *The Extragalactic Distance Scale*, ed. M. Livio, M. Donahue, and N. Panagia (Cambridge: Cambridge Univ. Press), 137.

Tonry, J. L., Blakeslee, J. P., Ajhar, E. A., and Dressler, A.: 1997, *ApJ* **475**, 399.

Tosi, M.: 1996, in *From Stars to Galaxies: The Impact of Stellar Physics on Galaxy Evolution*, ed. C. Leitherer, U. Fritze-von Alvensleben, and J. Huchra, ASP Conf. Series 98 (San Francisco: ASP), 299.

Tytler, D., Fan, X.-M., and Burles, S.: 1996, *Nature* **381**, 207.

Walker, T. P., Steigman, G., Schramm, D. N., Olive, K. A., and Kang, H.-S.: 1991, *ApJ* **376**, 51.

Weinberg, S. 1972, *Gravitation and Cosmology* (New York: Wiley).

White, M. and Silk, J. I.: 1996, *Phys. Rev. Lett.* **77**, 4704; erratum ibid., **78**, 3799.

White, M., Viana, P. T. P., Liddle, A. R., and Scott, D.: 1996, *MNRAS* **283**, 107.

White, S. D. M., Navarro, J. F., Evrard, A. E., and Frenk, C. S.: 1993, *Nature* **366**, 429.

Address for correspondence: Professor G. Steigman, Department of Physics, The Ohio State University, 174 West 18th Avenue, Columbus, OH 43210, USA

STARS AND STELLAR SYSTEMS AT $z > 5$: IMPLICATIONS FOR STRUCTURE FORMATION AND NUCLEOSYNTHESIS

MARTIN J. REES

Institute of Astronomy, Madingley Road, Cambridge, CB3 OHA, UK

Abstract. Enough UV radiation was generated before $z = 5$ to have ionized the intergalactic medium. If this comes from stars (probably aggregated in systems of subgalactic scale), one straightforwardly calculates that the associated nucleosynthesis would be sufficient to produce a universal abundance of order 1 percent of solar. The first 'pre-galaxies' may eventually be detectable by their direct UV emission, with characteristic spectral features at Lyman alpha; high-z supernovae may also be detectable . Other probes of the IGM beyond $z = 5$, and of the epochs of reheating and reionization, are discussed, along with possible links between the diffusion of pregalactic metals and the origin of magnetic fields.

1. Introduction

The structures in our present universe are the outcome of more than 10 billion years of evolution. Slight irregularities imprinted at very early eras led to increasing contrasts in the density from place to place, until overdense regions evolved into bound structures. Any acceptable theory must account for the present clustering properties of galaxies and dark matter; it must also match the actual universe at all past eras that can be probed observationally.

We have for some years known about quasars with redshifts up to 5. But these objects are associated with atypical (even exceptional) galaxies, so their intrinsic properties are hard to relate to the general trend of galaxy formation. What has been especially exciting about recent developments is that the morphology and clustering of ordinary galaxies can now be probed out to similar redshifts: the powerful combination of HST and the Keck Telescope has now revealed many galaxies at $z > 3$. Also, the absorption features in quasar spectra (the Lyman forest, etc), probe the history of the clumping and temperature of a typical sample of the universe on galactic (and smaller) scales.

The mystery lies at still higher redshifts, between (in round numbers) a million years ($z = 1000$) and a billion years ($z = 5$). When the primordial radiation cooled below a few thousand degrees, it shifted into the infrared. The universe then entered a dark age, which continued until the first bound structures formed, releasing gravitational or nuclear energy that lit up the universe again. How long did the 'dark age' last? We now know that at least some galaxies and quasars had already formed by a billion years. But how much earlier did structures form, and what were they like?

In most cosmological theories, especially those that postulate adiabatic gaussian irregularities on the early universe, quasars and large galaxies should thin out

Space Science Reviews **84**: 43–53, 1998.
© 1998 *Kluwer Academic Publishers.*

beyond $z = 5$, but subgalactic structures may exist even at redshifts exceeding 10. This paper will briefly discuss the effects of the earliest stars and supernovae – especially the production of UV radiation and synthesis of the first heavy elements

2. Clustering in Hierarchical Models

I will focus on the cold dark matter (CDM) model (cf. Couchman and Rees, 1986; Ostriker and Gnedin, 1996, and references cited therein). But this is just a 'template' for more general deductions, which essentially apply to any 'bottom up' model for structure formation. There is no minimum scale for the aggregation, under gravity, of cold non-baryonic matter. However the baryons constitute a gas whose pressure opposes condensation on very small scales. The gas therefore does not 'feel' the very smallest condensations, but can aggregate in clumps of dark matter for which the escape velocity exceeds the sound speed. This corresponds to a dark-matter mass $M_J \simeq 10^6 \left(T_g/T_{rad}\right)^{3/2} M_\odot$.

During the 'dark age' the gas, with temperature T_g, became even cooler than the microwave background, with temperature T_{rad}: if it had cooled adiabatically, with no heat input since recombination, T_g would, at $z = 10$, have been below 5 K. The smallest bound structures, with mass $\sim M_J$, would have virialised at temperatures of a few times larger than T_g. Larger masses would virialise at temperatures higher by a further factor $\left(M/M_J\right)^{2/3}$. This virial temperature would be reached not solely by adiabatic compression, but also because of a shock, which would typically occur before the radius had decreased by a factor of 2.

These virialised systems would, however, have a dull existence as stable clouds unless they could lose energy and deflate due to atomic or molecular radiative processes – clouds that couldn't cool would simply remain in equilibrium, being later incorporated in a larger scale of structure as the hierarchy builds up. On the other hand, clouds that can cool radiatively will deflate. Most cooling mechanisms are more efficient at higher temperatures, as well as at higher densities. Once collapse starts, it proceeds almost isothermally, so that the internal Jeans mass falls as the density rises. A virialised self-gravitating cloud that can cool radiatively would eventually go into free-fall collapse, and (perhaps after a disc phase) fragment into smaller pieces.

Three 'cooling regimes' are relevant during successive phases of the cosmogonic process, each being associated with a characteristic temperature.

(i) For a H-He plasma the only effective cooling at low temperatures ($< 10^4$ K) comes from molecular hydrogen. Even this process cuts off below a few hundred degrees; but above that temperature it allows contraction within the cosmic expansion timescale. The H_2 fraction is never high, and it is in any case not a very efficient coolant – indeed systems that collapse at $z < 10$ fail to form enough molecules for effective cooling (eg. Fig. 1 of Tegmark et al., 1997) but molecular cooling almost certainly played a role in forming the very first objects that lit up the universe.

(ii) But even at high redshifts, H_2 cooling would be quenched if there were a UV background able to dissociate the molecules as fast as they formed. Photons of $h > 11.18$ eV can photodissociate H_2, as first calculated by Stecher and Williams (1967). These photons can penetrate a high column density of HI and destroy molecules in virialised and collapsing clouds. (If the incident spectrum has a non-thermal component extending up to keV energies, due, for instance, to early supernova remnants, there is a counterbalancing positive feedback because the number of photoelectrons is increased, and this enhances molecule formation.) If molecular cooling is ineffective, then a H-He mixture behaves adiabatically unless T is as high as 8-10 thousand degrees, when excitation of Lyman α by the Maxwellian tail of the electrons provides efficient cooling whose rate rises steeply with temperature. Because of this steep temperature-dependence, gas in this regime contracts almost isothermally, so that its Jeans mass decreases.

(iii) The UV from early stars will photoionize some (and eventually almost all) of the diffuse gas. When this happens, the HI fraction is suppressed to a very low level, so there is is no cooling by collisional excitation of Lyman lines; moreover the energy radiated whenever a recombination occurs is quickly cancelled by the energy input from a photoionization, so the only net cooling is via bremsstrahlung. The cooling is, in effect, then reduced by a factor of ~ 100 (Efstathiou, 1992). The minimum temperature (below which there is a net heating from the UV) depends on the UV spectrum, and on whether the He is doubly ionized: it is in the range 20-40 thousand degrees. When this third phase is reached, the thermal properties of the uncollapsed gas will resemble those of the structures responsible for the observed Lyman-forest lines in high-z quasars spectra – these are mainly filaments, draining into virialised systems. Such systems have velocity dispersions of 50 km/sec, and are destined to turn into galaxies of the kind whose descendents are still recognisable.

(These three regimes refer to a H He plasma. When heavy elements are present they can dominate the low-T cooling; ionization is still important in suppressing the most efficient channels for cooling.)

There are thus three stages in the build-up of hierarchical structure, characterised by different masses and virial temperatures. They also occur at different epochs – however, the demarcation is unlikely to be sharp because the range of amplitudes (for gaussian fluctuations) translates into a broad spread of turnaround times for a given mass scale.

3. Detection of Pregalaxies at $z > 5$

The UV that reheated and photoionized the IGM before $z = 5$ came primarily from stars that formed in dark matter clumps with total mass of order, or greater than, 10^8 solar masses, and internal velocity dispersion $\sigma \gtrsim 15$ km/sec. (There could have been a few precursors of smaller mass, but these would have generated enough UV to inhibit molecular cooling – i.e. to cause the transition from stage (i) to stage (ii) above – before most of the medium had been ionized.) What is the chance of detecting these ancient stellar systems?

The total integrated UV production beyond $z = 5$ must have been enough to ionize the IGM and build up the UV background whose strength can be inferred directly out to $z = 5$ from models of the Lyman forest, etc., as discussed by Tytler and Sargent in these proceedings. We therefore have a firm lower limit to the ionizing UV generated. The total amount could be substantially above this limit, because much could 'go to waste' through reprocessing in dense clouds, local absorption in the sources, etc.

Corroboration that O and B stars formed at these early eras comes from the heavy element abundance in high-z absorption clouds. The diffuse gas in the Lyman α forest, even at redshift $z = 3$, is enriched to a heavy element abundance $Z \sim 10^{-2} Z_\odot$. The evidence comes from absorption lines of CIV and other species associated with Lyman α forest lines with $N_{\rm HI} \gtrsim 10^{14}\,{\rm cm}^{-2}$ (Songaila and Cowie, 1996, and references therein). Detailed calculations of the expected column densities of the observed absorption lines, using hydrodynamic simulations of the Lyman α forest and realistic models for the spectrum of the ionizing background (Rauch, Haehnelt, and Steinmetz, 1996; Hellsten *et al.*, 1997) have shown that the carbon abundance needed to reproduce the observations is $[{\rm C/H}] = -2.5$. However, the metal abundances are similar to those of Population II stars, where oxygen is the most abundant element and is overabundant by a factor ~ 2 relative to carbon. With $Z_\odot = 0.02$, the metallicity of the Lyman α forest is then $Z \simeq 10^{-4}$. This metallicity should be approximately the same as the mean metallicity of the universe at $z = 3$ if the Lyman α forest contains most of the baryons, as is found to be the case in models of structure formation for the Lyman α forest similar to those analyzed in Hernquist *et al.* (1996) and Miralda Escudé *et al.* (1996). The mean metallicity of the universe could be significantly lower only if most of the baryons were metal-free and in $N_{\rm HI} < 10^{14}\,{\rm cm}^{-2}$ absorption systems, or in a very diffuse, unobserved intergalactic medium.

The ratio of the mass of heavy elements ejected by a star to the energy in ionizing photons emitted over the lifetime of the star, as derived from models of stellar evolution and supernova explosions, turns out to be about constant over the relevant mass range $10 M_\odot \lesssim M \lesssim 50 M_\odot$, so given a mean metallicity $\bar{Z}$ we can predict the energy in ionizing photons that was emitted for each baryon in the universe. According to Madau and Shull (1996), this energy is $0.002 \bar{Z} m_p c^2$ per baryon. To produce the heavy elements needed to provide a universal abundance $\bar{Z} = 10^{-4}$, would require about 10 ionizing photons per baryon (assuming a mean energy of 20 eV per ionizing photon).

At first sight, this seems to imply more UV than is inferred from the early ionization of the IGM. However, this ionization didn't just require one UV photon per baryon. Extra photons were needed to balance recombinations in dense clouds where there can be several recombinations during the reionization epoch (as in, for instance, the observed Lyman limit systems), or in the 'pregalaxies' themselves. Others may be absorbed by dust. Overall, one should be impressed by the gratifying consistency between the inferred UV background and the high-z metallicity. (In more radical pictures where black holes are involved in the early energy input, there could of course be abundant UV production without associated nucleosynthesis, because the energy supply would then be gravitational).

AGN formation requires virialised systems with larger masses and deeper potential wells (cf. Haehnelt and Rees, 1993). AGNs may 'take over' as the dominant UV source at redshifts below 5 (and the second ionization of He may be delayed until AGNs can provide a power-law contribution to the spectrum), but at higher z the OB stars would dominate.

Jordi Miralda-Escudé and I (Miralda Escudé and Rees, 1997a) have recently considered whether the 'pregalaxies' could feasibly be detected. The above considerations fix the average surface brightness of the sky due to 'pregalaxies'. (The uncertainties arise from the unknown role of dust, and the uncertain efficiency with which ionizing photons are converted into Lyman α in the pregalaxies themselves.) This surface brightness is the product of the number of galaxies per solid angle and their individual, mean flux. But the ease of detection obviously depends crucially on whether the UV background comes from huge numbers of individually ultra-faint systems, or (more optimistically) at least in part from systems of higher luminosity. The uncertainty here is much greater – it is essentially tied to the efficiency of forming massive stars in low-mass dark matter halos with shallow potential wells.

If star formation were highly efficient in all collapsed halos with velocity dispersion $\sigma \gtrsim 15\,\mathrm{km\,s^{-1}}$, corresponding to the virial temperature $T = \mu\sigma^2/k \simeq 10^4$ K , then reionization would take place as soon as the highest density peaks on the scales of these low velocity dispersions collapse.

For a top-hat spherical region of mass M that turns around when the age of the universe is $t_f/2$, the relation between mass and velocity dispersion is $M = 2^{3/2}/(2\pi G)\sigma^3 t_f \sim 10^9 M_\odot (\sigma/20\,\mathrm{km\,s^{-1}})^3 (t_f/10^9\,\mathrm{yr})$. If all the baryons turned into stars over a timescale of order the free-fall time of the halo, the star formation rate would be $0.3 M_\odot\,\mathrm{yr}^{-1}(\sigma/20\,\mathrm{km\,s^{-1}})^3)$. For a normal IMF, this can yield a luminosity close to the peak luminosity of a supernova. If the sources that reionized the universe have these characteristics, then we can use the mean Lyman α surface brightness we obtained earlier to conclude that the number density of these sources in the sky should be ~ 1 per arcsec^2 for the fiducial numbers we have chosen, and a formation redshift $z_f \sim 10$. The redshift z_f should of course depend on the detailed model for the amplitude of the primordial fluctuations on small scales.

The main problem with this scenario is that supernova explosions in galaxies of such low velocity dispersions may well expel the gas before more than a small fraction has turned to stars. This leaves us with two other options regarding the sources of the reionization photons:

(i) If just a small fraction of the gas in each object turns to stars before the remainder is expelled in a wind, then each galaxy emits fewer photons. More therefore have to form, from lower amplitude peaks. The luminosity of each galaxy would be reduced. Or

(ii) Star formation may be so inefficient in the low-escape-velocity halos that reionization has to await the formation of more massive galaxies (with $\sigma \gtrsim 100$ km s^{-1}.

Some moderately massive galaxies are in any case expected to form before reionization, because the power spectrum at small scales in cold dark matter models flattens to a slope of about $n = -2.7$ (so the amplitude of fluctuations decreases very slowly with scale), implying that at the same epoch, halos with $\sigma = 70$ km s^{-1} (with a mass ~ 40 times larger than halos with $\sigma = 20$ km s^{-1}) would be collapsing from $3 - \sigma$ peaks on this larger scale. In a gaussian theory, the $3 - \sigma$ peaks should contain $\sim 10\%$ as much mass as the $2 - \sigma$ peaks at a fixed epoch, so we see that the mass distribution of the pregalaxies should probably extend well above the minimum mass for efficient atomic cooling.

The main spectral feature that should identify any such galaxies at $z \gtrsim 5$ is a sharp break of the UV continuum at the Lyman α wavelength, due to the Gunn-Peterson trough (Gunn and Peterson, 1965); in addition, the Lyman α emission line may be present, depending on dust absorption and scattering of the Lyman α photons. Notice that the redshift at which the IGM was reionized is not highly relevant regarding the presence of the Gunn-Peterson trough, because even if the medium was reionized at $z \gg 5$, we know that the flux decrement caused by the Lyman α forest reaches a factor of 2 at $z \simeq 4$ and grows rapidly with redshift. Thus, the technique of identifying galaxies at $z \simeq 3$ from the Lyman continuum break (Guhathakurta et al., 1990; Steidel et al., 1996) should be replaced by the Gunn-Peterson trough at $z \gtrsim 5$ (see Madau, 1995, and Madau et al., 1996, for a careful analysis of the effects of the Lyman α forest on galaxy colors).

If there is one galaxy per arcsec^2 at redshifts $z_f \sim 10$, each galaxy would have an AB magnitude of 32, i.e., about 3 magnitudes fainter than the detection limit in the *Hubble Deep Field* (HDF) for the I-band (Williams et al., 1996). However, as mentioned above, galaxies with a mass 40 times larger would be ~ 400 times rarer, which still implies a number density of a few galaxies per square arc minute with AB magnitudes of ~ 28 [with the optimistic assumption of a rapid and efficient starburst in these more massive galaxies].

Detection with ground-based telescopes should be possible at $z < 6$. The Keck telescope can detect point sources to AB magnitudes $R < 28$, and $I < 27$, in a night of observing time (M. Rauch, 1997, priv. communication; Cohen, 1995a,b). Galaxies up to $z = 6$ can be detected in the I band; for higher redshift, the magnitude limit from a ground-based telescope degrades rapidly due to the high atmospheric background at $\lambda > 8500$ Å. The magnitude limit could be improved significantly with adaptive optics. With space instruments, the faintest galaxies in the HDF reach to $I \simeq 28.5$ (Williams et al., 1996). The *New Generation Space Telescope* could image galaxies to AB magnitudes ~ 31 in the near-infrared, and should be able to detect galaxies to much higher redshift (Mather and Stockman, 1996), with a much higher number density than has been seen so far. The prospect for detecting the high-redshift galaxies responsible for the enrichment of the Lyman α forest has also been analyzed by Haiman and Loeb, (1997b).

In general, the lensing magnification in rich lensing clusters may be used to stretch the magnitude limit. As an example, a lensing cluster with an Einstein ring radius $b = 30''$ should magnify to $A > 10$ an area of ~ 30 arcsec^2 in the source plane. In the example used above, about 30 galaxies with $AB = 32$ could be in this area, which would be magnified to $AB = 29.5$. Magnified images of high-redshift galaxies should characteristically appear in pairs around the critical lines, in a region that can be predicted from lensing models (see Miralda-Escudé and Fort, 1993; Kneib et al., 1996, and references therein), so this should help in their identification. These numbers indicate that a new deep field (similar to the HDF) imaged with HST in a rich cluster, adding also the H and J filters in the near-infrared, might well identify several galaxies at $z > 5$. In fact, the largest redshift object known at present (at $z = 4.92$) is already a gravitationally lensed galaxy (Franx et al., 1997).

The foregoing discussion has assumed that these high-redshift galaxies would be sufficiently small to remain unresolved. Resolved objects would need to have higher fluxes to be detected, since the detection is limited by the sky background. The likely scales are 100 pc, corresponding to angular sizes ~ 0.01 arcsec – not resolvable even with NGST.

The Lyman α emission line would contain about 10 percent as many photons as the stellar UV continuum in the range 1000-1300 A. If the sensitivity for detecting galaxies were limited by the sky background for emission-line searches, line searches would have an advantage if $\lambda/\lambda < 0.01$. But the line is actually likely to be so broadened by multiple scattering that this offers little advantage over a continuum search. (However, the distinctive shape of the Gunn-Peterson edge at the redshifted Lyman α frequency would be an interesting probe of the IGM. When the surrounding IGM is still mostly neutral, even photons on the red side can be scattered due to the damping wing of the Gunn-Peterson trough (Miralda-Escud/'e and Rees, 1997a). The edge of the Gunn-Peterson trough should have a distinctive shape).

4. Very High Redshift Supernovae (and Gamma-Ray Bursts?)

The intergalactic gas was already highly photoionized by $z = 5$; also, the mean abundance of heavy elements had attained a level about 0.01 solar by that time, this degree of contamination being about what would be expected if the reheating and ionization were due to OB stars (cf. Haiman and Loeb, 1997a; Gnedin and Ostriker, 1997).

It is straightforward to calculate how many supernovae would have gone off, in each comoving volume, as a direct consequence of this output of UV and heavy elements: there would be one, or maybe several, per year in each square arc minute of sky (Miralda-Escudé and Rees, 1997b). These would be primarily of Type II. The typical observed light curve has a flat maximum lasting 80 days. One would therefore (taking the time dilation into account) expect each supernova to be near its maximum for nearly a year. It is possible that the explosions proceed differently when the stellar envelope is essentially metal-free, yielding different light curves, so any estimates of detectability are tentative. However, taking a standard Type II light curve (which may of course be pessimistic), one calculates that these object should be approx 27th magnitude in J and K bands even out beyond $z = 5$. The detection of such objects would be an easy task with the NGST. With existing facilities it is marginal. The best hope would be that observations of clusters of galaxies might serendipitously detect a magnified gravitationally-lensed image from far behind the cluster.

As a speculative addendum I note that a few percent of observed gamma-ray bursts may come from redshifts as large as 5: this would be expected if the burst rate, as a function of cosmic epoch, tracks the star formation rate. At the time of writing, data on optical and X-ray afterglows is still very sparse, but it is at least an exciting possibility (cf. Wijers *et al.*, 1997) that there may be occasional flashes, far brighter than supernovae. from very large redshifts.

5. The First Stars: Some Uncertainties

The gravitational aspects of clustering can all be modeled convincingly by computer simulations. So also, now, can the dynamics of the baryonic (gaseous) component – including shocks and radiative cooling. But the nature of the simulation changes as soon as the first stars (or other compact objects) form. The first stars exert crucial feedback – the remaining gas is heated by ionizing radiation, and perhaps also by an injection of kinetic energy via winds and even supernova explosions. The three uncertainties here are:

(i) What is the IMF of the first stellar population? The high-mass stars are the ones that provide efficient (and relatively prompt) feedback. It plainly makes a big difference whether these are the dominant type of stars, or whether the initial IMF rises steeply towards low masses, so that very many faint stars form before there is a significant feedback.

(ii) The influence of the early stars depends on whether their energy is deposited locally or penetrates into the medium that is not yet in contracting systems. The UV radiation could, for instance, be mainly absorbed in the gas immediately surrounding the first stars, so that it exerts no feedback on the condensation of further clumps – the total number of massive stars needed to build up the UV background, and the concomitant contamination by heavy elements, would then be greater.

(iii) Quite apart from the uncertainty in the IMF, it is also unclear what fraction of the baryons that fall into a clump would actually be incorporated into stars before being re-ejected. The retained fraction would almost certainly depend on the virial velocity: gas more readily escapes from shallow potential wells. Ejection is even easier in potential wells so shallow that they cannot confine gas at the photoionization temperature.

All these three uncertainties would, for a given fluctuation spectrum, affect the rate at which, for given initial conditions, nucleosynthesis and UV production would build up.

6. Heavy Elements and Magnetic Fields

It is interesting to consider how the processed gas would be distributed. Would it be confined in the virialised systems, or could it spread through the entire IGM?

The ubiquity of carbon features in intermediate and high ($N > 3.10^{14} cm^{-2}$) column density absorption systems implies that heavy elements are broadly enough dispersed to have a large covering factor. These absorption systems may be associated with the subgalactic sites of star formation that produce the first heavy elements. The nucleosynthesis sites cannot therefore be too sparse if these elements are, within the time available, to diffuse enough so that they are encountered somewhere along every line of sight through a typical high-column-density cloud. The absorption line data tell us, however, only the mean abundance through the relevant cloud. They are compatible with 99 percent of the material being entirely unprocessed, and the heavy elements being restricted to 1 percent of the material with a full 'solar' abundance – the early heavy elements need not be thoroughly mixed, but they must have spread sufficiently to have a large 'covering factor' in the intermediate- and high-N clouds.

The first stars are important for another reason: they may generate the first cosmic magnetic fields. Moreover, mass loss (via winds or supernovae permeated by magnetic flux) would disperse magnetic flux along with the heavy elements. This flux, stretched and sheared by bulk motions, can be the 'seed' for the later amplification processes that generate the larger-scale fields pervading disc galaxies. The diffusion of the first magnetic flux, and of the first heavy elements, may be linked phenomena.

7. Where Are the Oldest Stars?

The efficiency of early mixing is important for the interpretation of stars in our own galaxy that have ultra-low metallicity – lower than the $Z \simeq 10^{-4}$ that would have been generated in association with the UV background at $z > 5$. If the heavy elements were efficiently mixed, then these ultra-metal-poor stars would themselves need to have formed before galaxies were assembled. To a first approximation they would subsequently cluster non-dissipatively; they would therefore be distributed in halos (including the halo of our own Galaxy) like the dark matter itself. More careful estimates slightly weaken this inference. This is because the subgalaxies would tend, during the subsequent mergers, to sink via dynamical friction towards the centres of the merged systems. There would nevertheless be a tendency for the most extreme metal-poor stars to have a more extended distribution in our Galactic Halo, and to have a bigger spread of motions.

The number of such stars depends on the IMF. If this were flatter, there would be fewer low-mass stars formed concurrently with those that produced the UV background. If, on the other hand, the IMF were initially steeper, there could in principle be a lot of very low mass (macho) objects produced at high redshift. These could be distributed like the dark matter. They could provide a few percent of the halo if omega were 1; a larger proportion in a low-density universe.

The UV background comes primarily from high-mass stars. These have short lifetimes, and are essentially the same that produce the primary heavy elements.

Most of these galaxies should by now have merged into more massive systems (forming part of the halo population of stars in normal galaxies like the Milky Way), but some could survive until today in galactic halos, or even as isolated objects. This may be the explanation of present-day dwarf spheroidal galaxies.

Acknowledgements

I am especially grateful to Jordi Miralda-Escudé, and would also like to thank Zoltan Haiman, Avi Loeb, and Max Tegmark, for discussion and collaboration on some of the topics described here.

References

Cen, R., Miralda-Escudé, J., Ostriker, J. P., and Rauch, M.: 1994, *ApJ* **437**, L9.
Cohen, J. G.: 1995a, *Keck LRIS Quick Reference Guide* (Caltech).
Cohen, J. G.: 1995b, *The Efficiency of the LRIS in the Spectroscopic Mode* (Caltech).
Couchman, H.M.P. and Rees, M.J.: 1986, *MNRAS* **221**, 53.
Efstathiou, G.: 1992, *MNRAS* **256**, 43p.
Franx, M., Illingworth, G. D., Kelson, D. D.: 1997, *ApJ*, in press.
Guhathakurta, P, Tyson, J. A., and Majewski, S. R.: 1990, *ApJ* **357**, L9.
Gunn, J. E., and Peterson, B. A.: 1965, *ApJ* **142**, 1633.
Haehnelt, M., and Rees, M.J.: 1993, *MNRAS* **263**, 168.
Haiman, Z., and Loeb, A.: 1997a, *ApJ* **476**, 458.
Haiman, Z., and Loeb, A.: 1997b, to appear in the *Proceedings of Science with the Next Generation Space Telescope* (astro-ph 9705144).
Haiman, Z., Rees, M. J., and Loeb, A.: 1997, *ApJ* **476**, 458.
Hellsten, U., Dave, R., Hernquist, L., Weinberg, D. H., and Katz, N.: 1997, *ApJ*, submitted (astro-ph 9701043).
Hernquist, L., Katz, N., Weinberg, D. H., and Miralda-Escudé, J.: 1996, *ApJ* **457**, L51.

Kneib, J. P., Ellis, R. S., Smail, I. R., Couch, W. J., and Sharples, R.: 1996, *ApJ* **471**, 643.

Madau, P.: 1995, *ApJ* **441**, 18.

Madau, P., Ferguson, H. C., Dickinson, M. E., Giavalisco, M., Steidel, C. C., and Fruchter, A.: 1996, *MNRAS* **283**, 1388.

Mather, J., and Stockman, H.: 1996, NASA Report.

Miralda-Escudé, J., Cen, R., Ostriker, J. P., and Rauch, M.: 1996, *ApJ* **471**, 582. (MCOR)

Miralda-Escudé, J., and Fort, B.: 1993, *ApJ* **417**, L5.

Miralda-Escudé, J., and Rees, M. J.: 1997b, *ApJ* **478**, L57.

Ostriker, J. P., and Gnedin, N.: 1996, *ApJ* **472**, 630.

Rauch, M., Haehnelt, M. G., and Steinmetz, M.: 1997, *ApJ*, in press, (astro-ph 9609083).

Songaila, A., and Cowie, L. L.: 1996, *AJ* **112**, 335.

Stecher, T.P., and Williams, D.A.: 1967, *ApJ* **149**, L1.

Tegmark, M., Silk, J., Rees, M. J., Blanchard, A., Abel, T., and Palla, F.: 1997, *ApJ* **474**, 1.

Thoul, A., and Weinberg, D. H.: 1996, *ApJ* **465**, 608.

Tytler, D., Fan, X.-M., and Burles, S.: 1996, *Nature* **381**, 207.

Wijers, R.A.M.J., *et al.*: 1997, *MNRAS*, in press.

Williams, R. E., *et al.*: 1996, *AJ* **112**, 1335.

BIG BANG THEORY AND PRIMORDIAL NUCLEI

T.P. WALKER

Department of Physics, Department of Astronomy, The Ohio State University,
Columbus, Ohio, U.S.A.

Abstract. A summary of the discussions of Working Group 1 (Rees, Reeves, Schramm, Steigman, Tammann, and TPW) on Big Bang Theory as it relates to primordial nuclei and their galactic evolution.

1. Introduction

Let me start by thanking the organizers for setting up a very successful workshop and by thanking the members of my Working Group for their many insightful comments during our meetings. The Big Bang Theory input to the field of primordial nuclei can be viewed on two fronts: (1) the Big Bang nucleosynthesis (BBN) production of light elements and their subsequent evolution in stars and/or galaxies, and (2) other measures of the baryon density of the Universe. The general constraints that BBN places on the baryon density are in good agreement with independent measures of the baryon component of the Universe. Key to both of these issues are the abundances of primordial nuclei. In this review, I discuss the theoretical issues we believe to be of importance to those who measure the abundances of light elements (D, He, Li, Be, and B).

2. Big Bang Nucleosynthesis

The rules in BBN are well defined: armed with measurements of the MWB temperature (Fixsen *et al.*, 1996) and of the Hubble constant (see Tammann, 1998, or Freedman, 1997), measurements of the light elements D, ^{3}He, ^{4}He, and ^{7}Li can in principle be used to determine the baryon density of the Universe. This is because the predictions of standard BBN are uniquely determined by one parameter, η, the baryon-to-photon ratio: $\eta_{10} = 273\Omega_B h^2$ (Ω_B is the ratio of the baryon density to the critical density and the Hubble parameter is $H_0 = 100h$ km/s/Mpc; $\eta_{10} = 10^{10}\eta$). For standard BBN the early universe is assumed to have been homogeneous and expanding isotropically and the energy density at the time nucleosynthesis begins (about 1 second after the Big Bang) is described by the standard model of particle physics ($\rho_{tot} = \rho_\gamma + \rho_e + N_\nu\rho_\nu$, where ρ_γ, ρ_e, and ρ_ν are the energy density of photons, electrons and positrons, and massless neutrinos and anti-neutrinos (one species), respectively, and N_ν is the equivalent number of massless neutrino species which, in standard BBN is exactly 3). In Fig. 1 (Hata *et al.*, 1997) the primordial abundances predicted by standard BBN are shown as a function of η. The width of each curve reflects the 2σ uncertainty in the predictions. Nothing has significantly

Space Science Reviews **84**: 55–62, 1998.

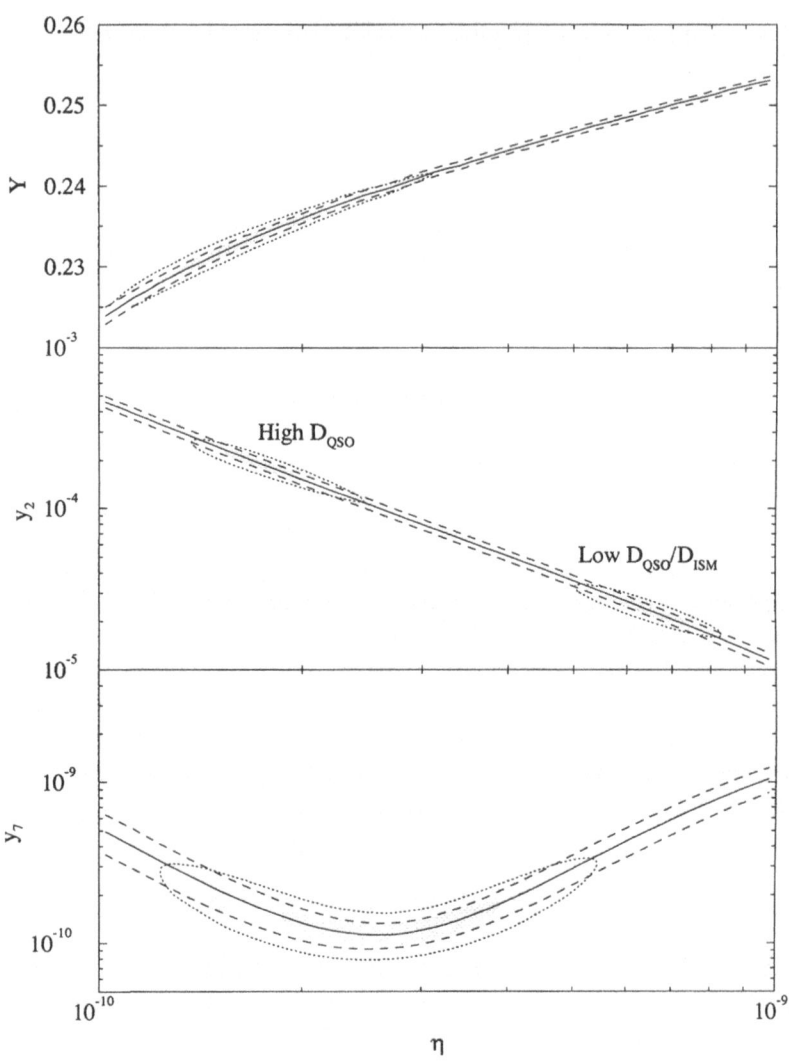

Figure 1. The predicted abundances of primordial nuclei as a function of the baryon-to-photon ratio η. Also shown are the inferred primordial abundances (see text for discussion).

changed in the theoretical predictions since Walker *et al.* (1991) and that is not surprising – the whole game in BBN is in extracting the primordial abundances (that have undergone astrophysical processing) rom observational data. In that arena there are advances on all fronts and all of them are positive except for ^{3}He, where our lack of understanding of stellar production and destruction has been backed up by the data (Rood *et al.*, 1998). So let's go through the current status of the primordial abundances of D, ^{4}He, and ^{7}Li.

2.1. DEUTERIUM

Any good measurement of primordial deuterium is a very accurate diagnostic of the baryon density of the Universe. Thus begat Schramm's "Deuteronomy" (see Schramm, 1998, and Fig. 1). In the not too distant past (see Walker *et al.*, 1991, for example), the primordial deuterium abundance was inferred from the pre-solar nebula and/or the ISM deuterium abundance and chemical evolution models which, for the purposes of deuterium destruction, need only give us the fraction of gas that has been processed through stars (for a certain set of chemical evolution models, this is shown as D_{ISM} in Fig. 1). Over the last year or so, methods for determining the primordial deuterium abundance have radically changed with the advent of deuterium measurements in QSO absorption line systems (QSOALS) (Tytler *et al.*, 1996; Burles and Tytler, 1996; Tytler *et al.*, 1996; Tytler *et al.*, 1997) (Songaila *et al.*, 1994; Carswell *et al.*, 1994; Rugers and Hogan, 1996a; Rugers and Hogan, 1996b; Songaila *et al.*, 1997; Webb *et al.*, 1997). Potentially, QSO deuterium provides the cleanest BBN abundance measurement in terms of the relative lack of contamination of the absorbing high redshift cloud. The only unfortunate bit is that the QSOALS observers are straddling the fence and, for the moment, theorists are stuck with (at least) two QSO deuterium abundances. It is quite difficult to live in a world where both abundances are consistent with a primordial origin(Copi *et al.*, 1996; Jedamzik and Fuller, 1996). In Fig. 1 I show the two QSO deuterium abundances. The consequence of either QSO deuterium being primordial has been much discussed, both before this meeting and within our Working Group – the politically correct conclusion is that BBN is "observationally challenged" and that's a good thing. Pinning down a single QSO deuterium abundance will allow us (via BBN) to predict the primordial abundances of ^{4}He and ^{7}Li, using BBN to say something about stellar and chemical evolution rather than the other way around. And clearly a matching set of deuterium abundances – primordial, pre-solar, and ISM – will provide constraints on models of Galactic chemical evolution (see Tosi, 1998).

So, with the importance of QSO deuterium, to what can we look forward? Under the right TAC conditions, the Sloan Sky Survey can provide 10–100 D/H systems and with a sample that large, we'll either know the primordial deuterium abundance quite well or we'll realize that the high redshift QSOALS are much more complicated than we thought.

2.2. HELIUM-4

Let's suppose, just for the sake of argument, that the primordial abundance of deuterium is the low one: D/H $\sim 2 \times 10^{-5}$. From Fig. 1 we see that this corresponds to a primordial ^{4}He mass fraction of $Y_p \approx 0.249$. Now we are "observationally challenged" because the combined data sets of Pagel *et al.* (1992) and Skillman *et al.* (1993; 1994; 1997), as extrapolated by Olive, Skillman, and Steigman (1996), yield $Y_p = 0.230 \pm 0.003$. Without extrapolating, Hogan, Olive, and Scully (1997) bayesianly obtain $Y_p \leq 0.243$ with the same data. That is, the predicted primordial ^{4}He is too large compared to this data set. The data of Izotov, Lipovetsky, and Thuan (1994; 1996) simply does not go to low enough metallicity and, as was shown by Olive, Skillman, and Steigman (1996), over the relevant metallicity range both data sets are consistent (Y_p by extrapolation with all the data is 0.234 ± 0.002). [*Note added post-Bern*: there is now additional data from Izotov and Thuan which, when added to their previous data, yields a data set comparable to that used by Olive, Skillman, and Steigman. The extrapolation of the Izotov and Thuan data set still yields $Y_p \sim 0.243$ (see Thuan, 1998)].

If the conflict between deuterium and helium persists we will be forced to alter standard BBN and, in particular, its predictions for primordial ^{4}He. A toy model which nicely contains ways to both increase and decrease ^{4}He production relative to standard BBN is the case of a massive ν_τ (Kawasaki *et al.* (1994) and references therein). The two relevant parameters are the ν_τ mass and lifetime. A ν_τ which is stable on BBN timescales (*i.e.*, $\tau_\nu \gtrsim 100$ s) and has a mass greater than a few MeV will increase Y_p relative to standard BBN. This is because such a neutrino still decouples when it is fairly relativistic and so its number density is comparable to that of a massless neutrino. However, its energy density at the onset of BBN is much greater than that of a massless neutrino since its mass is significantly greater than the temperature. Therefore, weak interactions decouple earlier, increasing the neutron-to-proton ratio at freeze out and thus the amount of ^{4}He. Just the opposite can occur if such a ν_τ decays rapidly compared to BBN timescales. The rapid decays and inverse decays keep the ν_τ's in equilibrium much longer than conventional weak interactions and their number density is exponentially suppressed, right along with their energy density. A typical example is a relative decrease in Y_p of about 0.01 for a ν_τ with a mass of 20 MeV and a lifetime ($\nu_\tau \rightarrow \nu_\mu + \phi$ where ϕ is a Majoron) of 0.1 s.

It is worth noting that establishing a bound on the energy density at the BBN epoch [the effective number of massless neutrinos à la Steigman, Schramm, and Gunn (1977)] requires two ingredients: (1) a lower bound to η (either from deuterium or perhaps ^{7}Li) and (2) an upper bound to the observed Y_p. Although some of my colleagues would disagree with me, I believe all bets are off in light of the problems between deuterium and ^{4}He outlined above. Olive *et al.* (1996), have proposed ignoring deuterium for the time being, which by the way is a dangerous thing to do, and use ^{7}Li to bound η. I discuss the primordial ^{7}Li abundance below.

2.3. LITHIUM-7

In order to determine the primordial abundance of ^{7}Li, we must understand the depletion of lithium in Pop II halo stars. Since there have been many talks on this in other sessions, I will simply summarize the relevant information here. The folklore on the lack of lithium depletion in Pop II halo stars is entirely based on the "Spite plateau" (Spite and Spite , 1982) – a nearly constant lithium abundance is observed for sufficiently warm metal-poor halo stars. Lithium is destroyed in stars by (p, α) reactions when the temperature is a few million degrees. As stars collapse to the main sequence their convection zones shrink as their core temperatures increase. The temperature at the base of the convective zone (as well as its final size) is a sensitive function of stellar mass (or effective temperature) and thus the presence of a plateau in lithium abundance vs. effective temperature indicates that convective burning was most likely inoperative for the plateau stars, *i.e.*, that plateau abundance is very close to the primordial abundance. This is precisely what the original Yale models (Deliyannis *et al.*, 1989) showed. This folklore is reflected in our choice for the primordial ^{7}Li abundance shown in Fig. 1. Note that the "valley" corresponds to the plateau abundance and the allowed upper slopes represent our (Hata *et al.*, 1995) best estimate for possible stellar depletion.

In addition, recall that the Yale group also claimed that by including rotational mixing as much as a factor of 10 lithium depletion could occur in Pop II halo stars (Pinsonneault *et al.*, 1992). The question was still whether or not such models could reproduce a tight plateau and the answer was that they could, but only if halo stars had very similar angular momenta histories. Recently Pinsonneault, Steigman, Narayanan, and myself (1997) have considered rotationally mixed models with more realistic angular momenta distributions as drawn from the Pleiades. If we adopt the same initial angular momenta distribution for halo stars as is *required* to understand lithium depletion in open clusters (where the lithium abundance decreases as a function of age) we reproduce a flat plateau with the observed dispersion and a factor of 2–3 lithium depletion. As can be seen from Fig. 1, the move out of the lithium valley further clouds the BBN consistency picture – primordial lithium now supports both high and low deuterium.

Of major importance to further constraining lithium depletion in Pop II stars are observations of ^{6}Li, Be, and B in these same stars (see Duncan, 1998). Each of the LiBeB isotopes are burned at slightly different temperatures narrowly bracketing the temperature of ^{7}Li burning. A complete set of all five LiBeB isotopes observed in a handful of stars will drastically enhance our search for the primordial lithium abundance. For example, the one Pop II star with a measured ^{6}Li abundance (HD 84937) (Hobbs and Thorburn, 1997) was used in the Pinsonneault *et al.* (1997) analysis to further delineate the amount of lithium destruction.

3. The $_B$ Inventory

In addition to BBN, there are several other methods which can measure the baryon density: dynamical probes (such as X-ray clusters and the shape parameter), the Ly-α forest, and the MWB, can all be used to probe $_B$. I'll briefly discuss them here and refer you to my Working Group co-member's workshop presentations for further details. The executive summary is that considerations other than BBN point

T. WALKER

to higher baryon densities which are consistent with the low D_{QSO} measurements but inconsistent with the helium-4 data and high D_{QSO}.

As discussed above, the best BBN can do is provide two estimates of $_B$: (1) $_B h^2 \sim 0.005$ consistent with high D_{QSO} measurements and the primordial helium-4 abundance, and (2) $_B h^2 \sim 0.02$ consistent with the low D_{QSO} measurements but inconsistent with the helium-4 data (however, see Thuan, 1998).

As discussed by Steigman (1998) and in Steigman et al. (1997), dynamical constraints on $_B$ can be obtained from five observables (for a review of cosmological parameters, see Tammann, 1998): the Hubble parameter, the age of the universe, the mass density as measured by clusters or large scale flows, the gas-mass fraction from X-ray clusters, and the shape parameter (required to match the initial perturbation spectrum to that observed in the galaxy correlation function). This method of constraining $_B$ and H_0 is consistent with the high η (low D_{QSO}) constraint from BBN but not with BBN's low η (high D_{QSO}) estimate.

The simulations of HeII absorption by the Ly-α forest by Croft, Weinberg, Katz, and Hernquist (1997) agree with observation provided $_B h^2 \gtrsim 0.0125$, again consistent with the high η estimate from BBN, but not the low. The utility of combining H Ly-α with He Ly-α is that HeII absorption can be important in regions where the HI density is too low to produce significant absorption and thus HeII absorption is sensitive to diffuse intergalactic gas. The mean opacities due to HI and HeII are a function of $\frac{2}{B}/$, where is the photoionization rate as calculated from the UV background due to QSOs. That is to say, the optical depth increases as you add more baryons or reduce the ionizing flux. The simulations allow one to predict the mean opacities for a given combination of $\frac{2}{B}/$. Croft et al. find that for $_B h^2 = 0.0125$ and the UV ionizing flux from QSOs alone, the predicted opacities are just marginally consistent with observation. Thus the lower bound on $_B h^2$ as quoted above. The Croft et al. results can be scaled to higher baryon densities – they find (with the minimal QSO UV flux) excellent agreement with observation for $_B h^2 = 0.023$, exactly what we would expect from BBN and low QSO deuterium.

Perhaps the definitive measurement of $_B h^2$ will come from from the next generation of microwave background (MWB) detectors: NASA's MAP (http://map.gsfc.nasa.gov) and ESA's PLANCK (a.k.a. COBRAS/SAMBA) (http://astro.estec.esa.nl:80/SA-general/ Projects/Cobras/cobras.html). With the assumption of an adiabatic, nearly power-law initial perturbation spectrum, the high order ($l \gtrsim 100$) multipole moments of the angular power spectrum of temperature fluctuations can yield accurated information on H_0, , , and $_B$ (see, for example Jungman et al., 1996). In particular, measurements of the temperature power spectrum can obtain $_B h^2$ to better than 0.001 (current MWB measurements (e.g., Saskatoon and CAT) find $_B h^2 \sim 0.04$ with large errors).

4. Conclusions

A succinct summary of Big Bang Theory as it relates to primordial nuclei is: $_Bh^2 \sim 0.02$. This is in good agreement with the low deuterium value obtained from QSOALS (and with chemical evolution models which constrain the primordial deuterium abundance to be no more than a factor of a few greater than the ISM abundance). It is also in good agreement with the lithium abundances in Pop II stars, provided there is modest destruction of lithium as suggested by models with rotational mixing. It is not in good agreement with the higher QSO observations nor with the primordial ^{4}He abundance as inferred from metal poor HII regions. The new ^{4}He data of Izotov and Thuan may suggest an underestimate of Y_p. Court is still out on QSO deuteria. We are still quite sure that $_B$ is small. Estimates for the mass density of the Universe (see Tammann, 1998, and Rees, 1998) indicate that the majority of the mass in the Universe is dark and non-baryonic.

Acknowledgements

I would like to thank my BBN collaborators, particularly Keith Olive, Gary Steigman, and David Schramm, for many useful discussions. I also again would thank the Workshop organizers and support staff for running such a stimulating meeting.

References

Burles, S. and Tytler, D.: 1996, *Science*, submitted, astro-ph/9603070.
Burles, S. and Tytler, D.: 1997, *Astron. J.* in press, astro-ph/9707176.
Carswell, R.F., Rauch, M., Weymann, R.J., Cooke, A.J., and Webb, J.K.: 1994, *M.N.R.A.S.* **268**, L1.
Copi. C.J., Olive, K.A., and Schramm, D.N.: 1996, astro-ph/9606156
Croft, R.A.C., Weinberg, D.H., Katz, N., and Hernquist, L.: 1996, *Ap. J.* in press, astro-ph/9611053.
Deliyannis, C.P., Demarque, P., Kawaler, S.D., Krauss, L.M., and Romanelli, P.: 1989, *Phys. Rev. Lett.* **62**, 1583.
Duncan, D.: 1998, *Space Sci. Rev.*, this volume.
Fields, B.D., Kainulainen, K., Olive, K.A., and Thomas, D.: 1996, *New Astronomy* **1**, 77.
Fixsen, D.J., Cheng, E.S., Gales, J.M., Mather, J.C., Shafer, R.A., and Wright, E.L.: 1996, *Ap. J.* **473**, 576.
Freedman, W.: 1997, in N. Turok (ed.), *Critical Dialogs in Cosmology*,World Scientific, Singapore, 92.
Hata, N., Scherrer, R.J. Steigman, G., Thomas, D., Walker, T.P., G., Bludman, S., Langacker, P.: 1995, *Phys. Rev. Lett.* **75**, 3977.
Hata, N., Steigman, G., Bludman, S., Langacker, P.: 1997, *Phys. Rev. D* **55**, 540.
Hobbs, L.M. and Thorburn, J.A.: 1997, preprint.
Hogan, C.J., Olive, K.A., Scully, S.: 1997, astro-ph/9705107.
Izotov, Y.L., Thuan, T.X., and Lipovetsky, V.A.: 1994, *Ap. J.* **435**, 647.
Izotov, Y.L., Thuan, T.X., and Lipovetsky, V.A.: 1996, preprint.
Jedamzik,K. and Fuller, G.M.: 1996, astro-ph/9609103
Jungman, G., Kaminokowski, M., Kosowsky, A., and Spergel, D.N.: 1996, *Phys. Rev. Lett.* **76**, 1007.
Kawasaki, M., Kang, H-S., Scherrer, R.J., Steigman, G., and Walker, T.P.: 1994, *Nucl. Phys.* **B419**, 105.

Olive, K.A., Skillman, E., and Steigman, G.: 1996, astro-ph/9611166.
Pagel, B.E.J., Simonson, E.A., Terlevich, R.J., and Edmunds, M.: 1992, *M.N.R.A.S.* **255**, 325.
Pinsonneault, M., Deliyannis, C.P., and Demarque, P.: 1992, *Ap. J. Suppl.* **78**, 181.
Pinsonneault,M., Walker, T.P., Steigman, G., and Narayanan, V.K.: 1997, astro-ph/9710035.
Rees, M.J.: 1998, *Space Sci. Rev.*, this volume.
Rood, R.T., Bania, T.M., Balser, D.S., and Wilson, T.L.: 1998, *Space Sci. Rev.*, this volume.
Rugers, M. and Hogan, C.J.: 1996, *Ap. J. Lett.* **459**, L1.
Rugers, M. and Hogan, C.J.: 1996, *Astron. J.* **111**, 2135.
Schramm, D.N.: 1998, *Space Sci. Rev.*, this volume.
Skillman, E. and Kennicutt, R.C.: 1993, *Ap. J.* **411**, 655.
Skillman, E., Terlevich, R.J., Kennicutt, R.C., Garnett, D.R., and Terlevich, E.: 1994, *Ap. J.* **431**, 172.
Skillman, E. *et al.*: 1997, unpublished.
Songaila, A., Cowie, L.L., Hogan, C.J., and Rugers, M.: 1994, *Nature* **368**, 599.
Songaila, A., Wampler, E.J., and Cowie, L.L.: 1997, *Nature* **385**, 137.
Spite, F. and Spite, M.: 1982, *Astron. Astrophys.* **115**, 357.
Steigman, G., Hata, N., Felten, J.E.: 1997, astro-ph/9708016.
Steigman, G., Schramm, D.N., and Gunn, J.E.: 1977, *Phys. Lett.* **B66**, 202.
Tammann, G.A.: 1998, *Space Sci. Rev.*, this volume.
Thuan, T.X.: 1998, *Space Sci. Rev.*, this volume.
Tosi, M.: 1998, *Space Sci. Rev.*, this volume.
Tytler, D., Fan, X.M., and Burles, S.: 1996, *Nature* **381**, 207.
Tytler, D., Burles, S., and Kirkman: 1996, *Ap. J.* astro-ph/9612121
Walker, T.P., Steigman, G., Schramm, D.N., Olive, K.A., and Kang, H-S.: 1991, *Ap. J.* **376**, 51.
Webb, J.K., Carswell, R.F., Lanzetta, K.M., Ferlet, R., Lemoine, M., Vidal-Madjar, A., Bowen, D.V.: 1997, *Nature* **388**, 250.

Address for correspondence: twalker@hoyle.mps.ohio-state.edu

II: EXTRAGALACTIC OBJECTS

ON THE MEASUREMENTS OF D/H IN QSO ABSORPTION SYSTEMS

Closing in on the primordial abundance of deuterium

S. BURLES

Department of Astronomy & Astrophysics, University of Chicago, 5640 S. Ellis Ave, Chicago, IL 60637

D. TYTLER

Center for Astronomy and Astrophysics, University of California, San Diego, 9500 Gilman Drive, La Jolla, CA 92093-0424

Abstract. We present our measurements of the deuterium to hydrogen ratio (D/H) in QSO absorption systems, which give D/H $= 3.40 \pm 0.25 \times 10^{-5}$ based on analysis of four independent systems. We discuss the properties of two systems which provide the strongest constraints on D/H. We outline the systematic effects involved in measurements of D/H and introduce a sophisticated method of analysis which properly accounts for these effects.

Key words: Cosmology: nucleosynthesis, abundances; Quasars: absorption lines

1. Introduction

The status of standard big bang nucleosynthesis (BBN) has been extensively reviewed in these proceedings and in the recent literature (Copi *et al.*, 1995; Sarkar, 1996; Hata *et al.*, 1997; Cardall and Fuller, 1996; Fuller and Cardall, 1996; Schramm and Turner, 1997). It has been emphasized that a determination of the primordial deuterium abundance has the potential to constrain the predictions of BBN, test the standard model, and give a precise measure of the present-day baryon density. Recent measurements of D/H towards bright, distant QSOs are now realizing this potential, and are already giving improved constraints on the models of BBN. Hogan (1998) gives a thorough overview of the present status of extragalactic D/H measurements; here we focus on measurements of the systems for which we have obtained high-quality spectra with the HIRES spectrograph (Vogt *et al.*, 1994) and Keck1 10-m telescope.

2. The Measurements

We have analyzed four high-redshift QSO absorption systems which place useful constraints on D/H. We show the limits on D/H in each of these systems in Fig. 1. Two of the systems, towards Q1251+3644 and Q1759+7539 have larger uncertainties for different reasons. Q1251+3644 is a faint QSO (V=19), and over 10 hours of observing time yields a spectrum with modest signal-to-noise ratio (SNR = 10) at high-resolution. The SNR drops quickly at lower wavelengths, and we

Space Science Reviews **84**: 65–75, 1998.

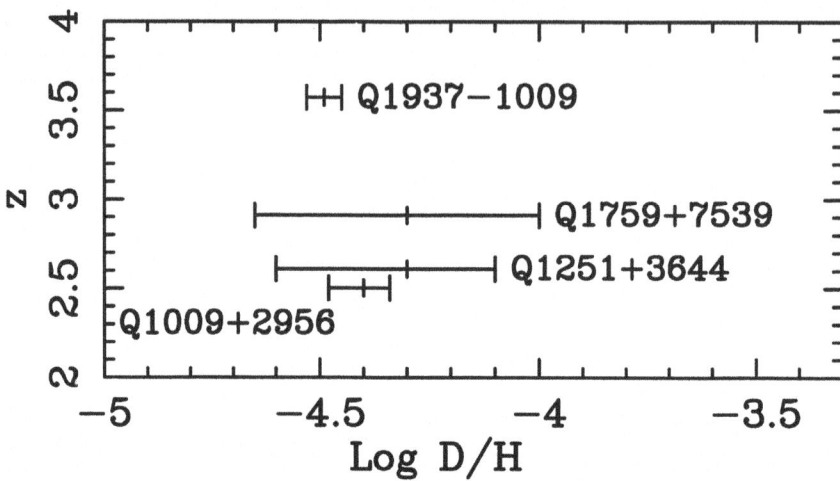

Figure 1. Constraints on D/H from four independent measurements in four separate QSO absorption systems. Shown are the most likely values and 68% confidence limits at the absorption redshift of the D/H systems.

cannot extract useful information from the highest order Lyman lines as a result. The system towards Q1759+7539 is a complex system of absorbers with a very high neutral hydrogen column density, log N(H I) > 10^{19} (all column densities expressed in cm^{-2}). The analysis of that system gives only an upper limit on D/H, at the 95% confidence level; the constraints are obtained from comparing the line profiles of Lyβ and Lyγ to the profile of Lyα. The method used to extract the limits on D/H will be described below. These results are preliminary, and improved constraints will require more data to allow a robust analysis.

Tytler *et al.* (1996) made the first measurement of low D/H in the absorption system at $z = 3.572$ towards Q1937–1009. We analyzed the high-resolution spectrum (8 hrs of exposure), which resolved the entire Lyman series up to Ly-19, as well as associated metal lines. By profile fitting the Lyman lines, with the position of the velocity components given by the metal lines, we find D/H = $2.3 \pm 0.3 \pm 0.3 \times 10^{-5}$ (statistical and systematic errors). The largest uncertainty in the measurement is the neutral hydrogen column density, log N(H I) = $17.94 \pm 0.06 \pm 0.05$, and the uncertainty stems from the saturated Lyman profiles (discussed in detail below). We then obtained a high quality low-resolution spectra from Keck with LRIS (Oke *et al.*, 1995), which gave better sensitivity shortward of the Lyman limit, to directly measure the total N(H I) in the system and therefore place better constraints on D/H. Utilizing both the high and low-resolution spectra, we find log N(H I) = 17.86 ± 0.02 by a direct measurement of the optical depth shortward of the Lyman limit at 4200 Å (Burles and Tytler, 1997a). With this constraint and a more sophisticated fitting procedure, we measure D/H = $3.3 \pm 0.3 \times 10^{-5}$ (Burles and Tytler, 1997b).

We discovered an absorption system at $z = 2.504$ towards Q1009+2956 ideal for a measurement of D/H. This system has a lower hydrogen column density, log N(H I) $= 17.39 \pm 0.06$. The highest order Lyman lines become unsaturated, which yields a precise measurement of N(H I) in both low and high resolution spectra (Fig. 2). Over twelve hours of Keck+HIRES produced a very high quality spectrum of the entire Lyman series, resolving the entire series up to Ly-22. We find strong evidence for contamination of the deuterium Lyα absorption feature, which unfortunately introduces the largest uncertainty in the measurement. With the contamination included in the analysis, we measure D/H $= 4.0 \pm 0.7 \times 10^{-5}$ (Burles and Tytler, 1997c).

The four independent systems support a low primordial abundance of deuterium, and together give D/H $= 3.4 \pm 0.25 \times 10^{-5}$. If this represents the primordial value, nucleosynthesis calculations from standard BBN models with three light neutrinos give the baryon-to-photon ratio, $\eta = 5.1 \pm 0.3 \times 10^{-10}$ and the baryon density, $\Omega_b h_{100}^2 = 0.019 \pm 0.001$. We arrive at very small statistical errors (10% at 95% confidence), so we must confront the existence of systematic effects. In the remainder of this paper we present a sophisticated method that we use in order to properly account for the systematic effects currently known.

3. An Improved Method to Measure D/H

Many groups have detected deuterium in absorption along the line of sight to distant QSOs (Songaila *et al.*, 1994; Carswell *et al.*, 1994; Tytler *et al.*, 1996, Wampler *et al.*, 1996; Rugers and Hogan, 1996a,b; Webb *et al.*, 1997; Burles and Tytler, 1997b,c). Here, we put forward an improved method to measure D/H, which has many advantages over the current, widely used, methods.

The goal is to extract from the spectrum of a QSO absorption system the most likely value of D/H, and the confidence intervals about that value. We are still required to model the D/H absorption system (DHAS) with a finite set of parameters (based on physical assumptions) to compare to the observed spectrum. With a given model, we can quickly minimize the difference between model and data, parameterized as χ^2, and use the final parameters to find the most likely value of D/H. But the uncertainty in D/H depends on correlations in the parameters, and is non-trivial to extract from the best fit solution.

The χ^2 function depends on the N parameters included to model the observed spectrum. Instead of mapping the χ^2 function in the full N dimensional parameter space, we choose to map χ^2 versus a single parameter, D/H (known as the "ridge" method). Towards Q1937–1009, for example, we compare a model with 234 free parameters to regions of the high-resolution spectrum which total 1298 pixels. We note the inventory of free parameters: 3 main components in Lyman limit system with 4 parameters each (N(H I), z_{abs}, T, b_{tur}) + 9 spectral regions with a total of 29 parameters for the continuum + 64 other H I lines with 3 parameters each (N(H I),

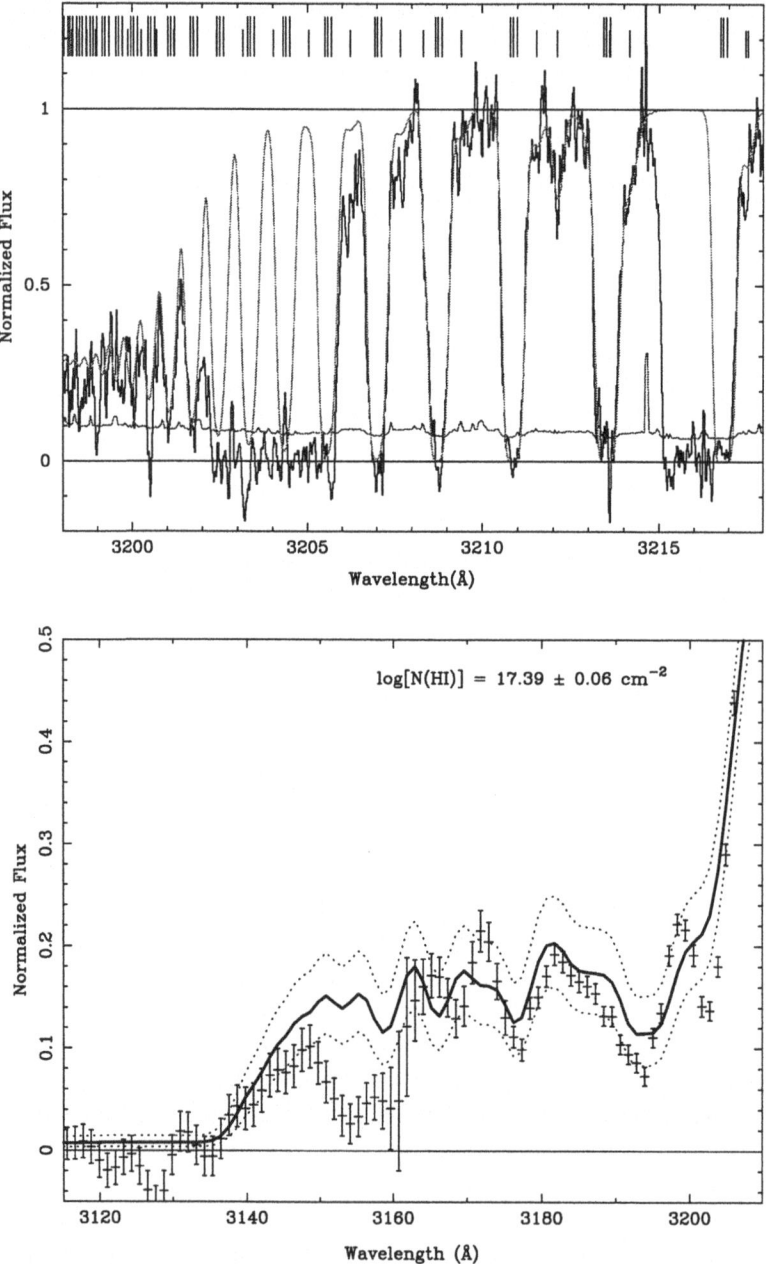

Figure 2. Lyman limit region of Q1009+2956 at $z = 2.504$.
Top: HIRES spectrum (FWHM = 8 km/s) showing Lyman lines 11 through 24, and model fit with three components and total H I column density, log N(H I) = 17.31.
Bottom: Lick spectrum (FWHM = 4 Å) showing residual flux and model fit shortward of Lyman limit. The solid line shows the best fit model with log N(H I) = 17.39, and dotted lines represent 68% confidence levels. Absorption shortward of 3135 Å is due to a lower redshift Lyman limit system.

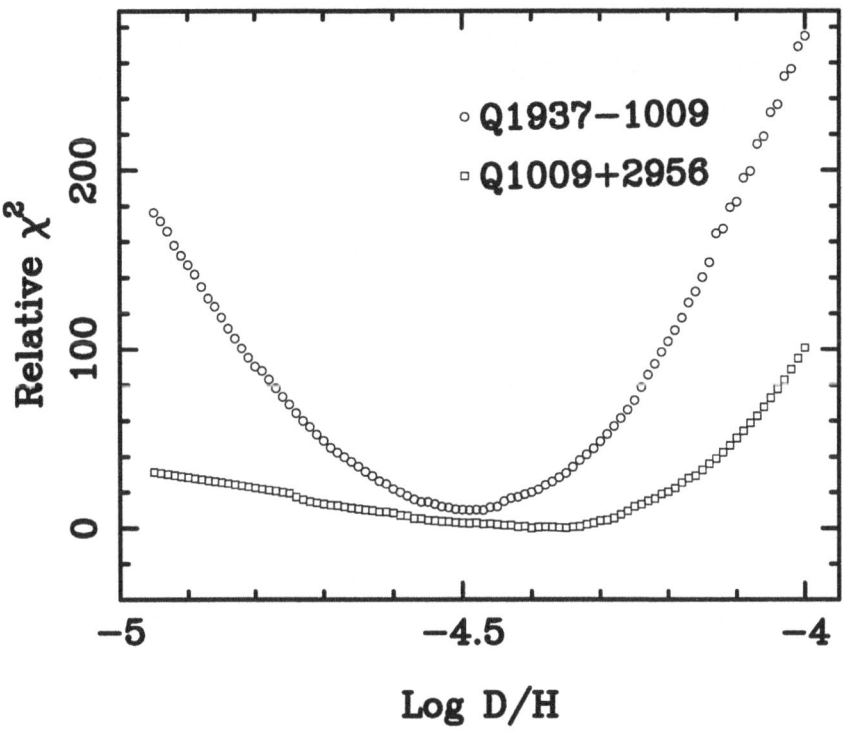

Figure 3. χ^2 functions of D/H in two separate absorption systems. The Relative χ^2 is shown to facilitate a direct comparison. The minima represent the most likely values of D/H in each system, and $\Delta\chi^2 = 4.0$ represents 95% confidence limits.

z_{abs}, b_{total}) which are near and affect the absorption lines of the D/H system + 1 parameter for D/H: $3 \times 4 + 29 + 64 \times 3 + 1 = 234$. To construct the minimized χ^2 function versus D/H alone, we choose a value for D/H, fix D/H for the iteration, and allow the other 233 parameters to freely change to minimize χ^2. We choose and fix a new value for D/H, and minimize χ^2 again. The results for both Q1937–1009 (this example) and Q1009+2956 are shown in Fig. 3. In Fig. 4 (bottom panel), note the large number (19) of H I lines which must be included in the Lyα region of Q1937–1009.

In this paper we present the difficulties inherent to measurements of D/H in QSO absorption systems, and discuss techniques we have implemented to overcome the systematic effects. We outline an improved method to measure D/H, which accounts for the total uncertainty, in contrast to the statistically uncorrelated errors of D and H column densities summed in quadrature. We show that the best two current measurements of D/H are complementary and give a robust measure of the primordial value, D/H = $3.4 \pm 0.25 \times 10^{-5}$.

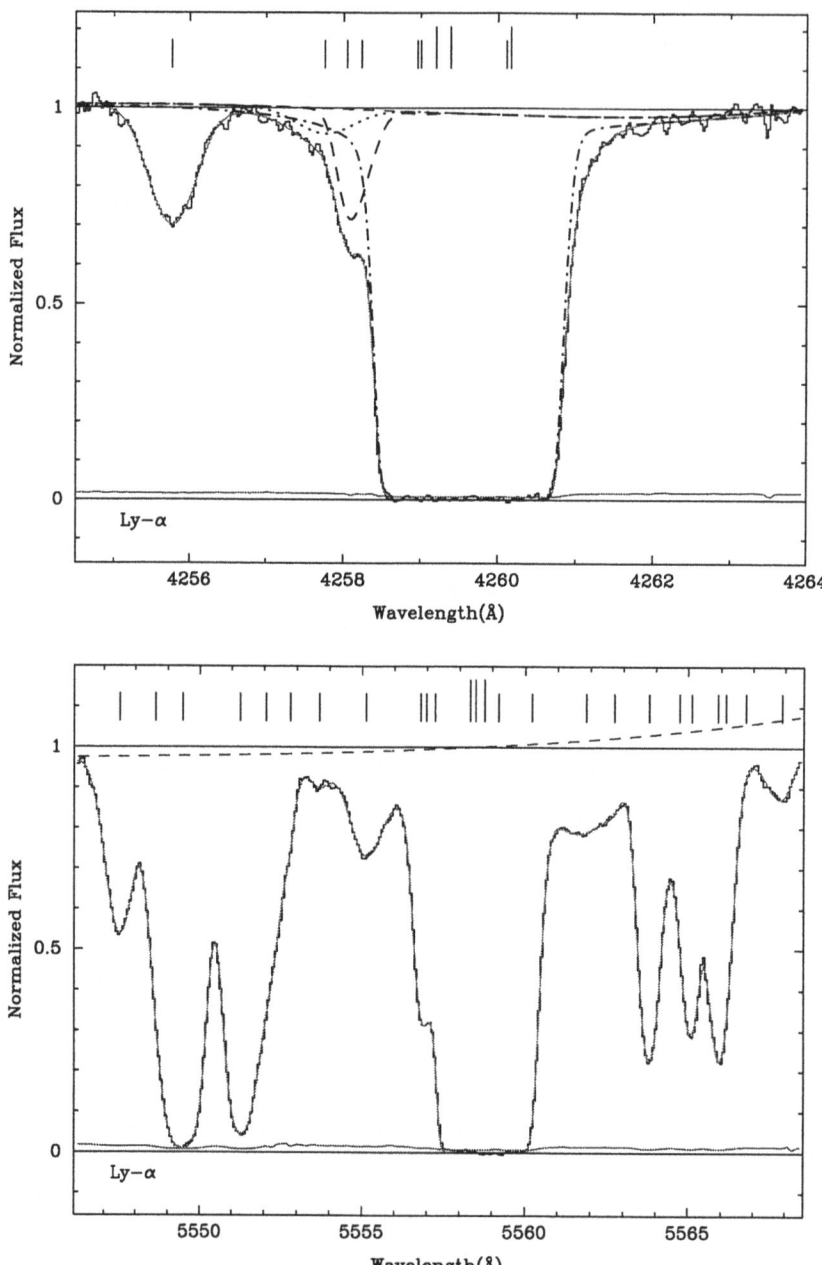

Figure 4. Top: Lyα region of $z = 2.504$ system towards Q1009+2956. The histogram shows the pixels of the HIRES spectrum normalized to the initial continuum estimate (solid line at unity). The 1σ errors per pixel is shown as the solid line near zero. The smooth dashed line near unity is the best fit continuum with five degrees of freedom. The model profile is composed of the main H I (dot-dashed) and D I (dashed), and contaminating H I (dotted).
Bottom: Lyα region of $z = 3.572$ system towards Q1937–1009. The dashed line shows the best fit continuum level.

4. Systematic uncertainties in D/H measurements

In this section, we describe the systematic uncertainties which are known to exist in measurements of D/H in QSO absorption systems. We explain how we account for the systematics, with the assumption that the nature of the effect is understood.

4.1. INTRINSIC QSO CONTINUUM

The observed QSO spectra are combinations of the intrinsic spectrum of the source with intervening absorption features. In the measurement of D/H, we are solely interested in the intervening features. We attempt and *require* the removal of the intrinsic QSO spectrum (continuum) to analyze the absorption features in the observed spectrum. The QSO continuum changes on much longer wavelength scales than the narrow absorption lines. A polynomial of low-order traces the smooth variations of the QSO continuum and gives the normalization of the observed spectrum to the QSO continuum. This procedure (called "continuum fitting") is not exact, and even worse, not well defined. However, most abundance measurements in QSO absorption systems have not taken into account the uncertainties introduced by subjectively fitting a fixed continuum to the observed spectrum. Once a continuum fit is complete, the solution is assumed, for all practical purposes, correct with no uncertainty. We have overcome this problem by assuming the continuum is not fixed or well determined. As a part of the abundance measurements, we include free parameters to account for the unknown shape and level of the continuum in regions of interest in the observed spectrum. We still fit an initial continuum (a first guess), but this is not assumed absolute or correct, only a likely estimate. The procedure has been outlined in Burles and Tytler (1997b), and should be adopted for all abundance measurements which could be influenced by continuum uncertainties.

As examples, we show the Lyα regions of Q1009+2956 (top) and Q1937–1009 (bottom) in Fig. 4. The initial continuum estimate is the solid line at unity and is used to normalize the spectrum before profile fitting. The smooth dashed line near unity is the resulting best fit continuum with five free parameters. The level of the final continuum lies near the initial estimate on the D/H Lyα feature, which confirms that the initial estimate was fairly well chosen. Allowing for a free continuum does not necessarily change the central value, but ensures that the final uncertainties take properly into account the uncertain nature of the intrinsic QSO continuum.

4.2. HYDROGEN CONTAMINATION OF DEUTERIUM

The absorption features of the Lyman series of H I and D I are modeled to measure D/H in QSO absorption systems. These features reside exclusively in the Lyα forest and are subject to blending, and therefore, contamination from other Lyman absorption. The line density of the Lyα forest rises steeply with redshift, so measurements at higher redshift, $z > 2$, are subject to more contamination from random, unrelated Lyα lines. Hogan (1998) discusses the *a posteriori* probabilities of unrelated H I contamination. On the other hand, structure in the absorption systems will increase with decreasing redshift, so contamination of measurements at lower redshift, $z \leq 1$, is dominated by Lyman absorption correlated on scales of the D-H velocity separation, 82 km s^{-1}. We must test for the presence of contamination in all measurements and ensure that the final results properly account for its existence. Contamination always gives lower values of D/H. A measurement which does not include contamination in the analysis gives an upper limit on D/H. In Fig. 4, we show the individual profiles of H I (dot-dashed) and D I (dashed) in the Lyman limit system and unrelated H I contamination redward of the D I profile (dotted line). Contamination was included in analysis towards both Q1937–1009 and Q1009+2956.

4.3. SATURATED ABSORPTION FEATURES

In measurements of D/H, one must select systems with high neutral hydrogen column densities, log N(H I) > 17, to be sensitive to D/H values less than 10^{-4}. In these systems, most of the lines of the Lyman series have saturated (optical depth, $\tau > 3$) line centers. And in systems with log N(H I) > 18, the entire Lyman series, as well as the continuum absorption of the Lyman limit, become saturated. In most systems, the measurement of the hydrogen column presents a greater difficulty than a measurement of the deuterium abundance itself.

To overcome the difficulty of determining N(H I), we obtain spectra with high signal-to-noise ratio (SNR) of the region shortward of the Lyman limit to independently constrain N(H I). It is possible to obtain a direct measurement of the total H I column in systems with $17 < $ log N(H I) < 18, $0.6 < \tau_{LL} < 6.0$, where τ_{LL} is the optical depth of Lyman continuum absorption at 912 Å (rest frame). The residual flux shortward of the Lyman limit provides a direct estimate of N(H I), independent of other characteristics of the system (e.g. velocity dispersion or deuterium column density). To obtain an accurate measurement, we must account for unrelated Lyman absorption from other systems and the uncertainty of the intrinsic QSO continuum near the Lyman limit (see Sec. 4.1). The majority of the Lyman absorption features can be specified by their related absorption features in high-resolution spectra (at higher wavelengths). The remaining features are statistically included from well determined Lyα forest distributions (Kirkman and Tytler 1997). The QSO continuum near the Lyman limit can be determined to better than 10% by utilizing high-resolution spectra (cf. Burles and Tytler 1997a). The uncertainty in the N(H I) measurement is proportional to the continuum uncertainty, Δ N(H I) $= \Delta$(cont)$/\tau_{LL}$. In Fig. '2, we show the Lyman limit region of the $z = 2.504$ D/H system in two separate spectra of Q1009+2956. The high resolution spectrum (top, FWHM = 8 km s^{-1}, 0.09 Å) shows the resolved high order Lyman lines up to Lyman-22. The Lyman lines become unsaturated at the highest orders, and give an optical depth in the continuum, $\tau_{LL} = 1.3$. The low resolution spectrum (bottom, FWHM = 4 Å) has better sensitivity and shows significant flux shortward of the Lyman limit at 3200 Å.

A one-parameter model, N(H I), is compared with the low resolution spectra to find the most probable value and confidence intervals of N(H I) (Burles and Tytler, 1997a). The features seen in the model fit in Fig. 2 are higher-order Lyman lines (Lyβ and above) associated with Lyman lines measured at longer wavelengths in the high resolution spectrum. Lyα absorption within $1.57 < z < 1.64$ is drawn from a distribution and included in the Monte Carlo analysis which gives the maximum likelihood of N(H I): log N(H I) = 17.39 ± 0.06, which includes the uncertainties from extrapolation of the unabsorbed QSO continuum.

4.4. THE CASE OF MESOTURBULENCE

Levshakov *et al.* (1998, and references therein) have studied the effects of correlations in turbulent velocity fields (mesoturbulence), on line formation in the Lyα forest. The standard line profiles (Voigt) are, in fact, the special case of mesoturbulence when the correlation length goes to zero. A substantial amount of work has been published to understand the impact of mesoturbulence on the abundance measurements in absorption systems, with emphasis on D/H (Levshakov and Kegel, 1997; Levshakov *et al.*, 1997; Levshakov *et al.*, 1998). We would like to point out a few observational constraints on mesoturbulence, and the limited role it plays (in our view) in measurements of D/H.

(1) The turbulent component should be consistent with all lines in the absorption system. Mesoturbulent profiles will significantly differ from Voigt profiles only when the turbulent component is comparable or larger than the thermal component. Both components contribute to the total velocity dispersion (b_{total}) which is measured in high resolution spectra: $b_{total}^2 = b_{meas}^2 = b_{turb}^2 + b_{therm}^2$. Comparisons of the metal line widths (e.g. C, Si) to the hydrogen and deuterium line widths place strict limits on the turbulent component. The heavy metal lines (Si and above) are dominated by turbulence at temperatures $T < 10^5$ K and their velocity dispersion usually gives a good measure of the turbulence in the system. In the two systems we have measured D/H, towards Q1937–1009 and Q1009+2956, the metal lines are very narrow, which give very low velocity dispersions, $b(Si) < 6$ km s^{-1}. The hydrogen lines are much wider, with typical widths $b(H) \approx 20$ km s^{-1}, which shows that thermal, uncorrelated, line formation is dominant in the optically thick Lyman lines.

(2) The line profiles of mesoturbulence can be distinguished from Voigt profiles by fitting the entire Lyman series (Levshakov and Kegel, 1996). The requirement for invoking mesoturbulence should be determined by the observational data. If the Lyman series lines show evidence for mesoturbulent profiles, then one should take it into account. In D/H systems to date, simple Voigt profiles can explain all of the profiles in the Lyman series and do not require the mesoturbulent model.

(3) Mesoturbulence is likely to play an important role in a wide variety of absorption line studies. The hydrogen lines in QSO spectra which do show deuterium belong to a class of absorbers with the narrowest intrinsic widths, and are therefore subject to the least amount of turbulent broadening. On the other hand, optically thick absorption lines of heavy metals (for instance, in damped-Lyα systems) are prime candidates for studies of mesoturbulence.

4.5. LINE BLENDING

The D/H absorption systems do not have simple, symmetric profiles which can be modeled as a single component. Analysis of the absorption profiles of the Lyman lines and associated metal lines requires multiple components. The components are usually separated by less than the intrinsic line widths of the individual components, which gives rise to blended profiles of the Lyman series. The severe blending represents a loss of information. The parameters describing individual components are poorly constrained, but those describing the entire system can be characterized and measured. Here lies the problem: the set of parameters describing the multiple components of the D/H system are not orthogonal when the components are blended. The uncertainties in the parameters are strongly correlated, and the variance in the parameters does not represent the total uncertainty. Therefore, one cannot measure the individual column densities of D and H accurately in each absorption component.

Acknowledgements

We are extremely grateful to the W. M. Keck foundation which made this work possible. We would like to thank our hosts at ISSI in Bern, J. Geiss and R. von Steiger, for their hospitality, and the editors of these proceedings, N. Prantzos, M. Tosi, and R. von Steiger for their help.

References

Burles, S. and Tytler, D.: 1997, *AJ* **114**, 1330.
Burles, S. and Tytler, D.: 1997, *ApJ*, in press, astro-ph/9712108.
Burles, S. and Tytler, D.: 1997, *ApJ*, submitted, astro-ph/9712109.
Cardall, C. Y. and Fuller, G. M.: 1996, *ApJ* **472**, 435.
Carswell, R. F., Rauch, M., Weymann, R. J., Cooke, A. J., and Webb, J. K.: 1994, *MNRAS* **268**, L1.
Carswell, R. F., Webb, J. K., Lanzetta, K. M., Baldwin, J. A., Cooke, A. J., Williger, G. M., Rauch, M., Irwin, M. J., Robertson, J. G., and Shaver, P. A.: 1996, *MNRAS* **278**, 506.
Copi, C.J., Schramm, D.N., and Turner, M.S.: 1995, *Science* **267**, 192.
Fuller, G. M., and Cardall, C. Y.: 1996, *Nucl. Phys. B* **51**, 71.
Hata, N., Steigman, G., Bludman, S., and Langacker, P.: 1997, *Phys. Rev. D,* **55**, 540.
Hogan, C. J.: 1998, *Space Sci. Rev.*, this volume.
Kirkman, D. and Tytler, D.: 1997, *ApJ* **484**, 848.
Levshakov, S. A. and Kegel, W. H.: 1996, *MNRAS* **278**, 497.

Levshakov, S. A. and Takahara, F.: 1996, *MNRAS* **279**, 651.

Levshakov, S. A. and Kegel, W. H.: 1997, *MNRAS* **288**, 787.

Levshakov, S. A., Kegel, W. H., and Takahara, F.: 1997, *MNRAS*, submitted.

Levshakov, S. A., Kegel, W. H., and Takahara, F.: 1998, *Space Sci. Rev.*, this volume.

Oke, J. B., Cohen, J. G., Carr, M., Cromer, J., Dingizian, A., Harris, F. H., Labrecque, S., Lucinio, R., Schaal, W., Epps, H., Miller, J.: 1995, *PASP* **107**, 375.

Rugers, M. and Hogan, C.: 1996, *ApJ* **459**, L1.

Rugers, M. and Hogan, C.: 1996, *AJ* **111**, 2135.

Sarkar, S.: 1996, *Rep. Prog. Phys.* **59**, 1493.

Schramm, D. N. and Turner, M. S.: 1997, *Rev. Mod. Phys.*, submitted.

Smith , M. S., Kawano, L. H., and Malaney, R. A.: 1993, *ApJS* **85**, 219.

Songaila. A., Cowie, L. L., Hogan C. J., and Rugers, M.: 1994, *Nature* **368**, 599.

Songaila, A., Wampler, E. J., and Cowie, L. L.: 1997, *Nature* **385**, 137.

Tytler, D., Fan, X-M., and Burles, S.: 1996, *Nature* **381**, 207.

Vogt, S. *et al.* 1994, Proc. SPIE, 2198, 362

Walker, T. P., Steigman, G., Schramm, D. N., Olive, K. A., and Kang, H. S.: 1991, *ApJ* **376**, 51.

Wampler, E. J.: 1996, *Nature* **383**, 308.

Wampler, E.J., Williger, G.M., Baldwin, J.A., Carswell, R.F., Hazard, C. and Mcmahon, R.G.: 1996, *A&A* **316**, 33.

Webb, J. K., Carswell, R. F., Lanzetta, K. M., Ferlet, R., Lemoine, M., Vidal-Madjar, A. and Bowen, D. V.: 1997, *Nature* **388**, 250.

A REVERSE MONTE CARLO STUDY OF H+D LYMAN ALPHA
ABSORPTION FROM QSO SPECTRA

S. A. LEVSHAKOV* and W. H. KEGEL

*Institut für Theoretische Physik der Universität Frankfurt am Main, Postfach 11 19 32,
60054 Frankfurt/Main 11, Germany*

F. TAKAHARA

Department of Physics, Tokyo Metropolitan University, Hachioji, Tokyo 192-03, Japan

Abstract. A new method based on a Reverse Monte Carlo [RMC] technique and aimed at the inverse problem in the analysis of interstellar (intergalactic) absorption lines is presented. The line formation process in chaotic media with a finite correlation length ($l > 0$) of the stochastic velocity field (*mesoturbulence*) is considered. This generalizes the standard assumption of completely uncorrelated bulk motions ($l \equiv 0$) in the *microturbulent* approximation which is used for the data analysis up-to-now. It is shown that the RMC method allows to estimate from an observed spectrum the proper physical parameters of the absorbing gas and simultaneously an appropriate structure of the velocity field parallel to the line-of-sight.

The application to the analysis of the H+D Lyα profile is demonstrated using Burles and Tytler [B&T] data for QSO 1009+2956 where the DI Lyα line is seen at $z_a = 2.504$.

The results obtained favor a *low* D/H ratio in this absorption system, although our upper limit for the hydrogen isotopic ratio of about 4.5×10^{-5} is slightly higher than that of B&T (D/H = $3.0^{+0.6}_{-0.5} \times 10^{-5}$). We also show that the D/H and N(HI) values are, in general, *correlated*, i.e. the derived D-abundance may be badly dependent on the assumed hydrogen column density. The corresponding confidence regions for an arbitrary and a fixed stochastic velocity field distribution are calculated.

1. Introduction

The measurement of deuterium abundance at high redshift from absorption spectra of QSOs is the most sensitive test of physical conditions in the early universe just after the era of Big Bang Nucleosynthesis [BBN]. The standard BBN model predicts strong dependence of the primordial ratio of deuterium to hydrogen nuclei D/H on the cosmological baryon-to-photon ratio η and the effective number of light neutrino species N_ν (e.g. Walker *et al.*, 1991). The accuracy of the theoretically calculated hydrogen isotopic ratio is rather high and for a given value of η, D/H is determined with a 15% precision (Sarkar, 1996). Therefore to be comparable with this small uncertainty in the predicted D/H value astronomical measurements of the deuterium and hydrogen column densities should be of a similar precision.

The analysis of interstellar (intergalactic) absorption lines is not, however, an easy and unambiguous task, especially for the case of optically thick lines. As demonstrated in a series of papers (Levshakov and Kegel, 1997; Levshakov, Kegel, and Mazets, 1997; Levshakov, Kegel, and Takahara, 1997; Papers I, II and III, hereafter, respectively) the main difficulty is connected with correlation effects

---

* On leave from A. F. Ioffe Physico-Technical Institute, St. Petersburg, Russia

Space Science Reviews **84**: 77–82, 1998.
© 1998 *Kluwer Academic Publishers*.

between different physical parameters that may occur in chaotic media along the line-of-sight. For instance, if one considers the line formation process in the light of a point source, then the observed spectrum reflects only one realization of the velocity field and, hence, large deviations from the expected average intensity $\langle I_\lambda \rangle$ may occur if the correlation length of the velocity field l is not very small compared with the cloud size L. Paper I clearly demonstrates that in general (i.e. when $l \neq 0$) the absorption line profile is asymmetric (skew) and may look like a barely resolved blend whereas homogeneous density and temperature have been adopted in these calculations.

This fact becomes crucial for the analysis of H+D Lyα spectra where hydrogen lines are always saturated. An example discussed in Paper II suggests that the apparent scatter of the D/H ratio of more than a factor of 4 revealed in the ISM may be caused by an inadequate analysis.

The study of extragalactic hydrogen spectra may turn out to be more complicated. Two limiting cases should be distinguished:

1. *Complete statistical ensemble.* If one observes UV-spectra of galaxies when the spectrograph aperture covers an essential part of the galactic surface, then $\langle I_\lambda \rangle$ should reasonably correspond to the observations since this case is a good approximation to the mathematical space averaging procedure. An example of the averaged mesoturbulent H+D Lyα spectra is considered by Levshakov and Takahara (1996). It has been shown that the standard Voigt-fitting procedure applied to these spectra may yield either *higher* or *lower* D/H value (up to a factor of 10) as compared with an adopted one.

2. *Poor statistical ensemble.* If a hydrogen spectrum is observed in an intervening cloud along a QSO line-of-sight, and the quasar itself may be considered as a point-like source, then the light beam intercepts the absorbing region in *only one* direction. To obtain an unambiguous result in this case the structure of the stochastic velocity field must be known exactly. The Voigt-fitting procedure may lead in this case to *underestimated* D/H values as will be shown below.

This report presents a self-consistent method which enables us to evaluate both the physical parameters of the gas cloud and the corresponding velocity field projection to the line-of-sight.

2. The RMC Method and Results

The main aim of the present study is the *inverse problem*, i.e. the problem to deduce physical parameters from an observed absorption spectrum in the light of a point-like source. For this we use the same mesoturbulent model specified in full detail in Paper II. The inverse problem is always an optimization problem in which an objective function is minimized. We use a χ^2 method to estimate a goodness of the minimization and a $\Delta(\chi^2)$ technique to draw confidence regions. Our model is fully defined by specifying the hydrogen column density $N(\text{HI})$, the kinetic temperature T_{kin}, the ratio of the rms turbulent velocity to the hydrogen thermal velocity σ_t/v_{th}, and the L/l ratio. In addition we have to specify the distribution of the velocity component parallel to the line of sight $v(s)$. Here $v(s)$ is a continuous random function of the coordinate s. In the numerical procedure it is sampled at evenly spaced intervals Δs. The number of intervals depends on the current values of σ_t/v_{th} and L/l, being typically ~ 100 (see Paper II, for details). Thus, to solve the inverse problem we have to optimize the objective function in the parameter space of very large and variable (depending on σ_t/v_{th} and L/l) dimension. The RMC method based on the computational scheme invented by Metropolis *et al.* (1953) proved to be adequate in this case.

Contrary to the standard Monte Carlo procedure in which random configurations of a given physical system are generated to estimate its average characteristics, the RMC takes an experimentally determined set of data and searches for a random parameter configuration which reproduces the observation. Applied to the analysis of absorption spectra, the computational procedure is described in detail in Paper III. Here we only note that the main idea was to divide the parameter space into two parts: (1) the subspace of constant dimension which contains the physical parameters like $N(HI)$, D/H, *etc.*, and (2) the subspace of variable dimension containing the velocity components $v(s_i)$.

We applied the procedure to the H+D Lyα line observed by B&T in the spectrum of QSO 1009+2956 at $z_a = 2.504$. This spectrum was selected since (i) it shows a well pronounced DI absorption, and (ii) it was obtained with high spectral resolution and signal-to-noise ratio.

Using a *two*-cloud microturbulent model B&T derived the following parameters: total hydrogen column density $N(HI) = 2.9^{+0.4}_{-0.3} \times 10^{17}$ cm^{-2}, D/H $= 3.0^{+0.6}_{-0.5} \times 10^{-5}$, $T_{kin} = 2.1^{+0.1}_{-0.1} \times 10^4$ K, and $2.4^{+0.7}_{-0.7} \times 10^4$ K for the blue and red subcomponents, respectively, the corresponding turbulent velocities 3.2 ± 0.4 km s^{-1} and 2.3 ± 1.4 km s^{-1}, and a difference in the radial velocity of 11 km s^{-1}.

We started with the calculation of a template H+D Lyα spectrum based on the parameters listed in Table 1 by B&T. To simulate real data, we added the experimental uncertainties to the template intensities which were sampled in equidistant bins as shown in Fig. 1 by dots and corresponding error bars. We adopted for the redshift the mean value of 2.504 and used a *one*-component mesoturbulent model with a homogeneous density and temperature.

Adequate profile fits for three different D/H values are shown in Fig. 1 by solid curves. The estimated parameters and corresponding values of χ^2 (per degree of freedom) are also listed for each solution. The velocity distributions $p(u)$ leading to these profiles [where $u = v(s)/\sigma_t$] are shown by histograms.

Although our model (*one* cloud with homogeneous density and temperature accounting for a *finite* correlation length in the stochastic velocity field) is quite different from that of B&T (*two* clouds, stochastic velocities in each one considered in the *microturbulent* approximation), we derive with the RMC method in this

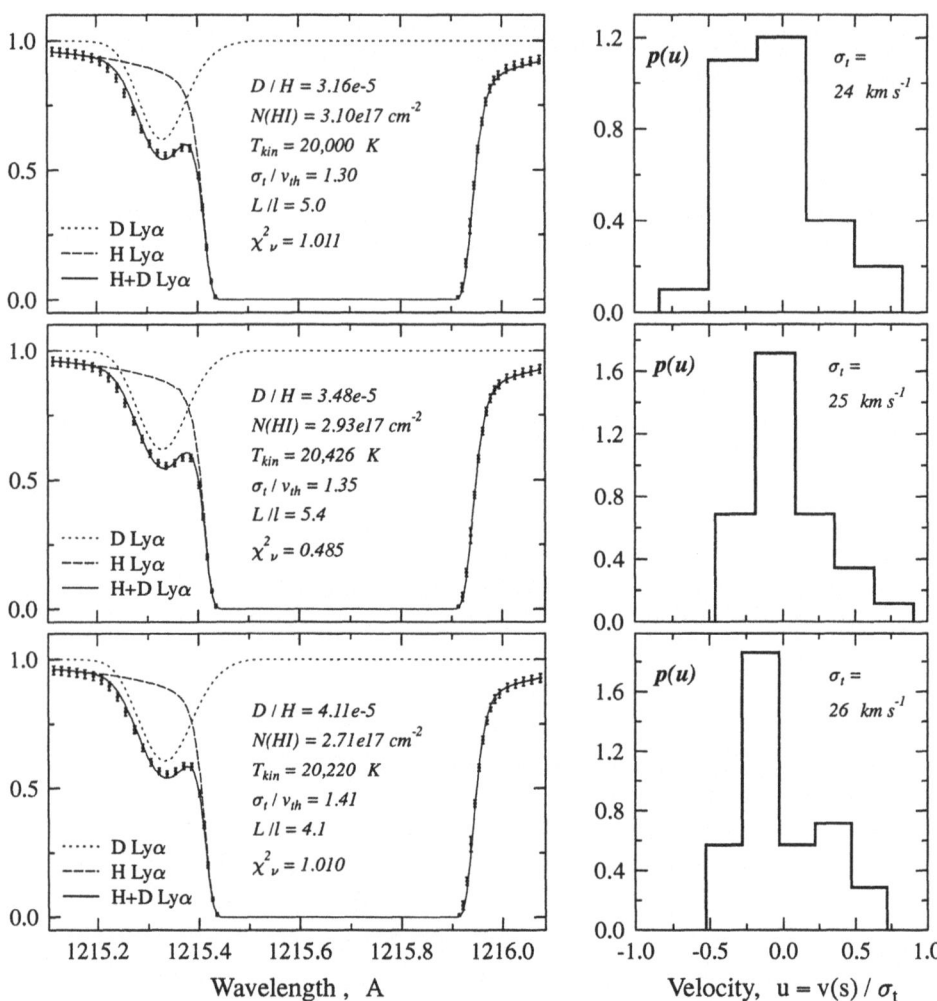

Figure 1. A template H+D Lyα profile (dots) representing the normalized intensities and their uncertainties in accord with the B&T data. The solid curves show the results of the RMC minimization. The corresponding projected velocity field distributions $p(u)$ are shown by histograms.

particular case values for N(HI), D/H, and T_{kin} which are not significantly different from those of B&T. The rms turbulent velocity σ_t, however, is substantially larger. For the additional parameter L/l we found a value of about 5, indicating a substantial variation of the large-scale velocity field along the line-of-sight.

The best RMC solution with the smallest $\chi^2 \simeq 0.5$ (the middle panel in Fig. 1) is labeled by the filled diamond in Fig. 2a whereas the filled circle is the B&T result. In Fig. 2a, we show confidence regions which are projected onto the plane "log N(HI) – log D/H" under the condition that the *parameters* T_{kin}, σ_t/v_{th} and L/l are fixed, but the velocity field *configuration* is free. The elongated and declined

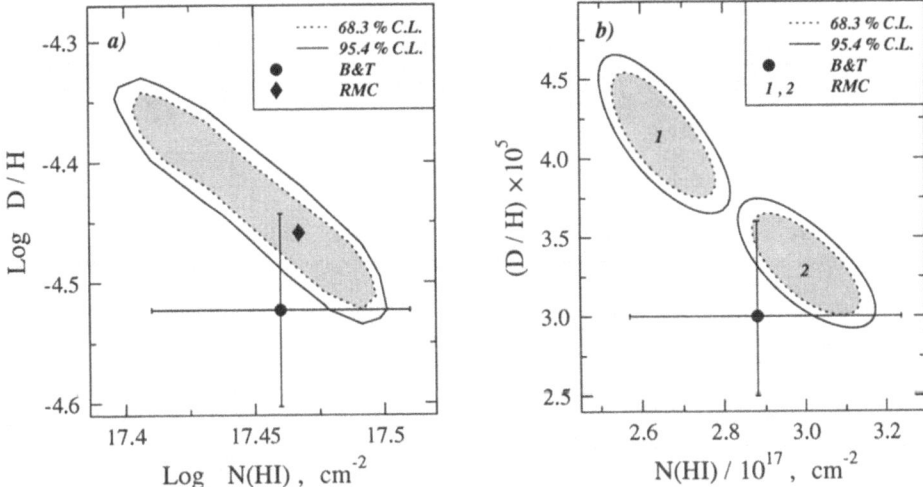

Figure 2. (a) - Confidence regions for the plane "log N(HI) – log D/H" when the other parameters T_{kin}, σ_t/v_{th} and L/l are fixed (listed in the middle panel of Fig. 1) but the configuration of the velocity field is free. The B&T and RMC best fitting parameters are labeled by the filled circle and diamond, respectively. The B&T error bars correspond to 1σ. (b) - Confidence regions 1 and 2 for the fixed configurations of the corresponding velocity fields. The fixed parameters T_{kin}, σ_t/v_{th} and L/l are listed in the lower and upper panels of Fig. 1, respectively.

shape of these confidence regions reflects a correlation between D/H and N(HI), i.e. the scatter of the D/H values cannot be simply determined by the projection of a given confidence region onto the corresponding axis.

Another important result obtained is the existence of possible multiple solutions with non-overlapping confidence regions. This may be a real problem, if the multiplicity scatters parameters significantly. To illustrate the problem we draw two confidence regions labeled by numbers 1 and 2 (Fig. 2b). These regions correspond to the RMC H+D profiles shown in the lower and upper panels of Fig. 1 which present the limiting D/H values for the mesoturbulent model. The regions 1 and 2 were calculated in this case under the condition that T_{kin}, σ_t/v_{th}, L/l and the corresponding configurations of the velocity field are fixed. Thus, if one succeeds to determine the velocity field distribution $p(u)$ by including additional absorption lines in the analysis, the accuracy of the D/H measurements may be improved significantly.

Acknowledgements

This work was supported by the Deutsche Forschungsgemeinschaft and by the RFBR grant No. 96-02-16905-a.

S. A. LEVSHAKOV ET AL.

References

Burles, S. and Tytler, D.: 1996, 'Cosmological deuterium abundance and the baryon density of the Universe', *preprint*, astro-ph 9603070 [B&T].

Levshakov, S.A. and Takahara, H.: 1996, 'The effect of spatial correlations in a chaotic velocity field on the D/H measurements from QSO absorption spectra', *Monthly Notices Roy. Astron. Soc.* **279**, 651–660.

Levshakov, S.A. and Kegel, W.H.: 1997, 'New aspects of absorption line formation in intervening turbulent clouds – I. General principles', *Monthly Notices Roy. Astron. Soc.* **288**, 787–801, [Paper I].

Levshakov, S.A., Kegel, W.H., and Mazets I.E.: 1997, 'New aspects of absorption line formation in intervening turbulent clouds – II. Monte Carlo simulation of interstellar H+D Lyα absorption profiles', *Monthly Notices Roy. Astron. Soc.* **288**, 802–816, [Paper II].

Levshakov, S.A., Kegel, W.H. and Takahara F.: 1997, 'New aspects of absorption line formation in intervening turbulent clouds – III. The inverse problem in the study of H+D profiles', *Monthly Notices Roy. Astron. Soc.*, submitted, [Paper III].

Metropolis, N. *et al.*: 1953, 'Equation of state calculations by fast computing machines', *J. Chem. Phys.* **21**, 1087–1092.

Sarkar, S.: 1996, 'Big bang nucleosynthesis and physics beyond the standard model', *Rep. Prog. Phys.* **59**, 1493–1609.

Walker, T.P. *et al.*: 1991, 'Primordial nucleosynthesis redux', *Astrophys. J.* **351**, 51–69.

Address for correspondence: S.A. Levshakov, Department of Theoretical Astrophysics,
A. F. Ioffe Physico-Technical Institute, 194021 St. Petersburg, Russia
e-mail: lev@astro.ioffe.rssi.ru

THE PRIMORDIAL HELIUM-4 ABUNDANCE FROM OBSERVATIONS
OF A LARGE SAMPLE OF BLUE COMPACT DWARF GALAXIES

TRINH XUAN THUAN

Dept. of Astronomy, University of Virginia, Charlottesville VA 22903

YURI. I. IZOTOV

Main Astronomical Observatory, Goloseevo, Kiev 252650, Ukraine

Abstract. We use a sample of 45 low-metallicity H II regions in blue compact dwarf (BCD) galaxies to determine the primordial helium abundance Y_P with a precision better than 5%. We have carefully investigated the physical effects which may make the He I line intensities deviate from their recombination values such as collisional and fluorescent enhancements, underlying He I stellar absorption and absorption by Galactic interstellar Na I. By extrapolating the Y vs. O/H and Y vs. N/H linear regressions to O/H = N/H = 0, we obtain $Y_P = 0.244\pm0.002$ and 0.245 ± 0.001, respectively, higher than previous determinations ($Y_P = 0.230 - 0.234$). Part of the difference comes from the fact that previous investigators have not taken into account underlying He I stellar absorption, especially in the NW component of the BCD I Zw 18 which, because of its extremely low metallicity plays a key role in the determination of Y_P. We derive a slope $dY/dZ = 2.3\pm1.0$, considerably smaller than those derived before. With this smaller slope and taking into account the errors, chemical evolution models with an outflow of well-mixed material can be built for star-forming dwarf galaxies which satisfy all the observational constraints. Our Y_P gives $\Omega_b h_{50}^2 = 0.058\pm0.007$, consistent with the lower limit set by dynamical measurements and X-ray observations of clusters of galaxies. It is also consistent, within the framework of standard big bang nucleosynthesis theory, with measurements of primordial [7]Li in galactic halo stars and with the D/H abundance measured in absorption systems toward quasars by Burles and Tytler (1997).

Key words: Helium-4, cosmic abundance, HII region, chemical evolution, dwarf galaxies

1. Introduction

Starting in 1993, we have embarked on a large-scale observational program to improve the present determination of the primordial helium abundance Y_P. We were motivated by several reasons. First, we had in hand a new large sample of low-metallicity blue compact dwarf galaxies (BCD) assembled from objective prism survey plates obtained with the 1m Schmidt telescope at the Byurakan Observatory during the Second Byurakan Survey (SBS, Markarian *et al.*, 1989). BCDs are galaxies with $M_B \geq -18$ presently undergoing an intense burst of star formation, as evidenced by their blue UBV colors and their HII region-like optical spectra with strong narrow emission lines superposed on a stellar continuum rising toward the blue (see Thuan, 1991, for a review). BCDs are ideal objects in which to determine Y_P. Their low metallicity ($Z_\odot/50 \leq Z \leq Z_\odot/3$) implies that their gas has not been processed through stars many times, so that the pollution of the gas in the HII regions by nonprimordial helium synthesized in stars is low, allowing us to bypass the chemical evolution problems which plague the determination of [3]He. Furthermore, the low metallicities of the HII regions in BCDs lead to low cooling

Space Science Reviews **84**: 83–94, 1998.
© 1998 *Kluwer Academic Publishers*

and high excitation, making the correction for unseen neutral helium negligible. Because the big-bang production of ^{4}He is relatively insensitive to the baryonic density of matter, Y_P needs to be determined to a very high precision (to better than 5%) in order to put useful constraints on the matter density. This precision can be achieved with very high signal-to-noise ratio emission-line spectra of BCDs. The theory of nebular emission is well understood enough to yield, in principle, the required accuracy. The SBS sample is particularly interesting as it goes ~ 2 mag deeper than the First Byurakan (or Markarian) Survey and contains significantly more low-metallicity BCDs than previous surveys. The SBS has uncovered about a dozen BCDs with $Z \leq Z_\odot/15$, more than doubling the number of such known low-metallicity objects. and filling in the metallicity gap between I Zw 18, the most metal-deficient BCD known with $Z \sim Z_\odot/50$, and other known BCDs (for a detailed description of the SBS BCD sample, see Izotov et $al.$, 1993, and Thuan et $al.$, 1994).

Second, the previous most careful determination of Y_P by Pagel et $al.$ (1992) using the most accurate and consistent data set up to that time gave $Y_P = 0.228 \pm 0.005$, below the limit set by the standard hot big bang model of nucleosynthesis (SBBN) and consistent with it only at the 2σ level. It was critical to obtain another measurement of Y_P with as high or better precision from an independent data set to test SBBN theory.

Third, there were new atomic data for the He atom: HeI recombination emissivities from Smits (1996) and collisional enhancement coefficients from Kingdon and Ferland (1995).

Fourth, we wished to improve on the methodology for determining Y_P. Pagel et $al.$ (1992) and previous authors based their Y_P determination on measurements of individual He I recombination lines and mainly of the He I $\lambda 6678$ line, known to be less subject to collisional enhancement as compared to other He I lines. We adopted an approach which made use of the maximum available information from the emission-line spectrum in order to achieve the very high precision required. There were several problems which we wished to address. In the range of high temperatures ($T_e \geq 12000$ K) found in our BCDs, there are two main mechanisms which may make the He line intensities deviate from pure recombination values. The first mechanism is self-absorption (or fluorescence) in the He I $\lambda 3889$ line. The second mechanism, also leading to the enhancement of some helium lines, is collisional excitation from the metastable $2\,^3$S and $2\,^1$S levels of He I. While collisional excitation is usually taken into account by previous investigators, fluorescence effects are not. There is also the problem of the electron number density N_e(He II). Previously, it was simply set to be equal to N_e(SII) as determined by the [S II] $\lambda 6717/\lambda 6731$ ratio. In the low-density regime (~ 100 cm^{-3}) which applies to BCDs, this ratio is a very uncertain indicator of N_e. This uncertainty is most important for hot H II regions with $T_e = 15000 - 20000$ K which are common in BCDs with the lowest metallicities. Moreover, the electron number density N_e(S II) so determined probably cannot be used because the S$^+$ and He$^+$ regions do not coincide. To address these two problems, we have adopted a self-consistent procedure where we have solved for the electron number density N_e(HeII) and the optical depth τ(3889) so that the He I $\lambda3889/\lambda4471$, $\lambda5876/\lambda4471$, $\lambda6678/\lambda4471$ and $\lambda7065/\lambda4471$ line ratios have their recombination values, after correction for collisional and fluorescent enhancement. The use of many He I lines is necessary to recognize which particular mechanism is responsible for changing the line intensity from its recombination value. The He I $\lambda3889$ and $\lambda7065$ lines play a particularly important role because they are especially sensitive to both unknown quantities. The detailed procedures are described in Izotov, Thuan and Lipovetsky (1994, 1997, hereafter ITL94 and ITL97) and Izotov and Thuan (1997a).

2. Underlying Stellar Absorption and Other Systematic Effects

Effects other than collisional and fluorescent enhancements can also change He I line intensities. An important effect is the underlying stellar absorption in He I lines caused by hot stars which can decrease the intensities of nebular He I lines. Model calculations of synthetic absorption line strengths in star forming regions by Olofsson (1995) show that the equivalent widths of He I absorption lines decrease as the starburst ages and are different for different helium lines. Furthermore, the dependence of He I equivalent widths on metallicity is small, and the equivalent width of the He I $\lambda4471$ absorption line can be as high as 0.35Å. Unfortunately, similar calculations for other important lines such as He I $\lambda5876$ and $\lambda6678$ are not yet available. The effect of underlying He I stellar absorption is most important for the emission lines with the smallest equivalent widths. Therefore, we expect that the He I $\lambda5876$ emission line which has the largest equivalent width should be the least affected by such absorption. This allows the use of the extinction-corrected He I $\lambda6678/\lambda5876$ line intensity ratio as an indicator of underlying stellar absorption. In galaxies where such absorption is important, this ratio should be smaller than the theoretical value given by Smits (1996).

The neglect of He I underlying stellar absorption can lead to a severe underestimate of Y_P. A case in point is that of the BCD I Zw 18 which, being the most metal-deficient object known, has played a key role in the determination of Y_P. I Zw 18 has two centers of star formation referred to as the NW and SE components. Most of the Y determinations in I Zw 18 have been for the brighter NW component either alone, or for the weighted mean of both NW and SE components which reduces to the first case, since the NW component is so much brighter. The many observations of the NW component in I Zw 18 have yielded uniformly the same low value of the helium mass fraction $Y\sim0.230$ (e.g. Pagel et al., 1992; Skillman and Kennicutt, 1993; ITL97; Izotov and Thuan, 1997b), mostly based on the He I $\lambda6678$ emission line. The good agreement between different sets of observers attests to the quality of the data. But there was always a puzzle concerning the helium mass fraction in the NW component: the Ys derived from the two other

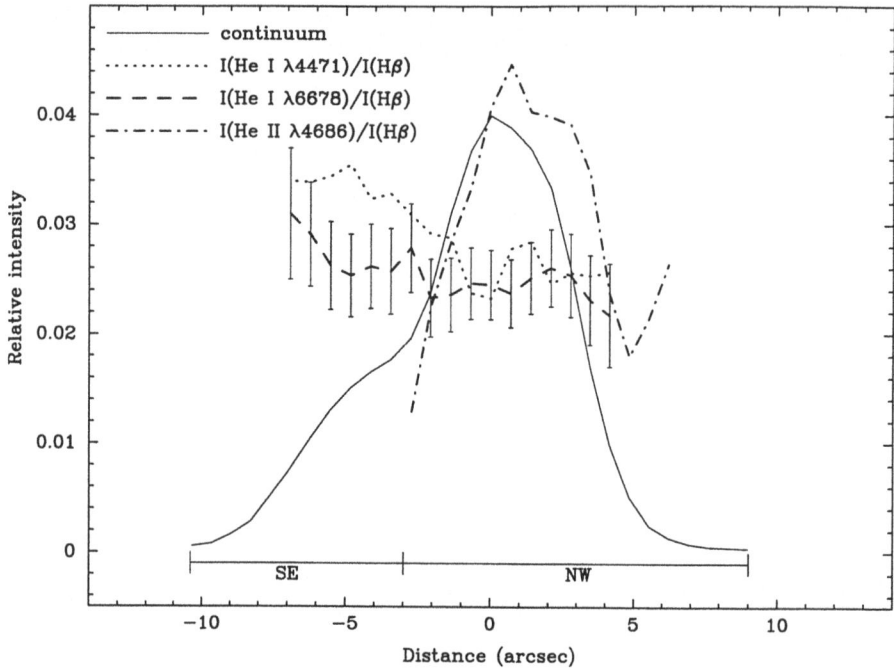

Figure 1. Intensity distributions of the He I λ4471, He I λ6678 and He II λ4686 emission lines, relative to Hβ and corrected for interstellar extinction, in the BCD I Zw 18. The continuum distribution normalized to a value 0.04, is shown for comparison. The decrease of the He I λ4471 and λ6678 line intensities from the SE to the NW component in I Zw 18 is evident. This decrease is mainly caused by underlying stellar absorption, which is most important in the NW component. Error bars are shown for the He I λ6678/Hβ intensity distribution. They are similar for the other line intensity distributions.

He I lines, λ4471 and λ5876, were always systematically below that derived from λ6678. While there is an explanation for the low intensity of the He I λ5876 line – absorption from Galactic interstellar Na I λ5890 – none was advanced for the abnormally low Y ≤ 0.200 derived from the He I λ4471 line, despite the nearly as high quality data for this line as for the He I λ6678 line. Izotov and Thuan (1997b) have argued that the He I λ4471 line in the NW component is subject to strong underlying stellar absorption, which diminishes its intensity. We plot in Figure 1 the spatial variations of the He I λ4471, He I λ6678 and He II λ4686 emission line intensities relative to that of Hβ in I Zw 18, all corrected for interstellar extinction. There is an evident systematic decrease of He I line intensities from the SE to the NW component. The difference between the two components can be as high as ∼30% for the He I λ4471 emission line intensity and ∼10% for the He I λ6678 emission line intensity. Only ∼5% of these differences can be explained by a higher fraction of He^{++} in the NW component. The remaining difference can plausibly be explained by significant underlying stellar absorption in both He I lines. If this explanation holds, all previous measurements of the helium abundance in the NW knot have systematically underestimated the true value. The He I $\lambda6678$ line is also affected, but because of its larger equivalent width, its intensity is reduced by only $\sim5\%$. Thus, as emphasized by ITL97 and Izotov and Thuan (1997a), unless underlying stellar absorption is properly understood and taken into account, the NW component of I Zw 18 cannot be used for primordial helium abundance determination. Such a use would give a Y_P which is systematically too low. Instead, Izotov and Thuan (1997b) argued that, to derive the true Y in I Zw 18, the SE component should be used. There, the Ys derived from the two He I $\lambda4471$ and $\lambda6678$ lines are in good agreement, suggesting that underlying stellar absorption is not as important. The derived helium mass fraction is higher, $Y = 0.242\pm0.009$, in excellent agreement with the Ys derived for other extremely metal-deficient BCDs. Izotov and Thuan (1997b) have suggested that very hot post main-sequence stars can be responsible for the presence of these strong He I absorption lines in the NW component. The population of post main-sequence stars is not as numerous in the SE component because it is younger by several 10^6yr than its NW counterpart, as deduced from its stellar populations and its larger Hβ equivalent width (128Å as compared to 82Å in the NW component). Assuming that the true Y in the NW component of I Zw 18 is equal to 0.243, Izotov and Thuan (1997b) have derived the considerable equivalent absorption widths of 0.4Å and 0.2Å for the He I $\lambda4471$ and $\lambda6678$ lines respectively.

In summary, we reiterate and stress the conclusion that to determine correct helium mass fractions, underlying He I stellar absorption has to be taken into account. To check the importance of absorption, it is essential to compare helium abundances derived from different He I emission lines. Possible problems with strong He I absorption cannot be detected when only one single He I line is considered. Fortunately, the strong He I absorption in the NW component of I Zw 18 constitutes the exception rather than the rule. As discussed by ITL97, underlying He I stellar absorption is relatively less important in other very metal-deficient BCGs such as SBS 0335–052, SBS 0940+544N or SBS 1159+545. This is mainly because the equivalent widths of their He I emission lines are considerably larger (~ 3 times) than those of the NW component of I Zw 18. Thus, given the same He I absorption equivalent widths in all BCGs, the $\sim5\%$ upward correction for the He I $\lambda6678$ line in the NW component of I Zw 18 reduces to a $\leq2\%$ correction in the case of other very metal-deficient BCGs. In practice, this correction for underlying stellar absorption is even smaller by a factor of ~3 as, to determine He abundances, we use several He I lines with the highest weight given to the He I $\lambda5876$ line which has an equivalent width ~ 3 times larger than that of the He I $\lambda6678$ line. In cases where the underlying He I absorption is evidently important, or may play a role (i.e when the He I equivalent width is small), we simply exclude the line from consideration.

The Galactic interstellar Na I $\lambda5890.0$ and $\lambda5895.9$ lines can also be a source of absorption for the He I $\lambda5876$ line in galaxies in the redshift range $z = 0.0023$ – 0.0035. As the velocity dispersion of the interstellar clouds responsible for Na I

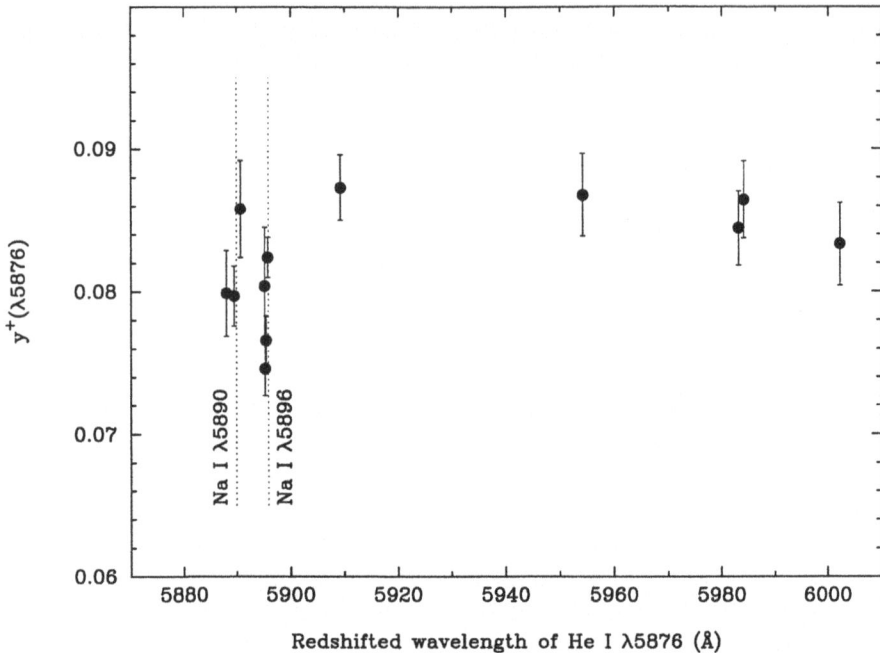

Figure 2. Ionized helium abundance $y^+(\lambda 5876)$ vs. redshifted wavelength of the He I $\lambda 5876$ emission line for galaxies in the Izotov and Thuan (1997a) sample. The rest wavelengths of the Galactic interstellar Na I $\lambda 5890$ and $\lambda 5896$ absorption lines are shown by vertical dotted lines. There is a clear drop of $y^+(\lambda 5876)$ in the region delimited by the dotted lines partly due to Galactic interstellar Na I absorption.

absorption can be as high as 30 km s^{-1}, red or blue shifts of $\sim$0.6Å of the Na I line have also to be taken into account. Davidson, Kinman and Friedman (1989) found the He I $\lambda 5876$ line intensity in the NW component of I Zw 18 to be attenuated by as much as 10% by Na I absorption, while ITL97 found the attenuation can be as high as 20%. Although the effect is smaller for the SE component because its slightly higher redshift ($z = 0.00271$ as compared to $z = 0.00254$ for the NW component) shifts the He I $\lambda 5876$ line more out of the Na I $\lambda 5890.0$ line, Izotov and Thuan (1997b) still found a $\sim$3% attenuation. Figure 2 shows the dependence of the ionized helium abundance y^+, as derived from the He I $\lambda 5876$ line, on the redshifted wavelength of the latter. The quantity y^+ has been corrected for collisional enhancement assuming an electron number density equal to N_e(S II). The rest wavelengths of the two Na I lines are shown by dotted vertical lines. It is evident from Figure 2 that the intensity of the He I $\lambda 5876$ line decreases appreciably when its redshifted wavelength is close to the rest wavelengths of the two Na I Galactic interstellar lines. Hence, we do not take into account that line in the calculations of helium abundance in those galaxies where Na I absorption is important. Again, the use of different He I lines permits a comparison of the ionized helium abundances calculated from each line and the detection of abnormally low

$y^+(\lambda 5876)$ caused by interstellar Na I absorption, which the use of a single He I line does not allow.

We have also examined critically other systematic effects which may influence the determination of Y_P (ITL97). We found the atomic data to be now in reasonably good shape, at least for the optical lines most frequently used to determine Y_P (He I $\lambda 4471$, $\lambda 5876$ and $\lambda 6678$), the best atomic data set being composed of Smits' (1996) He I emissivities and Kingdon and Ferland's (1995a) He I collisional enhancement correction factors. ITL97 also found the effects of corrections for neutral helium, possible deviations from case B recombination theory, temperature fluctuations in H II regions, Wolf-Rayet stellar winds, and supernova shock waves to be small.

3. Results

3.1. THE PRIMORDIAL HELIUM MASS FRACTION Y_P

As of this writing, we have assembled a sample consisting of 45 H II regions in 42 BCDs, 8 H II regions from ITL94, 18 H II regions from ITL97, another 18 H II regions from Izotov and Thuan (1997a) and the SE component of I Zw 18 from Izotov and Thuan (1997b). All these data were obtained and reduced in the same way, making it the largest homogeneous data set available for primordial helium abundance determination. It includes the two most metal-deficient BCGs known, I Zw 18 and SBS 0335–052, and all very metal-deficient BCGs (12+log(O/H)$\leq$7.7) listed by Kobulnicky and Skillman (1996), with the exception of CG 1116+51, Tol 65 and Tol 1214–277. We have performed linear regressions for the relations Y – O/H and Y – N/H with the helium mass fraction Y determined in two ways. First, the Ys are calculated by our preferred method, i.e. they are derived self-consistently, along with the electron number density, using the five brightest He I $\lambda 3889$, $\lambda 4471$, $\lambda 5876$, $\lambda 6678$ and $\lambda 7065$ emission lines in the optical range, and taking into account the collisional and fluorescent enhancements, and with lines subject to He I stellar and Galactic interstellar absorption excluded. Second, the Ys are determined from a single emission line, He I $\lambda 6678$, with the electron number density set equal to N_e(S II), and with only collisional enhancement taken into account. The second way is considered to compare our results with those of previous authors who usually used the latter method. Figure 3 shows both Y – O/H and Y – N/H regressions for the whole sample. The ITL94 and ITL97 data are shown by open circles and the galaxies in Izotov and Thuan (1997ab) by filled circles. We find for the Y – O/H relation $Y_P = 0.244\pm0.002$, greater by $\sim$0.4% than the value derived by ITL97 because we use here five-level instead of three-level atom calculations in the derivation of electron temperatures T_e(O III) and heavy element abundances. The value derived from the Y – N/H linear regression is slightly higher $Y_P = 0.245\pm0.001$. This is due to the fact that the very highest metallicity BCDs in our sample have a slightly higher N/O ratio than the lower

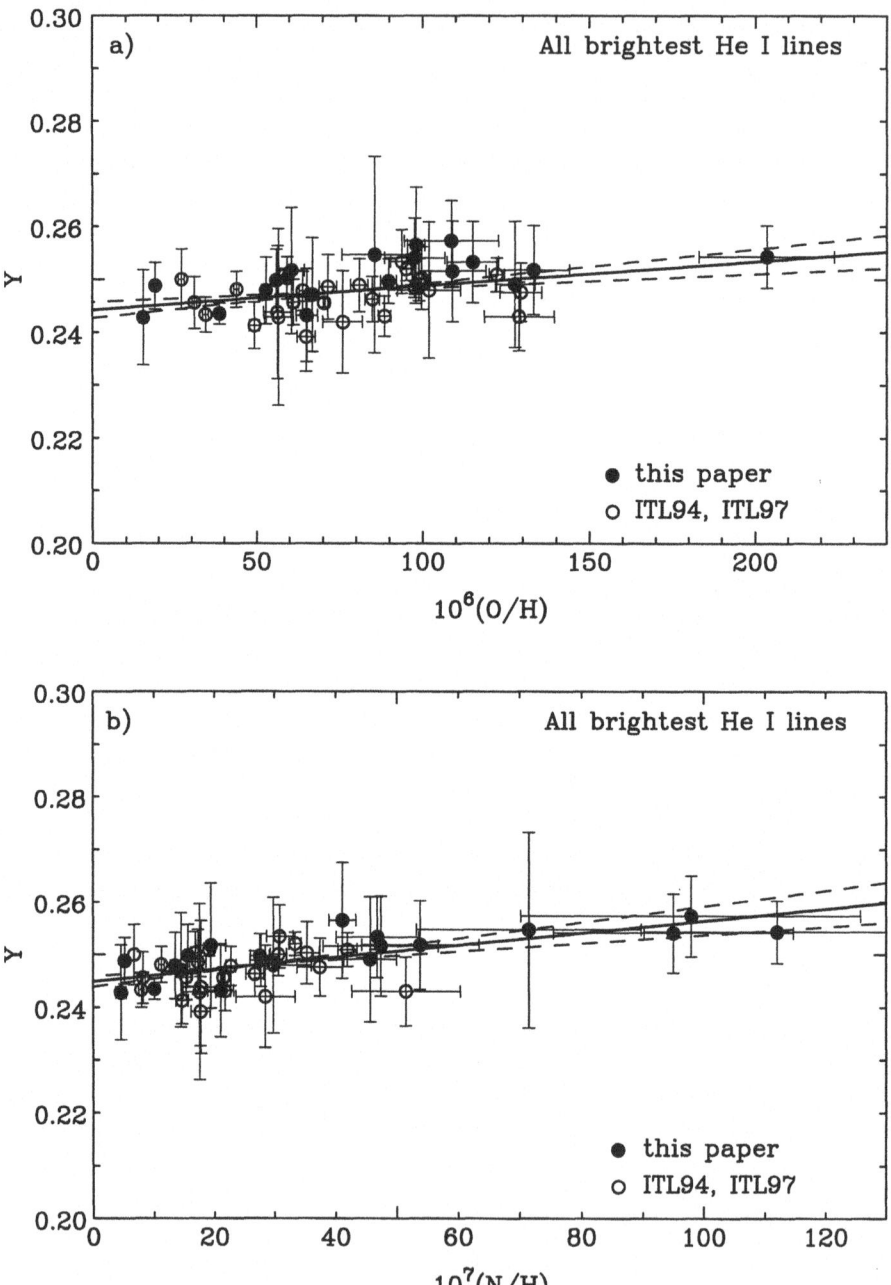

Figure 3. Linear regressions of (a) the helium mass fraction Y vs. oxygen abundance O/H and (b) the helium mass fraction Y vs. nitrogen abundance for our sample of 45 H II regions. The Ys are derived self-consistently by using the 5 brightest He I emission lines in the optical range. Collisional and fluorescent enhancements, underlying He I stellar absorption and Galactic Na I interstellar absorption are taken into account. Open circles denote data from ITL94 and ITL97 and filled circles are data from Izotov and Thuan (1997ab). 1σ alternatives are shown by dashed lines.

metallicity galaxies, thus making the Y – N/H relation slightly non-linear. This higher N/O value is caused by the presence of secondary nitrogen in the higher metallicity BCDs.

To compare our results with those of others we have also calculated Ys with the single line method. We set the electron number density to N_e (S II) to correct for collisional enhancement, but ignore the fluorescent enhancement correction and possible underlying He I stellar absorption. Figure 4 shows the Y – O/H and Y – N/H linear regressions for our sample when only the He I λ6678 emission line is used to calculate Y. The use of a single line does increase the formal error bars and the dispersion of the data points, but the results for Y_P remain similar to those obtained when all brightest He I lines are used. We obtain $Y_P = 0.247\pm0.004$ for the Y – O/H regression and $Y_P = 0.247\pm0.003$ for the Y – N/H regression.

3.2. MEAN Y OF THE LOWEST METALLICITY H II REGIONS

As the slopes for the regression lines in Figures 3 and 4 are relatively small, the mean Y of the most metal-deficient galaxies constitutes also a good estimate of the primordial helium abundance. Using the 5 brightest He I optical lines, the mean for the 2 most metal-deficient BCGs known, I Zw 18 SE and SBS 0335–052, is 0.246 ± 0.004 (standard deviation of the mean), while for the 10 lowest metallicity galaxies, it is $\bar{Y} = 0.246\pm0.002$. When only the He I λ6678 line is used, the mean for I Zw 18 SE and SBS 0335–052 is $\bar{Y} = 0.242\pm0.006$ and that for the 10 most metal-deficient BCGs is $\bar{Y} = 0.245\pm0.003$. The errors are larger because only one He I line is used. The mean Y value derived from the most metal-deficient BCGs in our sample is in excellent agreement with the Y_P derived from linear regressions.

We adopt as our best value $Y_P = 0.244\pm0.002$ as given by the Y – O/H linear regression. We do not average in the slightly higher Y_P given by the Y – N/H linear regression, as there may be some small contamination by secondary N. This is nearly the same value as obtained by ITL97.

3.3. THE dY/dZ SLOPE AND THE CHEMICAL EVOLUTION OF BCGS

The values of the slopes of the Y – O/H and Y – N/H linear regressions provide strong observational constraints on chemical evolution models of blue compact galaxies. We obtain dY/dO varying from 3.7 ± 1.6 to 1.4 ± 3.8 when both methods for calculating Ys are taken into consideration, with a corresponding variation of dY/dZ from 2.4 ± 1.0 to 0.9 ± 2.5. These slopes are consistent within the errors with $dY/dO = 2.6\pm1.4$ and $dY/dZ = 1.7\pm0.9$ obtained by ITL97. They are somewhat uncertain because our BCD sample does not include many very high metallicity objects, the majority of galaxies having $Z \leq Z_\odot/6$, and with only one data point at $Z \sim Z_\odot/4$. If we add to our sample the point for the well-studied Large Magellanic Cloud from Torres-Peimbert, Peimbert and Fierro (1989), we derive a slope $dY/dO = 3.5\pm1.5$ or $dY/dZ = 2.3\pm1.0$ when all He I lines are used and $dY/dO = 1.3\pm3.0$

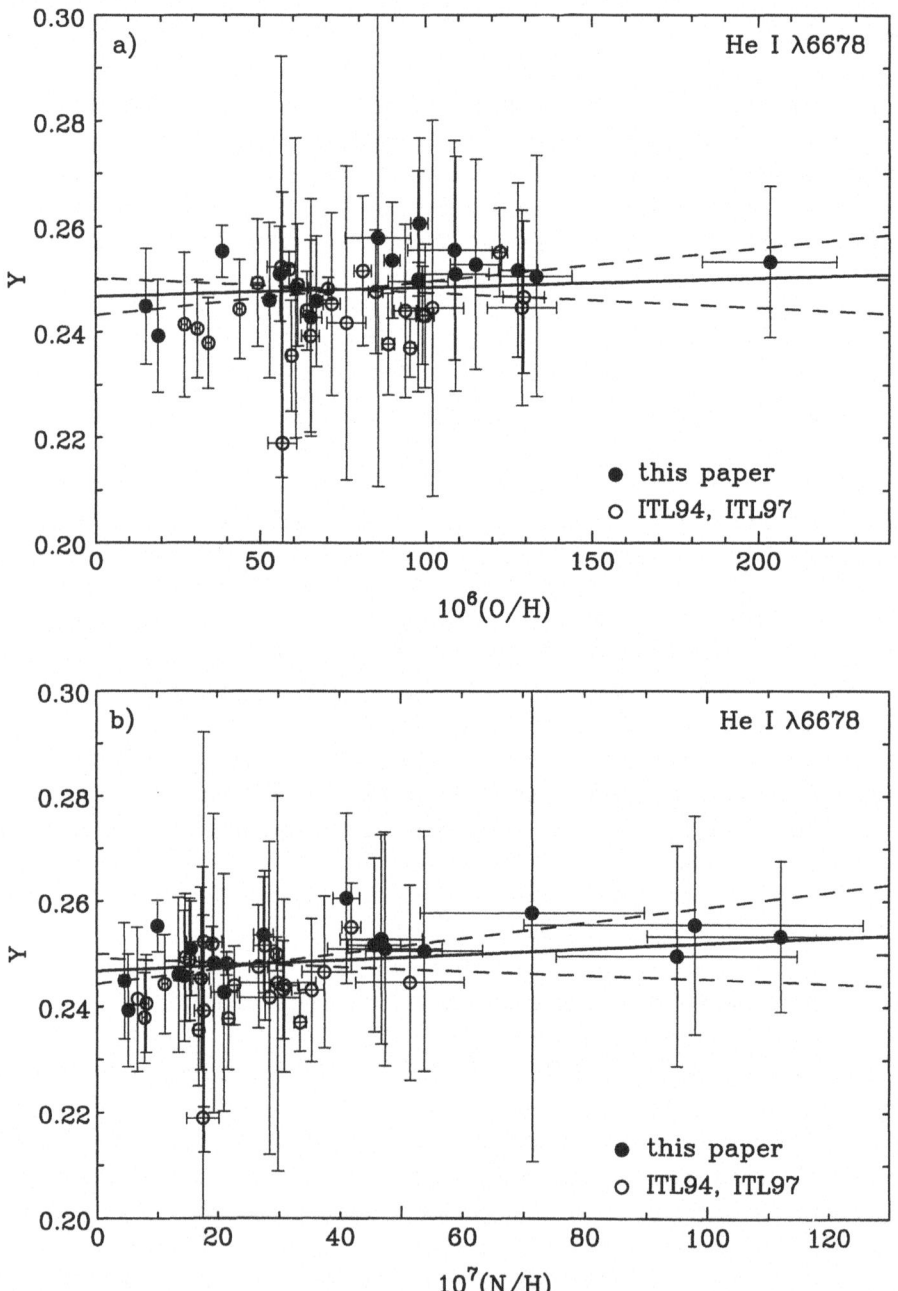

Figure 4. Same as in Fig. 3, except that the helium mass fraction Y is derived by using only the He I λ6678 emission line.

or $dY/dZ = 0.8\pm2.0$ when only He I $\lambda6678$ line is used. The derived Y_P do not change appreciably: $Y_P = 0.245\pm0.001$ in the first case and $Y_P = 0.247\pm0.003$ in the second case. This is to be compared to the values derived by previous authors: $dY/dO = 10.2\pm3.5$ by Pagel $et\ al.$ (1992), $dY/dO = 7.1\pm1.6$ by Carigi $et\ al.$ (1995). It is clear that, despite the uncertainties, our derived slopes are smaller than those obtained previously. A small dY/dO slope presents several advantages. First, it is more consistent with simple chemical evolution models of BCGs. A large slope implies an oxygen deficiency relative to helium, and previously derived large slopes have triggered the proposal of a slew of exotic chemical evolution models. For example, it has been suggested that an oxygen deficiency can be produced if the most massive stars collapse to black holes without enrichment of the interstellar medium with oxygen or if there exists a strong outflow of oxygen-enriched gas. But our smaller slope does not necessarily require such exotic solutions. Within the errors, simple chemical evolution models of BCGs and dwarf irregular galaxies with an outflow of well-mixed gas are sufficient to explain dY/dO, in addition to other observational constraints such as the gas to total mass fraction vs. oxygen mass fraction relation, the carbon-to-oxygen mass ratio and the heavy elements minus carbon and oxygen to oxygen mass ratio (Carigi $et\ al.$, 1995). Second, the extrapolation of the oxygen linear regression with $dY/dO = 3.7\pm1.6$ to solar metallicity gives $Y = 0.282\pm0.015$, in fair agreement with $Y = 0.277$ measured in the Sun.

3.4. COSMOLOGICAL IMPLICATIONS

For our adopted value of $Y_P = 0.244\pm0.002$, a number of neutrino families equal to 3 and a half-lifetime of the neutron equal ~887 s, we derive $\eta_{10} = 10^{10}\eta = 4.0\pm0.5$, where η is the baryon-to-photon number ratio and the error bars denote 1σ errors. This corresponds to a baryonic mass fraction $\Omega_b h_{50}^2 = 0.058\pm0.007$. For $H_0 = 70$ km s^{-1}Mpc^{-1}, $\Omega_b = 0.030\pm0.004$, consistent with the lower bound set by dynamical tests and with the baryonic mass density derived from X-ray observations of clusters of galaxies.

In the framework of SBBN, our derived value Y_P is consistent with measurements of the D + ^{3}He abundance in the local interstellar medium (ISM) and the Solar system, although the derived ISM abundances depend on the specifics of chemical evolution models (deuterium is destroyed and produced in stars, some of it ending up as ^{3}He) and solar system measurements of deuterium suffer from severe isotopic fractionation effects. More recently, D/H measurements in absorption systems toward quasars have been reported. These systems can provide in principle a direct measurement of the primordial deuterium abundance, since they sample metal-poor gas at early epochs where stellar destruction and production of deuterium is negligible. Burles and Tytler (1997) have obtained recently: D/H $= (3.4\pm0.3)\times10^{-5}$. This is to be compared with $(D/H)_p = (4.6\pm0.9)\times10^{-5}$ implied by our Y_P in the context of SBBN. Our value is consistent with that of Burles and Tytler at the 1σ level. It is also consistent with D/H = $(3.9\pm1.0)\times10^{-5}$ found by Chengalur *et al.* (1997) in the outer part of the Galaxy. Our value of Y_P = 0.244 ± 0.002 implies a primordial ^{7}Li abundance ^{7}Li/H = $(2.0\pm0.5)\times10^{-10}$ which is in good agreement with the value ^{7}Li/H = $(1.75\pm0.05_{1\sigma}\pm0.20_{sys})\times10^{-10}$ measured by Bonifacio and Molaro (1997) in halo stars.

References

Bonifacio, P., and Molaro, P.: 1997, *Monthly Notices of the RAS* **285**, 847.

Burles, S., and Tytler, D.: 1997, *Astrophysical Journal*, submitted.

Carigi, L., Colin, P., Peimbert, M., and Sarmiento, A.: 1995, *Astrophysical Journal* **445**, 98.

Chengalur, J. N., Braun, R., and Burton, W. B.: 1997, *Astronomy and Astrophysics* **318**, L35.

Davidson, K., Kinman, T. D., and Friedman, S. D.: 1989, *Astronomical Journal* **97**, 1591.

Izotov, Y. I., Lipovetsky, V. A., Guseva, N. G., Kniazev, A. Y., Neizvestny, S. I., and Stepanian, J. A.: 1993, *Astron. Astrophys. Trans.* **3**, 193.

Izotov, Y. I., Thuan, T. X., and Lipovetsky, V. A.: 1994, *Astrophysical Journal* **435**, 647. (ITL94)

Izotov, Y. I., Thuan, T. X., and Lipovetsky, V. A.: 1997, *Astrophysical Journal, Supplement Series* **108**, 1. (ITL97)

Izotov, Y. I., and Thuan, T. X.: 1997a, *Astrophysical Journal*, submitted.

Izotov, Y. I., and Thuan, T. X.: 1997b, *Astrophysical Journal*, submitted.

Kingdon, J. and Ferland, G. J.: 1995a, *Astrophysical Journal* **442**, 714.

Kobulnicky, H. A., and Skillman, E. D.: 1996, *Astrophysical Journal* **471**, 211.

Markarian, B. E., Lipovetsky, V. A., Stepanian, J. A., Erastova, L. K., and Shapovalova, A. I.: 1989, *Comm. Special Astrophys. Observ.* **62**, 5.

Olofsson, K.: 1995, *Astronomy and Astrophysics, Supplement Series* **111**, 57.

Pagel, B. E. J., Simonson, E. A., Terlevich, R. J., and Edmunds, M. G.: 1992, *Monthly Notices of the RAS* **255**, 325.

Skillman, E., and Kennicutt, R. C., Jr.: 1993, *Astrophysical Journal* **411**, 655.

Smits, D. P.: 1996, *Monthly Notices of the RAS* **278**, 683.

Thuan, T. X.: 1991, in *Massive Stars in Starbursts*, ed., C. Leitherer, N. R. Walborn, T. M. Heckman, and C. A. Norman (Cambridge: Cambridge Univ. Press), 183.

Thuan, T. X., Izotov, Yu. I., Lipovetsky, V. A. and Pustilnik, S. A.: 1994, in *Proc. ESO/OHP Workshop on Dwarf Galaxies*, ed. G. Meylan and P. Prugniel (Garching: ESO), 421.

Torres-Peimbert, S., Peimbert, M., and Fierro, J.: 1989, *Astrophysical Journal* **345**, 186.

TEMPERATURE FLUCTUATIONS, H II REGION ABUNDANCES, AND PRIMORDIAL HELIUM

GARY STEIGMAN
Departments of Physics and Astronomy
The Ohio State University, Columbus, OH 43210, USA

SUELI M. VIEGAS and RUTH GRUENWALD
Instituto Astronômico e Geofísico
Universidade de São Paulo, São Paulo, S.P. 04301-904, BRASIL

Abstract. There is evidence for temperature fluctuations in Planetary Nebulae and in some Galactic H II regions. If such fluctuations occur in the low metallicity, extragalactic H II regions used to probe the primordial helium abundance, the derived ^{4}He mass fraction, Y_P, could be systematically different from the true primordial value. Although this effect could be large, there are no data which allow us to estimate the size of the temperature fluctuations for the extragalactic H II regions. Therefore, we have explored this effect via Monte Carlo simulations of the data in which the abundances derived from a fiducial data set are modified by ΔT chosen from a distribution with $0 \leq \Delta T \leq \Delta T_{max}$ where ΔT_{max} is varied from 500 K to 4000 K.

Key words: Temperature Fluctuations, H II Region Abundances, Primordial Helium

1. Introduction

The primordial abundance of ^{4}He is key to testing the consistency of the standard hot big bang model of cosmology and to using primordial nucleosynthesis as a probe of particle physics (Steigman, Schramm, and Gunn, 1977). The availability of large numbers of carefully observed, low metallicity H II regions has permitted estimates of the primordial helium mass fraction, Y_P, whose statistical uncertainties are very small, $\approx 1\%$ (see, Olive and Steigman, 1995 (OS); Olive, Skillman, and Steigman, 1997 (OSS) and references therein). However, there remains the possibility that in the process of using the observational data to derive the abundances, contamination by unacknowledged systematic uncertainties has biased the inferred value of Y_P. Observers have identified many potential sources of systematic uncertainties (Davidson and Kinman, 1985; Pagel *et al.*, 1992 (PSTE); Skillman *et al.*, 1994; Izotov, Thuan, and Lipovetsky, 1994; 1997 (ITL); Peimbert, 1996) and, where possible, have designed their observing programs to minimize such uncertainties and/or to account for them. It is expected that the contributions from many of the potential sources of systematic uncertainty (if present at all) would vary from H II region to H II region and from observer (telescope/detector combination) to observer, introducing along with a systematic offset in the derived value of Y_P, an accompanying dispersion in the helium data. More insidious would be systematic errors in, for example, the atomic physics used to convert the observed equivalent widths to abundances, since an offset from the "true" value of Y_P would not be

Space Science Reviews **84**: 95–103, 1998.

accompanied by an enhanced dispersion in the fits to the data (e.g., Y vs. O/H or the weighted mean of Y for the lowest metallicity H II regions; see OSS). Recently, we have explored one particular source of potential systematic errors associated with deriving helium and heavy element abundances from emission-line data for extra-galactic H II regions: temperature fluctuations (Steigman, Viegas, and Gruenwald, 1997; hereafter SVG). This contribution to the proceedings of the ISSI Workshop on Primordial Nuclei and Their Galactic Evolution (Bern, Switzerland, 6 – 10 May 1997) is a summary of the results presented in SVG.

In low density regions, such as in H II galaxies, permitted-line emissivities (like those of H and He) decrease (slowly) with temperature (T) and are usually independent of electron density (n_e), unless collisional effects are important. In contrast, the forbidden-line emissivities (like those of O, N, S) are strongly dependent on T, and the dependence on the density may also be important. In H II galaxies, the [S II] line ratio is typically suggestive of low electron densities ($n_e < 500 \, \text{cm}^{-3}$), and except for high temperature regions, collisional effects are negligible. Thus, the derived heavy element abundances depend crucially on a good temperature determination. For these regions, the temperature used for the high-ionization lines is obtained from the [O III] line ratio, T_{OIII}, while that for the low-ionization lines is derived from T_{OIII} using results from photoionization models (see for example, PSTE).

A similar method is usually applied to planetary nebulae. In contrast to H II regions, the Balmer temperature T_B (obtained from the observed Balmer disconti-nuity) is also determined for several PNe. In many cases T_B is found to be lower than T_{OIII} (Peimbert, 1971; Liu and Danziger, 1993). As pointed out by Peimbert (1971), this discrepancy could be due to temperature fluctuations, which, however, are not predicted by the standard photoionization models for these nebulae (Liu and Danziger, 1993). If, indeed, the gas temperature is overestimated by T_{OIII}, with the true temperature given by T_B, the heavy element abundances derived from for-bidden lines using T_{OIII} are underestimated (Viegas and Clegg, 1994). Indeed, the abnormal chemical abundances inferred for some H II galaxies with WR features may result from just such an overestimate of the gas temperature (Esteban and Peimbert, 1995). Although one well-observed giant H II region, NGC 2363, shows just such a discrepancy between the Paschen temperature and T_{OIII} in two knots (González-Delgado et al., 1994), temperature fluctuations appear to be absent in the M33 giant H II region NGC 604 (Terlevich et al., 1996). It must be stressed that absence of evidence for a Paschen discontinuity is not evidence for absence of temperature fluctuations since a 10% difference (not uncommon) corresponds to $\approx 2000 \, \text{K}$ (E. Terlevich, Private Communication).

The H II galaxies used to derive the primordial helium abundance are anything but "classical", homogeneous Stromgren spheres and winds from pockets of very hot, young stars may well shock the gas producing temperature fluctuations which, if neglected, would yield an underestimate of the true oxygen abundance. This may introduce a systematic bias into the inferred value of Y_P derived from the helium versus oxygen correlation. Furthermore, temperature fluctuations will also have a direct effect on the helium abundance determined from the recombination lines (Esteban and Peimbert, 1995). Although the ratio of He and H emissivities is very nearly independent of temperature, the helium line intensities are modified from the pure recombination case by collisional excitation (Cox and Daltabuit, 1971; Ferland, 1986) which is temperature-sensitive. If collisions are important, and if the "true" H II region temperature has been overestimated, then the collisional correction will have been overestimated and the "true" helium abundance under-estimated. This effect, if present, will be largest for the hottest, most metal poor H II regions. When collisions are negligible it is the small temperature dependence of the hydrogen and helium recombination lines which matters and Y_P will be slightly overestimated (Peimbert, 1995). Thus the combined effects, which tend to increase the derived oxygen abundance in the metal rich H II regions and the helium abundance in the metal poor regions, may "tilt" the inferred Y versus O/H relation, flattening the slope and increasing the intercept, Y_P. In SVG we analysed these effects in order to quantify the resulting systematic uncertainty in Y_P. Below our approach is described and our results summarized. For more details and references, see SVG.

2. The Method

Given the lack of data on the Balmer (or Paschen) temperatures for the key, low metallicity, extragalactic H II regions used in the derivation of the primordial helium abundance, we are forced to statistical estimates of the magnitude of the possible corrections. First, we adopted a "fiducial" data set using 43 low metallicity (O/H $\leq 1.5 \times 10^{-4}$) H II galaxies from PSTE, ITL, Skillman and Kennicutt (1993) and Skillman *et al.* (1994) whose role is merely to provide a comparison, in order to quantify the *changes* induced by temperature differences. The restriction to low metallicity regions is to minimize the contamination of primordial helium by that produced in stars. For each H II region in our fiducial data set we adopt the same algorithms used by these authors to derive the temperature and the fractional abundances in the low-ionization zones. The electron density is obtained from the [S II] line ratio and we use the Brocklehurst (1972) recombination coefficients for the He and H lines along with the collisional corrections for the He lines from Clegg (1987). The more recent computations of the collisional corrections by Kingdon and Ferland (1995) are in excellent agreement with those of Clegg (1987). The fractional abundance of He^+ we adopt is the unweighted average of the abundances obtained from He I $\lambda4471$, He I $\lambda5876$ and He I $\lambda6678$ lines. Ignoring any statistical uncertainties, we use these abundances to find a "standard", or fiducial, linear fit to Y versus O/H (unweighted) in the absence of temperature fluctuations. We emphasize that we are not so much interested in the "best fit" Y versus O/H relation for our fiducial data set but, rather, the *differences* in the Y versus O/H relations between ΔT zero and nonzero ($\Delta T \equiv T_{OIII} - T$). As may have been anticipated, the higher metallicity regions tend to be cooler which has the effect that the oxygen abundance derived for the high metallicity regions is very sensitive to temperature fluctuations.

Since for planetary nebulae the observed temperature difference $\Delta T = T_{OIII} - T_B$ is typically 2000 K (Liu and Danziger, 1993) with one PN having $\Delta T = 6000$ K, and for the giant H II region NGC 2363, $\Delta T \geq 3000$ K (González-Delgado *et al.*, 1994), we have recalculated the helium and oxygen abundances for each H II region with ΔT chosen from a distribution which ranges from zero to ΔT_{max}, according to a probability $P(\Delta T)$ to be described below. ΔT_{max} is varied in the range $500 \leq \Delta T_{max} \leq 4000$ K. We repeated this procedure 10,000 times for each choice of ΔT_{max}. For most H II regions Y does not decrease with decreasing temperature as might have been expected from the effective recombination coefficients alone (Esteban and Peimbert, 1995) provided we account for the collisional effect on the He lines; for H II galaxies with high gas temperature, the collisional effect dominates, leading to an increase of Y. For each of the 10,000 realizations (for each choice of ΔT_{max}) of our H II region data set we fit a linear Y versus O/H relation to derive Y_P and the slope $\Delta Y / \Delta Z$, where the heavy element mass fraction $Z \approx 20(O/H)$ (see PSTE).

3. Results: Y Versus Z

3.1. FULL DATA SET

The results for our full data set of 43 H II regions, selected to have O/H $\leq 1.5 \times 10^{-4}$, are shown in Figure 1. In the top panel is plotted the offset in Y_P as a function of ΔT_{max}. As anticipated, the intercept, Y_P, is systematically increased by an amount which increases with ΔT_{max}. For the not unreasonable choice of $\Delta T_{max} = 2000$ K, the increase in the inferred value of Y_P is significant, comparable to some estimates of the upper bound to the systematic uncertainty in Y_P (OS; OSS). The middle panel displays the change in the slope of the Y vs. Z relation with ΔT_{max}. As expected, the increase in the intercept of the Y vs. Z relation is accompanied by a flattening of the slope.

It might have been anticipated (see OSS) that such large systematic offsets would be accompanied by an increase in the dispersion of the abundances around the best fit Y versus Z relation. To test for this, for each realization we have computed σ, the variance of the residuals between the data (Y) and those values predicted by the corresponding linear fit for that realization (Y_{fit}). The bottom panel shows the change with ΔT_{max} of the dispersion about the best fit linear Y vs. Z relation. Surprisingly, this effect of added dispersion, if present at all, is very small compared, for example, to the typical uncertainty in the individual H II region Y determinations which are of order 0.010 (OSS). Thus, until there are observations

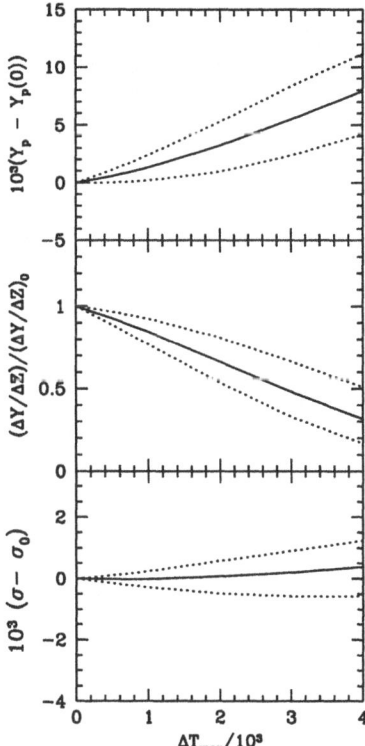

Figure 1. In the three panels the solid curves show the variation of the mean values of the offsets in Y_P (top panel), the ratio of slopes (middle panel) and the dispersion around the best fit linear Y vs. Z relation (bottom panel) as a function of ΔT_{max}. The dotted curves show the 95%CL ranges.

of H II region temperatures determined from hydrogen lines, it is difficult to set a firm upper bound to the magnitude of the systematic offset in Y_P due to the possible presence of temperature fluctuations. In the meanwhile it is, therefore, important to consider how best to analyze current data so as to minimize the potential importance of such temperature fluctuations.

3.2. Truncated Data Set

A significant contribution to the systematic offset in Y_P in the presence of temperature fluctuations comes from the flattening of the slope of the Y versus Z relation due to the large increase in oxygen abundance for the cooler, higher metallicity H II regions. This suggests that the uncertainty in Y_P derived from a linear fit to Y versus O/H might be minimized if the data set were restricted to the hotter, lowest metal abundance H II regions. We explored this by identifying for each realization of our Monte Carlos the subset of H II regions with low metallicity: O/H $\leq 0.9 \times 10^{-4}$. For these subsets we fit linear Y versus Z relations and compared the slopes and intercepts (as well as σ) to those found for the corresponding low

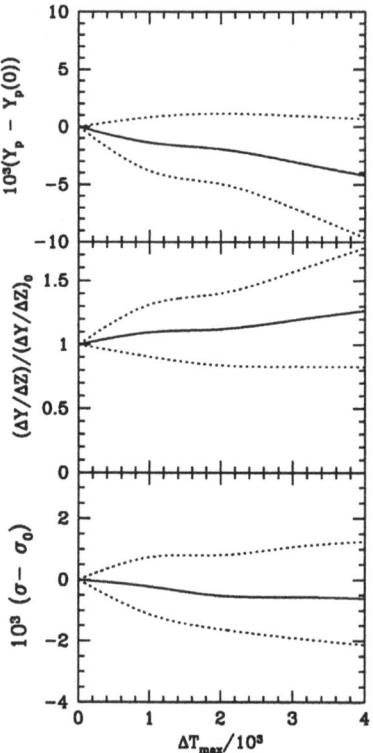

Figure 2. Similar to Figure 1, but for the low-metallicity subset of H II regions (O/H $\leq 0.9 \times 10^{-4}$).

metallicity (fiducial) subset in the absence of temperature fluctuations. The results are shown in Figure 2. As expected, the changes in slope and intercept for this low metallicity set are much smaller. Indeed, the trend is even opposite that for our full data set: *lower* intercept, *higher* slope (although the effect is so small as to be only marginally significant). The lesson is clear. To minimize the uncertainty in Y_P derived from a fit of Y versus Z, due to possible temperature fluctuations, we should restrict our attention to the most metal poor H II regions.

3.3. LOWEST Y H II REGIONS

Since it is generally accepted that the helium abundance has only increased since the big bang, an alternate approach to using H II regions observations to infer the primordial abundance is to compute the mean of Y for those regions with the lowest helium abundances; then, $Y_P \leq \langle Y \rangle$ (OS; OSS). We note here that the lowest oxygen abundance H II regions may not necessarily be those with the least stellar pollution of pristine, primordial helium. For example, oxygen produced by short-lived massive stars during a prior epoch of star formation may have been blown out of the H II galaxy, while helium produced by longer-lived low mass stars

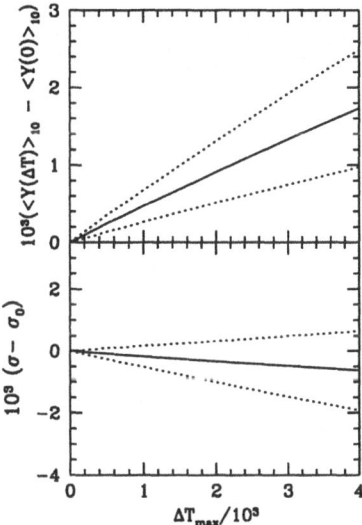

Figure 3. Similar to Figures 1 and 2 for the means of the ten lowest Y H II region helium abundances as a function of ΔT_{max}.

was retained yielding a (relatively) high Y, low O/H H II region. Thus, the lowest helium abundance regions are likely to provide a less biased upper bound to Y_P than that derived from the lowest O/H H II regions. Therefore, we identified the ten lowest Y H II regions in our fiducial data set ($\Delta T = 0$) and found the (unweighted) average of Y, $\langle Y \rangle_{10}(0)$. Then, from our Monte Carlo simulations of the data we identified for each realization (for various choices of ΔT_{max}) the ten lowest Y H II regions (which are often, but not always, the same regions as those in the fiducial set) and we computed $\Delta \langle Y \rangle_{10} \equiv \langle Y \rangle_{10}(\Delta T) - \langle Y \rangle_{10}(0)$.

The trend of $\Delta \langle Y \rangle_{10}$ with ΔT_{max} is shown in Figure 3 where the corresponding change in the dispersion around the mean (σ) is also shown. As expected, although temperature fluctuations tend to increase $\langle Y \rangle_{10}$ systematically (mainly due to the reduced correction for collisional excitation), the effect is quite small.

4. Discussion

Until data on temperature fluctuations in these key extragalactic H II regions become available, the best strategy is to analyze the current data in a manner which minimizes this potential systematic error. As we have seen, one possibility is to restrict attention to the lowest metallicity regions. OS and OSS have done this for the subset with O/H $\leq 0.9 \times 10^{-4}$ for which OSS derive from a linear fit to Y versus O/H: $Y_P = 0.230 \pm 0.007$ (95%CL). From our Monte Carlo simulations of the data we find that the systematic correction for this subset is likely to be very small (indeed, it may even be negative!); for $\Delta T_{max} = 4000\,\mathrm{K}$, $\Delta Y_P = -0.004 \pm 0.005$

(see Figure 2). If, perhaps naively, we apply this correction to the OSS result, we infer: $Y_P = 0.226 \pm 0.009$ (95%CL). In this case the "tension" between primordial helium and low primordial deuterium (Hata *et al.*, 1995) is exacerbated.

Another "safe" approach to using the existing data to estimate Y_P is to ignore any metallicity information and simply take a mean ($\langle Y \rangle$) of the helium abundance for the lowest abundance H II regions. Since helium is expected to only increase after BBN, this should provide an upper bound to Y_P. In general, the H II regions with the lowest values of Y tend to be the lowest metallicity regions which are also the hottest. For such regions there is often a non-negligible correction for collisions when Y is derived from the helium line intensities. If, in fact, the gas is cooler, this correction will have been overestimated and the "true" value of Y is larger. Thus, temperature fluctuations will increase $\langle Y \rangle$. From our Monte Carlos we selected, for each realization, the ten H II regions with the lowest Y values and we found, for $\Delta T_{max} = 4000\,K$, a small systematic increase: $\Delta \langle Y \rangle_{10} = 0.002 \pm 0.001$ (see Figure 3). For their ten lowest Y regions, OSS find $\langle Y \rangle_{10} = 0.230 \pm 0.006$ (95%CL), so that even with our largest correction we infer a revised mean of 0.232 ± 0.006 suggesting that $Y_P \leq 0.238$ (95%CL). Here, too, the "tension" between D and ^{4}He (Hata *et al.*, 1995) fails to be relieved.

5. Summary

Temperature fluctuations in low metallicity, extragalactic H II regions may significantly effect the determination of the primordial abundance of helium. If present, they may increase the metallicity of the higher metallicity, relatively cooler regions and increase the helium abundance of the more metal poor, hotter regions flattening the inferred Y versus Z relation which leads to a higher zero-metallicity intercept (Y_P). We have found that such a systematic offset is not necessarily accompanied by a significant increase in the dispersion of the data around the best fit linear Y versus Z relation and, therefore, may remain invisible in the absence of direct data on temperature differences in these regions. It is clear that temperature fluctuation data is crucial to constraining the uncertainty in Y_P. In the hopes of obtaining these important data we (S.M.V. & R.G. and colleagues) have proposed to make such observations. In the absence of such data, we have noted that restricting attention to the lowest metallicity regions ($Z \leq Z_\odot/10$) will tend to minimize this systematic offset. Alternatively, the mean of the helium abundances of the lowest Y H II regions is also likely to be robust against the effect of temperature fluctuations.

Acknowledgements

The work of S.M.V. and R.G. is partially supported by FAPESP and CNPq in Brasil; that of G.S. is supported by the DOE at Ohio State. G.S. wishes to acknowledge

ISSI, and especially Professor J. Geiss and R. von Steiger, for the friendly and efficient organization of a timely and fruitful workshop.

References

Brocklehurst, M.: 1972, *MNRAS* **157**, 211.
Clegg, R. E. S.: 1987, *MNRAS* **229**, 31P.
Cox, D. P., and Daltabuit, E.: 1971, *ApJ* **167**, 257.
Davidson, K., and Kinman, T. D.: 1985, *ApJS* **58**, 321.
Esteban, C., and Peimbert, M.: 1995, *A&A* **300**, 78.
Ferland, G. J.: 1986, *ApJ* **310**, L67.
González-Delgado, R. M., Pérez, E., Tenorio-Tagle, G., Vilchez, J. M., Terlevich, E., Terlevich, R., Telles, E., Rodriguez-Espinosa, J. M., Mas-Hesse, M., García-Vargas, M. L., Díaz, A. I., Cepa, J., and Castañeda, J.: 1994, *ApJ* **437**, 239.
Hata, N., Scherrer, R. J., Steigman, G., Thomas, D., Walker, T. P., Bludman, S., and Langacker, P.: 1995, *Phys. Rev. Lett.* **75**, 3977.
Izotov, Y. I., Thuan, T. X., and Lipovetsky, V. A.: 1994, *ApJ* **435**, 647 (ITL).
Izotov, Y. I., Thuan, T. X., and Lipovetsky, V. A.: 1997, *ApJS* **108**, 1 (ITL).
Kingdon, J., and Ferland, G. J.: 1995, *ApJ* **442**, 714.
Liu, X., and Danziger, J.: 1993, *MNRAS* **263**, 256.
Olive, K. A., Skillman, E., and Steigman, G.: 1997, *ApJ* **483**, 788 (OSS).
Olive, K. A., and Steigman, G.: 1995, *ApJS* **97**, 49 (OS).
Osterbrock, D. E. 1989, 'Astrophysics of Gaseous Nebulae and Active Galactic Nuclei', (Mill Valley, University Science Books)
Pagel, B. E. J., Simonson, E. A., Terlevich, R. J., and Edmunds, M.: 1992, *MNRAS* **255**, 325 (PSTE).
Peimbert, M.: 1971, *Bol. Obs. Tonantzintla y Tacubaya* **6**, 29.
Peimbert, M.: 1995, in 'The Analysis of Emission Lines', ed. R. E. Williams and M. Livio (Cambridge; Cambridge Univ. Press), 165.
Peimbert, M.: 1996, *Rev. Mex. Astr. Astrofis., Serie de Conferencias* **4**, 55.
Skillman, E., and Kennicutt, R. C.: 1993, *ApJ* **411**, 655.
Skillman, E., Terlevich, R. J., Kennicutt, R. C., Garnett, D. R., and Terlevich, E.: 1994, *ApJ* **431**, 172.
Steigman, G., Schramm, D. N., and Gunn, J.: 1977, *Phys. Lett.* **B66**, 202.
Steigman, G., Viegas, S. M., and Gruenwald, R.: 1997, *ApJ* **490**, In Press (Nov. 20, 1997) (SVG).
Terlevich, E., Diaz, A. I., Terlevich, R., González-Delgado, R. M., Pérez, E., and García-Vargas, M. L.: 1996, *MNRAS* **279**, 1219.
Viegas, S. M., and Clegg, R.: 1994, *MNRAS* **271**, 993.

Address for correspondence: Professor G. Steigman, Department of Physics, The Ohio State University, 174 West 18th Avenue, Columbus, OH 43210, USA

TOWARD AN UNDERSTANDING OF THE SYSTEMATIC UNCERTAINTIES IN DERIVING THE PRIMORDIAL HELIUM ABUNDANCE FROM H II REGION OBSERVATIONS

EVAN D. SKILLMAN
Astronomy Department, University of Minnesota,
116 Church Street, SE, Minneapolis, MN, 55455, USA

ELENA TERLEVICH
Institute of Astronomy, Madingley Road,
Cambridge CB3 0HA, England

ROBERTO TERLEVICH
Royal Greenwich Observatory, Madingley Road,
Cambridge CB3 0EZ, England

Abstract. Recent determinations of the primordial He abundance have given significantly different results. We are attempting to identify some of the causes of these differences and propose observational solutions. Here we identify a systematic difference in how the data are interpreted (differences in corrections for the presence of neutral helium) and the importance of a systematic bias towards lower derived helium abundances (underestimating the presence of underlying stellar absorption).

Key words: H II Region Abundances, Big Bang Nucleosynthesis, Primordial Helium

> *"He was also one of the first to realise fully the importance of repeating the same observation many times under different conditions, in order that the various accidental sources of error in the separate observations should as far as possible neutralise one another."*
> Berry (1898) regarding Tycho.

1. Introduction and Motivation

Since the original suggestion by Peimbert and Torres-Peimbert (1974; 1976) the preferred method for determining the primordial helium abundance consists of measuring the He/H ratio in HII regions of ever decreasing metallicity and extrapolating this relationship back to zero metallicity. Over the last two decades, improvements in this technique have come from increasing the sample of low metallicity HII regions, decreasing the statistical uncertainties associated with the emission line ratios measured from the HII region spectra, improvements in the understanding of the atomic physics and radiative transfer necessary to convert the emission line ratios into a helium abundance, and a better appreciation of possible systematic effects and their limits.

The work of Pagel *et al.* (1992, PSTE) represents a landmark study in this field. They presented new data, reanalyzed and homogenized data from the literature, and

Space Science Reviews **84:** 105–114, 1998.

provided a detailed description of the methodology required. They found that the relationships between He/H and metallicity (using both O/H and N/H) were linear to within the uncertainties. They then derived a value of the primordial helium abundance of 0.228 ± 0.005, or an upper limit of 0.242 (95% confidence interval) accounting for "reasonably likely" systematic effects.

PSTE also noted that HII regions with Wolf-Rayet features in their spectra may have their abundances altered by the mass loss processes associated with the stars in the exciting cluster. However, Olive and Steigman (1995) found no statistical evidence (from a study of the data set of PSTE) in support of this hypothesis. Recently, Kobulnicky and Skillman (1996) pointed out that the absence of low metallicity HII regions with Wolf-Rayet features in their spectra skews the comparison with normal HII regions. Constraining the comparison to the metallicity range in which Wolf-Rayet features are found, there is no evidence of a difference.

While the desired precision for the primordial helium abundance is better than 2% (because of the mild dependence of He/H on the baryon-to-photon ratio in the early universe, see, e.g., Walker *et al.*, 1991), the He/H determinations for the individual HII regions observed by PSTE typically carry uncertainties in the range of 3 to 8%. Skillman *et al.* (1994) obtained multiple observations of a very low metallicity HII region in the nearby dwarf galaxy UGC 4483 and showed that it is possible to determine the relevant emission line ratios with accuracies of order 2%, and that spatially resolved spectra showed no evidence of the presence of neutral helium.

In the past few years, estimates of the primordial helium abundance, key to the consistency of SBBN, have come under great scrutiny. There have now been several discussions of the uncertainties in converting observed spectra into He/H abundances and attention has focussed on both minimizing statistical uncertainties and understanding systematic uncertainties (e.g., Davidson and Kinman, 1985; Dinerstein and Shields, 1986; Skillman *et al.*, 1994).

One goal which has remained salient is the need for high quality spectra of very low metallicity objects. Although there is no secure evidence for a change in slope of the $\Delta Y/\Delta Z$ relationship for H II regions, having accurate He/H abundances at very low metallicities minimizes any extrapolation to the primordial He abundance.

Recently Izotov *et al.* (1997, ITL97) have added new observations of HII regions over a large range in metallicity and have derived a new value of the primordial helium abundance, using exclusively their own data. The resultant value of 0.243 ± 0.003 is significantly higher than that of PSTE (see Figure 1). ITL97 attribute the bulk of the difference to the use of the improved He emissivities of Smits (1996) and the collisional excitation rates calculated by Kingdon and Ferland (1995; based on the calculations of Sawey and Berrington, 1993).

Olive, Skillman, and Steigman (1997, OSS) have shown that the new data of ITL97 are in excellent agreement with the previous abundance determinations of PSTE. They further show that the changes in the atomic data are negligible (as had been stated previously in the literature, Smits, 1996; Kingdon and Ferland,

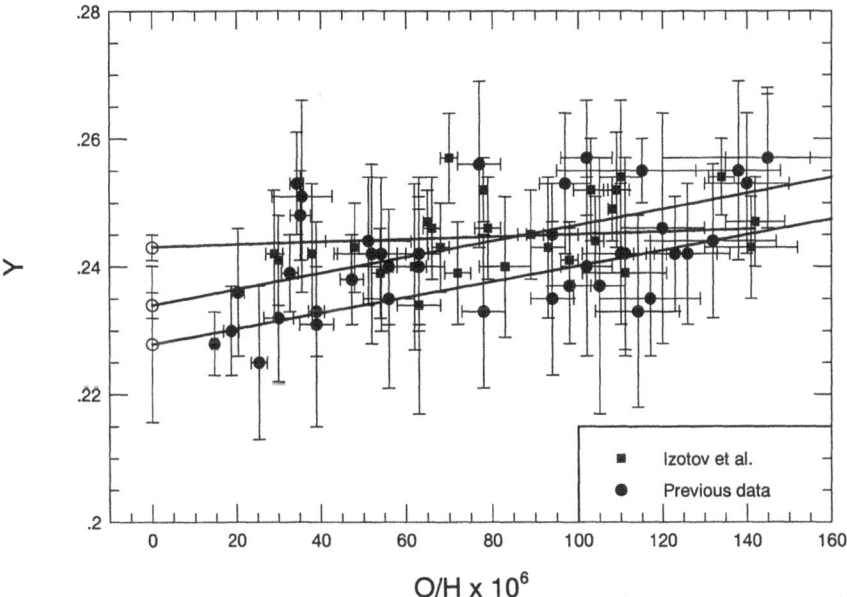

Figure 1. The helium (Y) and oxygen (O/H) abundances for the extragalactic H II regions from ITL97 (filled squares) and from other observations (filled circles – mostly from the compilation by PTSE). Regressions (using oxygen data only) from PSTE, OSS, and ITL97, are shown as solid lines. The intercepts for these regressions are shown with open circles plus error bars (from lowest to highest: PSTE, OSS, and ITL97).

1995). The main reason for the increased value derived by ITL97 is the rejection of observations of I Zw 18 (the HII region with the lowest measured metallicity, Skillman and Kennicutt, 1993) from their analysis and a resulting lack of low metallicity data points.

Given the quality of the data obtained so far and the attention to statistical and systematic uncertainties, one might consider that the primordial helium abundance determination is in good shape. In fact, there appears to be considerable concerns in the community about the currently published values. Since there is a lack of concordance with the recently obtained *low* values of D/H (Tytler *et al.*, 1996), it has been inferred that helium abundance determinations suffer from rather large systematic uncertainties. It has further been suggested that the best upper limit is derived by taking the extrapolated value, adding 2σ of statistical uncertainty, and adding (linearly!) every imaginable systematic uncertainty at its maximum possible value.

OSS have attempted to determine the best estimate given all the available data (0.234 ± 0.002). A thorough discussion of the systematic effects and their known limits is given therein, and a semi-empirical analysis supports the estimate of 0.005 given by PSTE (resulting in a "firm" 2σ upper bound of 0.244). Both of the above approaches to systematic errors leave something to be desired. To best estimate the total systematic error, we will need to estimate not only the amplitude, but also the *distribution*, of each of the various uncorrelated sources of systematic errors.

2. The Neutral Helium Correction

The degree to which the hydrogen and helium ionization zones in an HII region coincide is generally determined by the hardness of the ionizing radiation field, and may be governed, in part, by geometry (e.g., Osterbrock, 1989). Thus, there is always concern that in a specific observation of an HII region that neutral helium is co-existent with ionized hydrogen along the line of sight (see, e.g., discussion in Dinerstein and Shields, 1986).

Historically, a correction has been applied to the helium abundances in order to correct for unobserved neutral helium. Vílchez and Pagel (1988) used the models of Stasińska (1990) to demonstrate that ratios of ionization fractions of sulfur and oxygen provided an accurate measure of the hardness of the radiation field. PSTE used this technique to determine whether such a correction was necessary. The proposed methodology consisted of a simple test: if the radiation field was soft enough that a significant correction for neutral helium was implied, this correction was probably too uncertain for the proposed candidate to be useful for a helium abundance measurement.

Izotov, Thuan, and Lipovetsky (1994; ITL94) applied neutral helium corrections based on Stasińska (1990), without adopting the methodology of PSTE. Unfortunately, the correction derived in ITL94 is based only on the neutral helium fraction and does not take into account the neutral hydrogen fraction (see Vílchez, 1989). The result is that the derived helium abundances are too high. This correction accounts for about a one percent upward bias in the dataset of ITL97, or about 25% of the difference between ITL97 and OSS.

3. The Importance of Underlying Absorption

A second uncertainty and potential source of systematic error is the need to compensate for stellar absorption underlying the helium emission lines. We have been analyzing detailed spectroscopic observations of SBS0335-052 in order to address this concern. Izotov *et al.* (1997; I97) have independently identified SBS0335-052 as a region in which underlying stellar absorption affects the He abundance determination. The equivalent width of the $\lambda6678$ line is 10 Å, similar to that in I Zw 18, and, as discussed in SK93, previously it had been thought that this correction would be negligible at these equivalent widths (cf. Kunth and Sargent, 1983; Olofsson, 1995).

In order to test that assumption, in Figure 3 we have constructed a diagnostic diagram consisting of a plot of the He abundance derived from a single He emission

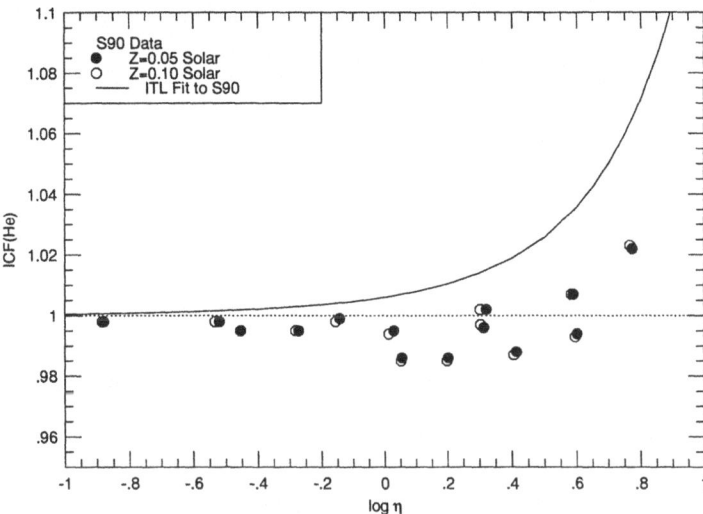

Figure 2. The ionization correction factor for neutral helium as a function of the radiation hardness parameter η (as defined in Vílchez and Pagel, 1988). The filled and open circles are the results of models calculated by Stasińska (1990). The solid line is the fit to these models given by ITL94.

line (assuming no underlying absorption) versus the equivalent width of the He emission line. We have plotted data taken from I97 and each of our three independent spectra. In all four cases one can clearly see the effects of He absorption lines in the underlying stellar spectrum. The weaker lines produce values of the He abundance which are significantly less than the values produced by the stronger lines. We have not included corrections for collisional enhancement (i.e., the plot is consistent with a low density), but the pattern of decreasing He abundance with decreasing EW cannot be corrected for by increasing an assumed density (due to the different density dependences of the four He lines).

The four lines plotted should have nearly equal values of the EW of underlying absorption. Observations of individual Galactic B supergiants (Lennon, Dufton, and Fitzsimmons, 1993) show that the EW of the absorption lines of $\lambda 6678$, $\lambda 4471$, and $\lambda 4026$ are all of approximately equivalent strength and share the same dependency on stellar effective temperature. The models by Auer and Mihalas (1972) show relatively good agreement for all four lines for temperatures in excess of 35,000 K and surface gravity values values of $\log g = 4$ and 4.5.

Plotted in the same diagram are the results of a simple model which predicts the amplitude of the effect under the assumption that the EW of the absorption underlying the plotted emission lines is equal for all lines. That is, we have plotted:

$$\left(\frac{He^+}{H^+}\right)_{derived} = \left(\frac{He^+}{H^+}\right)_{intrinsic} \times \frac{EW(\text{He I ems.}) - EW(\text{He I abs.})}{EW(\text{He I ems.})} \tag{1}$$

110   SKILLMAN ET AL.

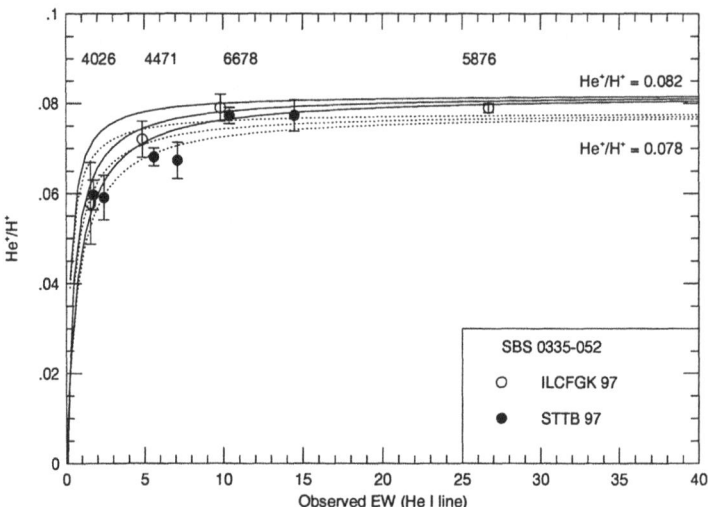

Figure 3. A plot of the He abundance derived from various He I emission lines versus the equivalent width of those lines as observed in SBS0335-052. The data come from I97 and our new study. The decrease in derived He/H with decreasing EW is interpreted as due to underlying stellar absorption. The dashed lines indicate models for two different values of He/H, and a range of EW of underlying absorption (0.25, 0.5, 0.75 Å). For the emission lines plotted, the assumption of equal EW of underlying absorption is consistent with models and observations of single stars (see text). Note that corrections for collisional enhancement of the emission lines have not been applied.

The models are shown for two different values of He/H and three values of EW of underlying absorption (0.25 Å, 0.50 Å, 0.75 Å). The two models presented demonstrate that the data are fairly well described by this simple model. From these models it is clear that for all values of the EW(He I abs.) the λ5876 line (EW ≈ 25) will suffer from only negligible effects of underlying absorption. However for values of EW(He I abs.) of 0.5 to 0.75 Å, there is already some effect on the λ6678 line and clearly λ4471 and λ4026 are strongly affected.

At this point, it is very reasonable to ask whether the effects of underlying absorption have been underestimated in the past, and, if so, why. Kunth and Sargent (1983) proposed the very simple test of looking for a trend in derived He abundance with EW(ems.). They found no evidence for this effect in their data (which span approximately the same range in EW(ems.) as the rest of the observations in the literature). In Figure 4, we have replotted their data as the ratio of He abundances derived from the λ4471 and λ5876 lines and found that there is, indeed, a slight trend with EW(Hβ) implying that underlying absorption may be present at a detectable level (note that IIZw40 should be ignored, as the λ5876 line is affected by Galactic Na I absorption as noted by PSTE).

In the past, we also relied on the modeling by Olofsson (1995). In Figure 5, we have plotted some of the results from those models. The result that was normally taken from this work was that the EW of λ4471 in absorption was generally of

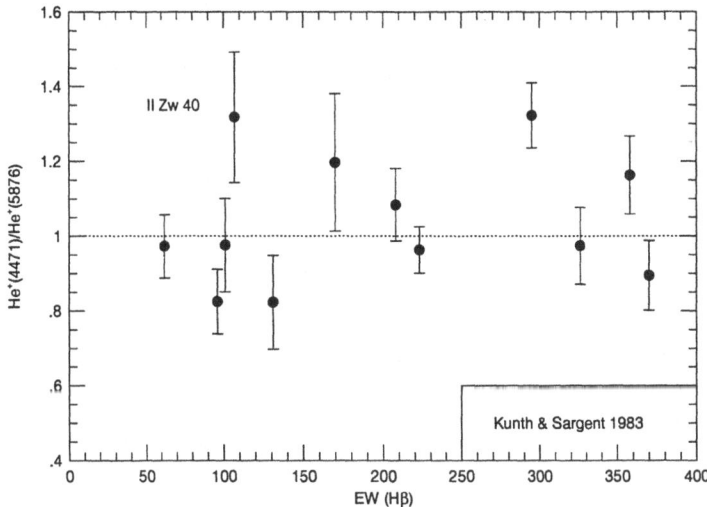

Figure 4. The ratio of He abundances derived from the $\lambda 4471$ and $\lambda 5876$ lines as a function of EW(Hβ) for the data of Kunth and Sargent (1983).

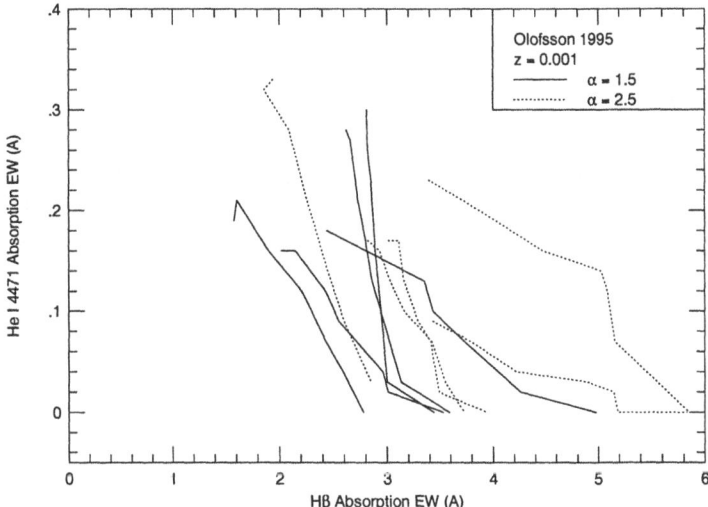

Figure 5. The EW of the stellar absorption in the $\lambda 4471$ line as a function of the EW of the absorption in the Hβ line from models calculated by Olofsson (1995). Note that none of the models produce EW(Hβ) $\leq$ 2 Å.

order 0.1 Å or less. As can be seen from Figure 5, the trend is for lower EW(4471) with increasing EW(Hβ). Since most observed HII regions have EW(Hβ) $\leq$ 2 Å, this implies that the underlying absorption values may be much larger than 0.1. We have also discovered an inconsistency in the relative strengths of the modeled He absorption lines. That is, in observed stars (e.g., Lennon *et al.*, 1993) and in numerical models (e.g., Auer and Mihalas, 1972), the strengths of the $\lambda 4471$ and

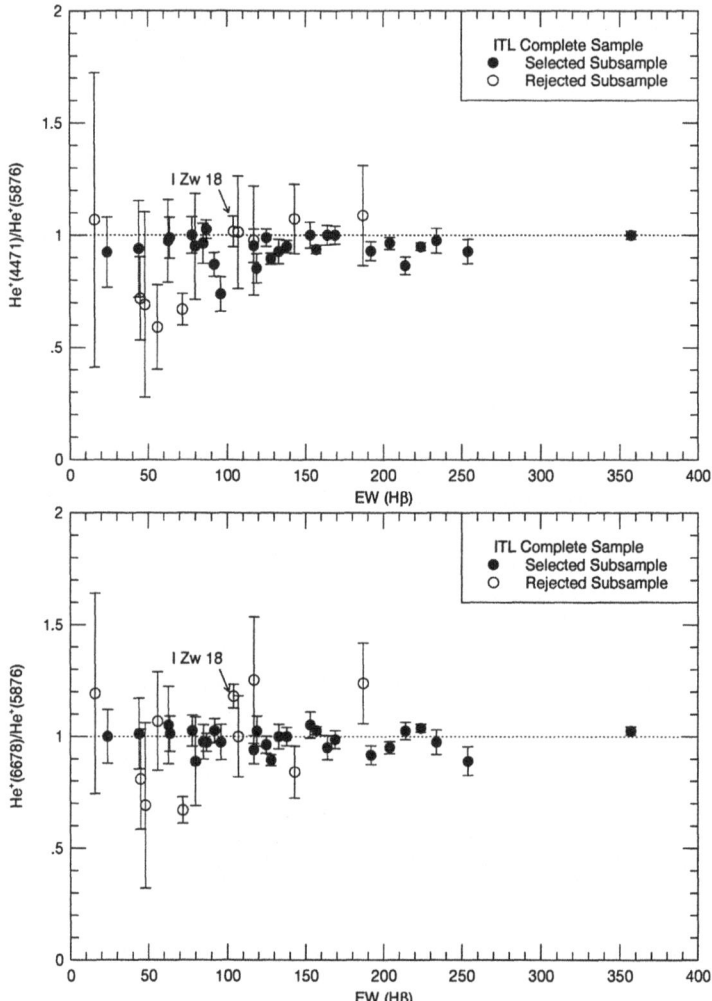

Figure 6. A comparison of the He abundances derived from the λ4471, λ5876, and λ6678 lines for the observations by ITL97. Note the systematic difference between the λ4471 and λ5876 measurements and the relatively good agreement between the λ6678 and λ5876 results.

λ4026 lines are about a factor of two stronger than λ4387 and λ4922, while in the models of Olofsson, the opposite is true. This potentially implies that the underlying absorption in λ4471 and λ4026 could have been underestimated by a factor of 4 in Olofsson's models (EWs for λ5876 and λ6678 are not calculated).

Clearly, we need to be much more careful in our treatment of this effect in the future. As an exercise, we looked at the data in ITL97 to see if we could find the signature of the effect of underlying absorption. In Figure 6, we plot the ratios of the He abundances derived from the three strongest lines. In the top panel, we see that the He abundances derived from λ4471 are systematically lower than those derived from $\lambda 5876$. Some of the "rejected" points are low EW(ems.) points with anomalously low He($\lambda 4471$) which points directly to strong underlying absorption. The $\lambda 6678$, $\lambda 5876$ comparison looks much better in this regard. Probably the lesson to be learned from this diagnostic diagram is that $\lambda 4471$ should rarely be used in determining emission line He abundances in these objects. During the workshop we presented other diagnostic diagrams for showing the effects of underlying absorption. While we lack space to reproduce them all, the basic points are made in Figure 6.

What are the greater implications for this realization that the effects of underlying absorption could have been underestimated in the past? Thuan (these proceedings) has proposed that underlying absorption is important in the NW component of I Zw 18. From a re-inspection of the observations of Skillman and Kennicutt (1993), we agree. OSS demonstrated that a large part of the difference between the estimates of the primordial helium mass fraction Y_P of about 0.23 (PSTE; OSS) and 0.243 (ITL97) was due to the inclusion/exclusion of I Zw 18. Since $\lambda 5876$ cannot be used due to contamination by Na I absorption, this makes measurements of He in I ZW 18 particularly vulnerable. At this point we can safely say that the uncertainty in the He abundance in I Zw 18 has been underestimated in the past.

The unfortunate result is that there are few extremely low metallicity objects with reliable He abundances. Thus, the derivation of Y_P is more reliant on the assumption of a perfectly linear relationship between He and Z. This is unfortunate, because it is precisely in the region of very low metallicity where popular theories of galactic wind dominated evolution imply a deviation from a linear relationship. Clearly there is more work to be done.

4. Summary and Future Outlook

It is a bit disturbing that we now have significantly different estimates of the primordial He abundance. It is important to identify the sources of the discrepancy. A first step will be to ensure that the observations are treated identically (e.g., a consistent treatment for neutral helium). We have also learned to be more wary in the identification of underlying absorption.

As a result, in order to significantly improve on the current status, we must: (1) Increase the number of measurements of new regions with extremely low metallicities. This will require very large initial samples, as the number of criteria that the candidates must meet is increasing (low metallicity, high surface brightness, hard radiation field, weak underlying absorption). (2) Use only measurements with uncertainties of order 3% or less. With these measurements we will be able to establish the intrinsic dispersion in the He/H vs. O/H, N/H relationships and test for several classes of systematic errors. Note that very few of the measurements of individual HII regions in the literature meet this criterion. (3) Continue to improve

114   SKILLMAN ET AL.

our knowledge of the basic atomic data for He and the relevant collisional and radiative transfer effects.

References

Auer, L. H., & Mihalas, D.: 1972, *ApJS* **24**, 193.
Berry, A.: 1898, *A Short History of Astronomy From the Earliest Times Through the Nineteenth Century*, Dover.
Davidson, K. and Kinman, T. D.: 1985, *ApJS* **58**, 321.
Dinerstein, H. L., and Shields, G. A.: 1986, *ApJ* **311**, 45.
Izotov, Y. I., Lipovetsky, V. A., Chaffee, F. H., Foltz, C. B., Guseva, N., and Kniazev, A. Y.: 1997, *ApJ* **476**, 698. (I97)
Izotov, Y. I., Thuan, T. X., and Lipovetsky, V. A.: 1994, *ApJ* **435**, 647. (ITL94)
Izotov, Y. I., Thuan, T. X., and Lipovetsky, V. A.: 1997, *ApJS* **108**, 1. (ITL97)
Kingdon, J., and Ferland, G.: 1995, *ApJ* **442**, 714.
Kobulnicky, H. A., and Skillman, E. D.: 1996, *ApJ* **471**, 211.
Kunth, D., and Sargent, W. L. W.: 1983, *ApJ* **273**, 81.
Lennon, D. J., Dufton, P. L., and Fitzsimmons, A.: 1993, *A&AS* **97**, 559.
Olive, K., and Steigman, G.: 1995, *ApJS* **97**, 49.
Olive, K. A., Skillman, E. D., and Steigman, G.: 1997, *ApJ* **483**, 788. (OSS)
Olofsson, K.: 1995, *A&AS* **111**, 57.
Osterbrock, D. E.: 1989, *Astrophysics of Gaseous Nebulae and Active Galactic Nuclei*, University Science Books.
Pagel, B. E. J., Simonson, E. A., Terlevich, R. J., and Edmunds, M. G.: 1992, *MNRAS* **255**, 325. (PSTE)
Peimbert, M., and Torres-Peimbert, S.: 1974, *ApJ* **193**, 327.
Peimbert, M., and Torres-Peimbert, S.: 1976, *ApJ* **203**, 581.
Sawey, P. M. J., and Berrington, K. A.: 1993, *Atomic Data Nucl. Data Tables* **55**, 81.
Skillman, E. D., and Kennicutt, R. C., Jr.: 1993, *ApJ* **411**, 655.
Skillman, E. D., Terlevich, R. J., Kennicutt, R. C., Garnett, D. R., and Terlevich, E.: 1994, *ApJ* **431**, 172.
Smits, D.: 1996, *MNRAS* **251**, 316.
Stasińska, G.: 1990, *A&AS* **83**, 501.
Tytler, D., Fan, X.-M., and Burles, S.: 1996, *Nature* **381**, 207.
Vílchez, J. M.: 1989, *Ap&SS* **157**, 9.
Vílchez, J. M. and Pagel, B. E. J.: 1988, *MNRAS* **231**, 257.
Walker, T. P., Steigman, G., Schramm, D. N., Olive, K. A., and Kang, H.: 1991, *ApJ* **376**, 51.

PRIMORDIAL HELIUM AND $\Delta Y/\Delta Z$ FROM H II REGIONS AND FROM FINE STRUCTURE IN THE MAIN SEQUENCE BASED ON HIPPARCOS PARALLAXES *

E. HØG

University Observatory, NBIfAFG, Juliane Mariesvej 30, DK-2100 Copenhagen Ø, Denmark

B.E.J. PAGEL, L. PORTINARI** and P.A. THEJLL***

NORDITA, Blegdamsvej 17, DK-2100 Copenhagen Ø, Denmark

J. MACDONALD

Department of Physics and Astronomy, University of Delaware, Newark DE 19716, USA

L. GIRARDI‡

Dipartimento di Astronomia, Vicolo dell' Osservatorio 5, 35122 Padova, Italy

Abstract. The primordial helium abundance Y_P is important for cosmology and the ratio $\Delta Y/\Delta Z$ of the changes relative to primordial abundances constrains models of stellar evolution. While the most accurate estimates of Y_P come from emission lines in extragalactic H II regions, they involve an extrapolation to zero metallicity which itself is closely tied up with the slope $\Delta Y/\Delta Z$. Recently certain systematic effects have come to light in this exercise which make it useful to have an independent estimate of $\Delta Y/\Delta Z$ from fine structure in the main sequence of nearby stars. We derive such an estimate from Hipparcos data for stars with $Z \leq Z_\odot$ and find values between 2 and 3, which are consistent with stellar models, but still have a large uncertainty.

Key words: galaxies: abundances, ISM: abundances, stars: abundances, stars: evolution

1. Introduction

One success of Big-Bang cosmology has been to predict primordial abundances of light elements (e.g. Copi, Schramm and Turner, 1995); but precise determination of each primordial abundance involves difficulties, both from measurements and in extrapolating through evolution that has taken place in the meantime. In particular, it has been pointed out by Hata *et al.* (1995) that with some widely accepted figures for primordial abundances, specifically with D/H $\leq 3 \times 10^{-5}$ (Tytler, Fan and Burles, 1996) and $Y_P \leq 0.242$ or 0.243 with 95% confidence (Pagel *et al.* 1992, hereinafter PSTE; Olive, Skillman and Steigman, 1997), there is actually inconsistency with the 'standard' Big-Bang nucleosynthesis theory assuming 3 massless neutrino types.

This conflict can be resolved if there are even quite small deficiencies in the data. The primordial deuterium abundance could be higher (e.g. Songaila, Wampler and

---

* Based on data from the ESA Hipparcos astrometry satellite
** On leave from: Dipartimento di Astronomia, Vicolo dell'Osservatorio 5, 35122 Padova, Italy
*** Present address: Danish Meteorological Institute, Lyngbyvej 100, DK-2100 Copenhagen Ø, Denmark
‡ On leave from: Instituto de Fisica, UFRGS, Caixa Postal 15051, 91501-970 Porto Alegre, RS, Brazil

Space Science Reviews **84**: 115–126, 1998.

Cowie, 1997) or the primordial helium abundance could be higher, or both. Recently several results have emerged favouring the higher helium abundance: Hata *et al.* (1996) have refined the arguments from D + ^{3}He (cf. Yang *et al.*, 1984) to derive a 95% confidence bound on deuterium, D/H $\leq 6.2 \times 10^{-5}$ implying $Y_P \geq 0.243$; allowance for ionized hydrogen associated with Lyman-α forest clouds indicates $1 \leq \eta_{10}/4 \leq 1.5$ implying $0.243 \leq Y_P \leq 0.247$ (Weinberg *et al.*, 1997); and a new study of extragalactic H II regions by Izotov, Thuan and Lipovetsky (1997, hereinafter ITL) has given $Y_P = 0.243 \pm 0.003$ (s.e.), which is more consistent with the above bounds on baryonic density. The origin of the difference in this latter value from that of PSTE does not lie in the use of different atomic data, nor in any significant difference in the helium abundance for objects in common with PSTE or with Skillman and Kennicutt (1993) and Skillman *et al.* (1994) who also derived values close to 0.23, but is largely the result of a lower $\Delta Y/\Delta Z$ slope resulting from their rejection of I Zw 18 which is still the least heavy-element enriched H II galaxy known. The justification for that rejection is discussed below. Anyway, ITL find a slope $\Delta Y/\Delta Z = 1.7 \pm 0.9$, in contrast to the value of about 4 found by PSTE and this translates directly into the discrepancy between the two values of Y_P. Now an estimate of $\Delta Y/\Delta Z$ (although not of Y_P itself, because of degeneracy with the mixing length) can also be obtained from fine structure in the main sequence of nearby stars, and so we have attempted to make such an estimate as will be described later.

2. Problems with extragalactic H II regions

Olive, Skillman and Steigman (1997) have presented a plot showing ITL's helium data in comparison with the others and suggest that the difference arises mainly from the fact that ITL's data do not go to the lowest metallicities available, after rejection of IZw 18. However, in the meantime, Izotov *et al.* (1997) have published a new analysis of another low-metallicity H II galaxy, SBS 0335–052, which fits the regression of ITL, whereas an earlier study of the same object by Melnick, Heydari-Malayeri and Leisy (1992) had fitted that of PSTE. This gives an opportunity to look for the cause of the difference. SBS 0335 has two major components, of which one has λ 4471 strongly affected by an underlying absorption line, while the other one looks clean. Some results from the two sets of authors for the latter component are shown in Table I.

The table shows excellent agreement for the raw measurements of the atomic He^+/H^+ ratio y^+, before correcting for collisional excitation from the 2^3S level. This correction is proportional to electron density n_e, which was estimated by Melnick *et al.* from the [S II] line ratio and by Izotov *et al.* by demanding consistency between the three lines in the table and λ 7065, resulting in smaller corrections, especially for λ 5876 which carries the most weight in the final average. Both methods are subject to errors, which should have been included in the error budgets.

Table I

Helium abundances in SBS 0335–052B

	y^+ (raw)		y^+ (corr.), y^{++}	
	Melnick *et al.* 92	Izotov *et al.* 97	Melnick *et al.* 92 n_e(SII) = 220	Izotov *et al.* 97 n_e(HeI) = 150
4471	.078	.076	.073 ± 8	.072 ± 4
5876	.084	.085	.075 ± 2	.079 ± 1
6678	.082	.081	.080 ± 4	.079 ± 3
mean y^+			.075 ± 5	.078 ± 1
y^{++}			.001	.003
y			.076 ± 5	.081 ± 2
Y			.233 ± 16	.245 ± 6

Table II

Weak helium lines in IZw 18

	theor.	meas.
$\frac{4388}{4471}$	0.12	0.10 ± .03
$\frac{4922}{4471}$	0.26	0.23 ± .03

In the Izotov *et al.* results, there is a significant discrepancy between 5876 and 4471, which can be explained in one of two ways: either 4471 is affected by underlying absorption or their collisional correction is an underestimate.

As was pointed out by PSTE, the effect of underlying absorption can be estimated by measuring the intensities of weak helium emission lines that are nevertheless strong in absorption, e.g. $\lambda\lambda$ 4026, 4388 and 4922 and comparing with theoretical intensities (Brocklehurst, 1972; Smits, 1996). Such measurements are not available for SBS 0335, but some are available for IZw 18, both from PSTE and from ITL, the latter being apparently somewhat more precise. Their results are shown in Table II.

It can be seen from the table that there is indeed a 1σ deficiency in the line ratios, which provides some evidence for absorption underlying λ 4471. For example, given that 4922 and 4471 have absorption-line strengths in a ratio of about 1:1.5 (Auer and Mihalas, 1972), the 10 per cent deficiency in their emission ratio would correspond to about a 4 per cent underestimate of 4471. A good quantitative estimate would need spectra with better resolution and signal:noise than are available at present, but the correction could amount to several per cent. Since λ 5876 is unusable for IZw 18 because of absorption by Galactic sodium and λ 6678 is not guaranteed to be free from underlying absorption, it is indeed quite possible that the current estimates of helium in IZw 18 (and possibly other very low-metallicity systems) are a few per cent too low and that the ITL solution is closer to reality. In the next sections we investigate the $\Delta Y/\Delta Z$ slope on the basis of stellar data.

3. Fine structure in the main sequence

As has been discussed previously by Perrin *et al.* (1977) and Fernandes, Lebreton and Baglin (1996), the location of the stellar main sequence as a function of metallicity is affected by $\Delta Y/\Delta Z$ because concomitant changes in Y and Z push the sequence in opposite directions. This can be seen from quasi-homology relations of the form (Cox and Giuli, 1968; Fernandes, Lebreton and Baglin, 1996)

$$\frac{L}{f(T_{\text{eff}})} \propto \epsilon_0^{0.32} \kappa_0^{0.35} \mu^{-1.33}, \tag{1}$$

where the energy generation constant $\epsilon_0 \propto X^2$, the opacity constant $\kappa_0 \propto (1 + X)(Z + Z_0)$ with $Z_0 \simeq 0.01$ and the molecular weight $\mu \propto (3 + 5X - Z)^{-1}$ leading to a magnitude offset above the zero-age zero-metallicity main sequence where $X = X_0 \simeq 0.76$

$$
\begin{aligned}
-\Delta M_{\text{bol}} = {}& 1.6 \log\left[1 - \frac{Z}{X_0}\left(1 + \frac{\Delta Y}{\Delta Z}\right)\right] \\
& + 0.87\left[\log\left\{1 - \frac{Z}{1 + X_0}\left(1 + \frac{\Delta Y}{\Delta Z}\right)\right\} + \log\left(1 + \frac{Z}{Z_0}\right)\right] \\
& + 3.33 \log\left[1 - \frac{5Z}{3 + 5X_0}\left(1.2 + \frac{\Delta Y}{\Delta Z}\right)\right].
\end{aligned}
\tag{2}
$$

For high metallicities, around $0.7Z_\odot \leq Z \leq 1.5Z_\odot$, the effects of Y and Z cancel out for $\Delta Y/\Delta Z \simeq 5.5$ (Fernandes *et al.*, 1996), but this is not the case for lower metallicities (e.g. Cayrel, 1968). Perrin *et al.* (1977) and Fernandes *et al.* (1996) tested stellar main sequences derived from ground-based parallaxes for relatively metal-rich stars against theoretical main sequences roughly corresponding to Eq. (2). Perrin *et al.* found no correlation with metallicity, suggesting quite a high $\Delta Y/\Delta Z$, but with a large uncertainty, whereas Fernandes *et al.*, lacking metallicity data and examining just the scatter, deduced only that $\Delta Y/\Delta Z \geq 2$. In this investigation we use Hipparcos parallaxes (ESA, 1997) to investigate a sample of mainly metal-poor stars based on a proposal submitted by one of us (BEJP) in 1982.

In the case of an old stellar population, direct application of Eq. (2) is not very useful in practice (as well as not being very accurate) because the effects of stellar evolution increase sharply with luminosity above, say, $M_V \simeq 5.5$, so that the sequences cannot be expected to run straight and parallel over a wide range of luminosities. It is more useful to translate Eq. (2) into a range of $\log T_{\text{eff}}$ at a fixed absolute magnitude using the slope of the evolved main sequence, which is about

20 magnitudes per dex in T_{eff}. We thus derive

$$-\Delta \log T_{\text{eff}} = 0.08 \log \left[1 - \frac{Z}{X_0} \left(1 + \frac{\Delta Y}{\Delta Z} \right) \right]$$
$$+ 0.0435 \left[\log \left\{ 1 - \frac{Z}{1 + X_0} \left(1 + \frac{\Delta Y}{\Delta Z} \right) \right\} + \log \left(1 + \frac{Z}{Z_0} \right) \right]$$
$$+ 0.167 \log \left[1 - \frac{5Z}{3 + 5X_0} \left(1.2 + \frac{\Delta Y}{\Delta Z} \right) \right]. \tag{3}$$

Since the terms in Z are small, apart from $(1 + Z/Z_0)$, we can expand the logarithms, also making use of $X_0 \simeq 0.765$, to give

$$\Delta \log T_{\text{eff}} \simeq 0.01Z + 0.11Z(1 + \Delta Y/\Delta Z) - 0.044 \log(1 + Z/Z_0). \tag{4}$$

The older quasi-homology relation given by Faulkner (1967):

$$L \propto (X + 0.4)^{2.67}(Z + Z_0)^{0.455} f(T_{\text{eff}}) \tag{5}$$

leads to a very similar relation

$$\Delta \log T_{\text{eff}} \simeq 0.124Z(1 + \Delta Y/\Delta Z) - 0.057 \log(1 + Z/Z_0). \tag{6}$$

These relations provide a rough guide to the behaviour of numerically computed isochrones for stars fainter than $M_V \simeq 5.5$ as a function of $\Delta Y/\Delta Z$; qualitatively, the spread in the main sequences is a decreasing function of this parameter.

4. The data

From the initial sample of about 1000 stars proposed for this programme, we have selected a subsample for which the Hipparcos parallaxes have errors better than 6 per cent, and effective temperatures and metallicities are given in either or both of the catalogues by Cayrel de Strobel et al. (1992) and Carney et al. (1994). Where available, the catalogue effective temperatures ('old' T_{eff}'s with an estimated rms error ± 100 K) have been replaced with temperatures derived using the infra-red flux method of D.E. Blackwell by Alonso et al. (1996); these 'new' temperatures do not seem to differ systematically from the 'old' ones, but are very much more accurate; consistency among three different infra-red colours indicates a precision close to ± 50 K. Table III gives the data for those of the subsample that are fainter than $M_V = 5.5$; effective temperatures and metallicities from Alonso et al. are denoted by T_{eff}^A and $[\text{Fe/H}]^A$ respectively.

An important factor in comparing stellar data with theoretical isochrones is the relationship between the metallicity [Fe/H] and the heavy-element mass fraction Z. In most cases we have used the formula by Salaris, Chieffi and Straniero (1993)

$$Z = Z_0(0.638 f_\alpha + 0.362), \tag{7}$$

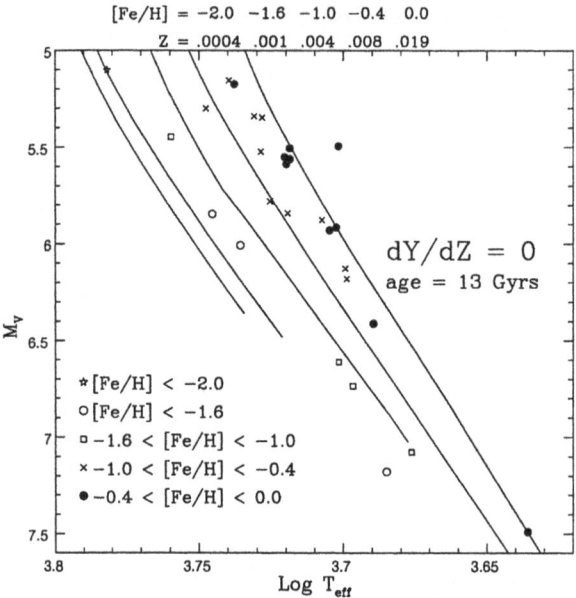

Figure 1. 13 Gyr isochrones deduced from Padova evolutionary tracks shifted by -0.009 in log T_{eff} for $\Delta Y/\Delta Z = 0$. Stellar data are plotted with 'new' effective temperatures where available. The 'star' symbol represents HD 19445. The values of Z for the 5 isochrones, together with the corresponding [Fe/H], are indicated on top of the plot.

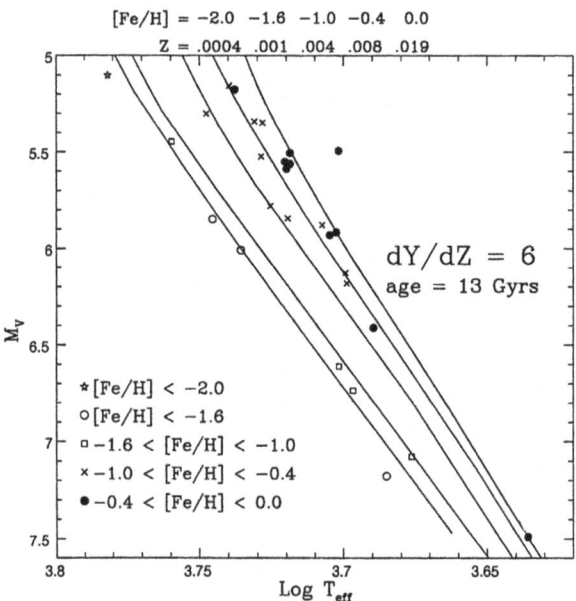

Figure 2. 13 Gyr isochrones from Padova evolutionary tracks as in Fig. 1, but for $\Delta Y/\Delta Z = 6$. Stellar data as in Fig. 1.

Table III

The sample of lower main-sequence stars

HIC	HD	V	M_V	±	T_{eff}	T_{eff}^A	[Fe/H]	[Fe/H]^A	log Z	
5031	6348	9.15	6.18	0.10	4998		-0.67		-2.31	
5336	6582	5.17	5.78	0.03	5210	5315	-0.61	-0.67	-2.31	
10138	13445	6.12	5.93	0.01	5067		-0.24		-1.99	
18915	25329	8.51	7.17	0.04	4828	4842	-1.56	-1.64	-3.04	
19849	26965	4.43	5.91	0.01	5040	5040	-0.28	-0.17	-1.94	
23080	31501	8.19	5.59	0.08	5254		-0.33		-2.06	
38541	64090	8.31	6.05	0.07	5369	5441	-1.78	-1.82	-3.22	
39157	65583	6.99	5.87	0.03	5305	5242	-0.61	-0.60	-2.27	
57939	103095	6.44	6.63	0.02	5032	5029	-1.36	-1.35	-2.75	
58949	104988	8.16	5.59	0.07	5247		-0.23		-1.98	
62607	111515	8.15	5.55	0.07	5354		-0.81		-2.37	
73005	132142	7.77	5.88	0.03	5090	5098	-0.55	-0.55	-2.22	
74234	134440	9.45	7.08	0.11	4756	4746	-1.52	-1.57	-3.32	a
74235	134439	9.07	6.73	0.09	4883	4974	-1.57	-1.52	-3.32	b
94931	—	8.84	6.10	0.07	4859	5004	-0.87	-0.42	-2.13	c
98020	188510	8.83	5.85	0.10	5418	5564	-1.78	-1.80	-3.20	
99461	191408	5.32	6.41	0.01	4893		-0.32		-2.06	
104214	201091	5.20	7.49	0.03	4364	4323	-0.05	-0.05	-1.80	
106122	204814	7.93	5.56	0.05	5232		-0.28		-2.02	

a $[Z] = [\text{Fe/H}]$

b $[Z] = [\text{Fe/H}]$

c BD +41 3306

where Z_0 is the solar Z ($Z_\odot = 0.019$) scaled according to [Fe/H] and f_α is the factor by which oxygen and α-particle elements are enhanced relative to iron, taking f_α from Pagel and Tautvaišiene (1995). Thus for [Fe/H] < -1, $Z = 2Z_0$. However, in the case of HD 134439 and 134440 we take $Z = Z_0$, following King (1997).

5. Comparison with theoretical isochrones

We have used two sets of isochrones, one computed from Padova evolutionary tracks normalized to the Sun and shifted by -0.009 in $\log T_{\text{eff}}$ to fit solar-metallicity stars, the other computed by one of us (JM) normalized to low-metallicity stars and shifted by -0.01 in $\log T_{\text{eff}}$ to fit the low-metallicity stars of our sample. For our purpose, anyway, it is only the spread of the isochrones that counts, not their absolute position.

In Figures 1 to 4, we compare these isochrones for the lower main sequence with the stellar data, based on 'new' effective temperatures where available. Comparison

E. HØG ET AL.

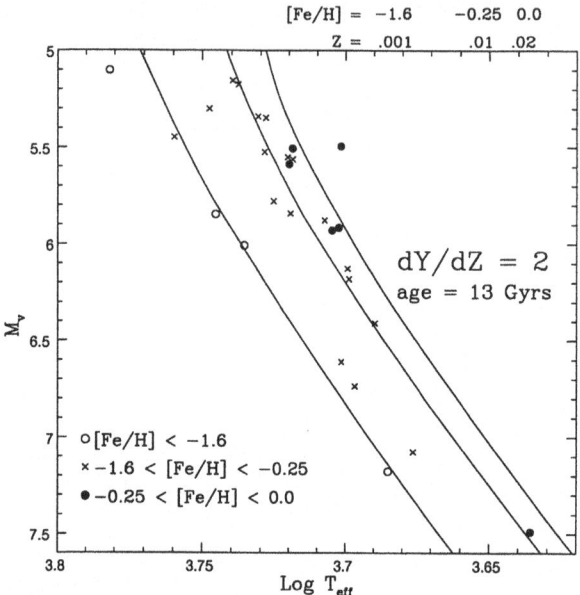

Figure 3. 13 Gyr isochrones from MacDonald evolutionary tracks and shifted by -0.01 in log T_{eff} for $\Delta Y/\Delta Z = 2$. Stellar data as in Fig. 1.

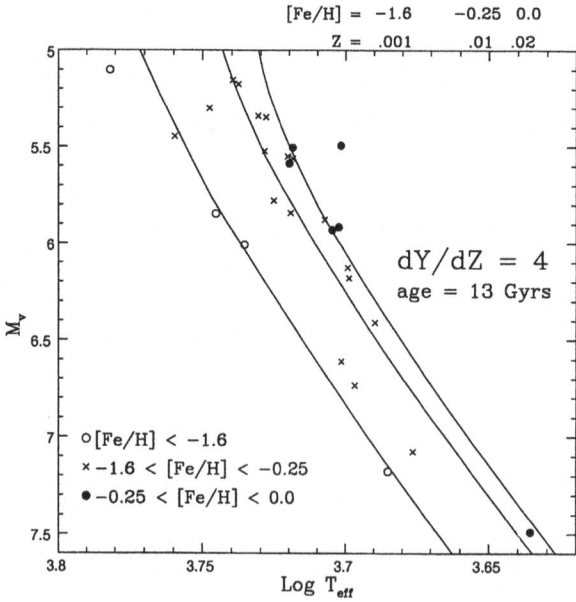

Figure 4. As Fig. 3, but for $\Delta Y/\Delta Z = 4$,

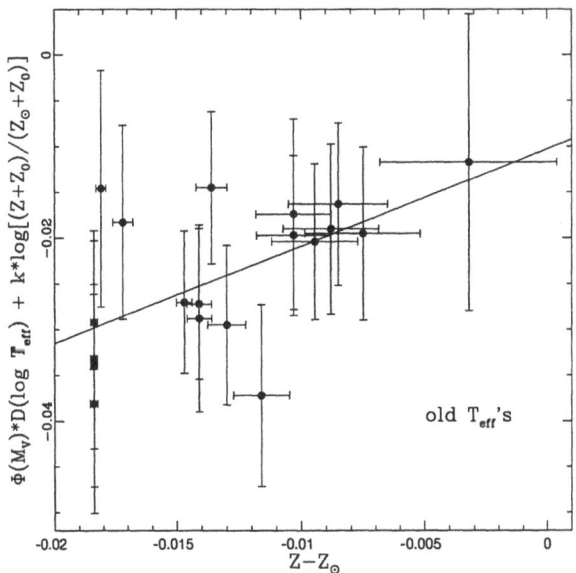

Figure 5. Maximum-likelihood regression for Padova 13 Gyr isochrones and 'old' effective temperatures for stars with $5.5 \le M_V \le 7.5$ assuming $\sigma_{T\text{eff}} = 100$ K and $\sigma_{[\text{Fe}/\text{H}]} = 0.1$. The slope corresponds to $\Delta Y/\Delta Z = 6.2 \pm 2.8$.

of the Padova isochrones with the data suggests that $\Delta Y/\Delta Z = 0$ is excluded, because the spread of the isochrones is too great, whereas $\Delta Y/\Delta Z = 6$ is too large. (With the 'old' temperatures, $\Delta Y/\Delta Z = 6$ actually gives the best fit.) The MacDonald isochrones are more widely spaced in metallicity and cover a smaller range in $\Delta Y/\Delta Z$; a value of 4 seems to give a slightly better fit than a value of 2. However, these visual impressions are largely based on the extremes in the abundance range, and the intermediate-metallicity stars are too scattered to allow any choice of $\Delta Y/\Delta Z$ from inspection. In the next section we shall try to improve on these qualitative impressions by applying a maximum-likelihood calculation based on an extension of the idea of quasi-homology.

6. Statistical analysis using quasi-homology relations

In order to be able to apply maximum-likelihood linear regression techniques to the data, we have carried out numerical experiments using *Mathematica* to find quasi-homology relations analogous to Eq. (6) applicable to the theoretical 13 Gyr isochrones in a limited range of absolute magnitude, specifically $5.5 \le M_V \le 7.5$, of the form

$$\phi(M_V)\Delta \log T_{\text{eff}} + k \log(1 + Z/Z_0) = aZ(1 + \Delta Y/\Delta Z), \qquad (8)$$

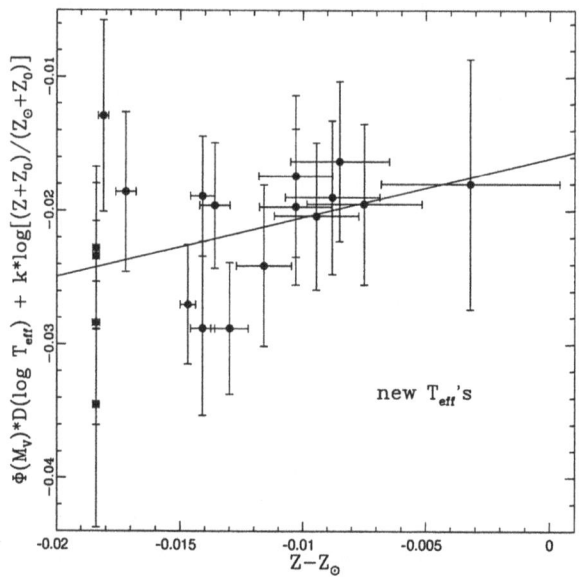

Figure 6. Maximum-likelihood regression for Padova 13 Gyr isochrones and 'new' effective temperatures for stars with $5.5 \leq M_V \leq 7.5$ assuming $\sigma_{T_{\text{eff}}} = 50$ K and $\sigma_{[\text{Fe/H}]} = 0.1$. The slope corresponds to $\Delta Y/\Delta Z = 2.0 \pm 1.7$.

where $\phi(M_V)$ is a normalization to allow for the convergence of the Padova isochrones towards low luminosities. The specific relations that we found are

$$\frac{\Delta \log T_{\text{eff}}}{1 - 0.234(M_V - 6.0)} + 0.054 \log(1 + Z/Z_0) = 0.147Z(1 + \Delta Y/\Delta Z)$$
$$Z_0 = 0.0015 \qquad (9)$$

for Padova isochrones, and

$$\Delta \log T_{\text{eff}} + 0.059 \log(1 + Z/Z_0) = 0.128Z(1 + \Delta Y/\Delta Z)$$
$$Z_0 = 0.0025 \qquad (10)$$

for MacDonald isochrones. The coefficients are quite similar to those in Eq. (6), but Z_0 turns out to be much lower than the widely quoted value of 0.01.

Figs. 5 and 6 show plots of the left-hand side of Eq. (9), normalized to solar metallicity, against Z for the 'old' and 'new' effective temperatures respectively, with the maximum-likelihood fit computed using a version of the program of Pagel and Kazlauskas (1992). From the figures it appears that the 'old' temperatures give a poorly defined regression with a large slope (i.e. large $\Delta Y/\Delta Z$), while the new ones give a more tidy regression with a smaller slope, but still a large error (see Table IV). The values of $\Delta Y/\Delta Z$ derived respectively from the 'old' and 'new'

Table IV
$\Delta Y/\Delta Z$ from regression analysis with 'new' temperatures

	Padova isochrones	MacDonald isochrones ($\alpha = 1.0$)
$5.5 \leq M_V \leq 6.5$	3.1 ± 2.2	0.2 ± 2.5
$5.5 \leq M_V \leq 7.1$	3.3 ± 2.0	1.7 ± 2.1
$5.5 \leq M_V \leq 7.5$	2.0 ± 1.7	2.6 ± 1.8

temperatures are consistent within their combined errors, but we consider the one from the 'new' temperatures to be much more reliable.

7. Discussion

Table IV gives some solutions derived from different ranges of absolute magnitude. The MacDonald isochrones give a trend with absolute magnitude which is absent in the case of the Padova isochrones and all the results are consistent with $\Delta Y/\Delta Z$ between 2 and 3, but with a disappointingly large error. To reduce the error one needs a more complete sample of stars with known metallicities, effective temperatures and parallaxes, which will become available once the complete Hipparcos catalogue is published.

On the face of it, our stellar value of $\Delta Y/\Delta Z$ is intermediate between those claimed by ITL and by PSTE for extragalactic H II regions with similar metallicities, assuming that in this respect the two kinds of objects have undergone similar chemical evolution. Our results tend to support that assumption and to remove the higher estimates by PSTE which have inspired chemical evolution models of dwarf galaxies involving strong effects of metal-enhanced galactic winds (e.g. Pilyugin, 1993). Our results also suggest that the primordial helium abundance has been somewhat underestimated by Olive, Skillman and Steigman (1997) using data from PSTE and other preceding investigations, probably because of underlying absorption lines in some of the lowest-abundance objects, and that ITL's result $Y_P \simeq 0.24$ is currently the best estimate available.

Chieffi, Straniero and Salaris (1995) have suggested that the mixing-length parameter α might vary systematically with metallicity. If so, this could introduce systematic errors in the estimation of $\Delta Y/\Delta Z$ from shifts in the main sequence. However, such an effect was suspected from small deviations in the main-sequence slope which are not apparent in our data, at least when compared to Padova isochrones. Values of $\Delta Y/\Delta Z$ expected from stellar models are in the neighbourhood of 2 (van den Hoek, 1997), not inconsistent with our result.

References

Alonso, A., Arribas, S., and Martinez-Roger, C.: 1996, *A&A Suppl.* **117**, 227.
Auer, L.H. and Mihalas, D.: 1972, *Ap. J. Suppl.* **24**, 193.
Bell, R.A. and Gustafsson, B.: 1989, *MNRAS* **236**, 263.
Brocklehurst, M.: 1972, *MNRAS* **153**, 471.
Carney, B.W., Latham, D.W., Laird, J.B., and Aguilar, L.: 1994, *Astr. J.* **107**, 2240.
Cayrel, R.: 1968, *Ap. J.* **151**, 997.
Cayrel de Strobel, G., Hauck, B., François, C., Thévenin, F., Friel, E., Mermilliod, M., and Borde, S.: 1992, *A&A Suppl.* **95**, 273.
Chieffi, A., Straniero, O. and Salaris, M.: 1995, *Ap. J. Lett.* **445**, L39.
Copi, C., Schramm, D.N., and Turner, M.S.: 1995, *Science* **267**, 192.
Cox, J.P., and Giuli, R.T.: 1968, *Principles of Stellar Structure*, Gordon and Breach.
ESA: 1997, The Hipparcos and Tycho Catalogues, ESA SP-1200.
Faulkner, J.: 1967, *Ap. J.* **147**, 617.
Fernandes, J., Lebreton, Y., and Baglin, A.: 1996, *A&A* **311**, 127.
Hata, N., Scherrer, R.J., Steigman, G., Thomas, D., Walker, T.P., Bludman, S., and Langacker, P.: 1995, *Phys. Rev. Lett.* **75**, 3977.
Hata, N., Scherrer, R.J., Steigman, G., Thomas, D., and Walker, T.P.: 1996, *Ap. J.* **458**, 637.
Izotov, Y.I., Lipovetsky, V.A., Chaffee, F.H., Foltz, C.B., Guseva, N.G., and Kniazev, A.Y.: 1997, *Ap. J.* **476**, 698.
Izotov, Y.I., Thuan, T.X., and Lipovetsky, V.A.: 1997, *Ap. J. Suppl.* **108**, 1.
King, J.R.: 1997, *Astr. J.*, **113**, 2302.
Melnick, J., Heydari-Malayeri, M., and Leisy, P.: 1992, *A&A* **253**, 16.
Olive, K., Skillman, E.D., and Steigman, G.: 1997, preprint astro-ph/96 11166.
Pagel, B.E.J. and Kazlauskas, A.: 1992, *MNRAS* **256**, 49P.
Pagel, B.E.J., Simonson, E.A., Terlevich, R.J., and Edmunds, M.G.: 1992, *MNRAS* **255**, 325.
Pagel, B.E.J. and Tautvaišiene, G.: 1995, *MNRAS* **276**, 505.
Perrin, M.-N., Hejlesen, P.M., Cayrel de Strobel, G., and Cayrel, R.: 1977, *A&A* **54**, 779.
Pilyugin, L.: 1993, *A&A* **277**, 42.
Salaris, M., Chieffi, A., and Straniero, O.: 1993, *Ap. J.* **414**, 580.
Smits, D.P.: 1996, *MNRAS* **278**, 683.
Songaila, A., Wampler, E.J., and Cowie, L.L.: 1997, *Nature* **385**, 137.
Tytler, D., Fan, X.-M., and Burles, S.: 1996, *Nature* **381**, 207.
van den Hoek, R.: 1997, *Thesis*, Amsterdam University.
Weinberg, D.H., Miralda-Escudé, J., Hernquist, L., and Katz, N.: 1997, astro-ph/97 01012.
Yang, J., Turner, M.S., Steigman, G., Schramm, D.N., and Olive, K.: 1994, *Ap. J.* **281**, 493.

Address for correspondence: E. Høg, University Observatory, NBIfAFG, Juliane Mariesvej 30, DK-2100 Copenhagen Ø, Denmark, e-mail erik@astro.ku.dk
B.E.J. Pagel, NORDITA, Blegdamsvej 17, DK-2100 Copenhagen Ø, Denmark, e-mail pagel@nordita.dk
L. Portinari, Dipartimento di Astronomia, Vicolo dell'Osservatorio 5, 35122 Padova, Italy, e-mail portinari@astrpd.pd.astro.it
P.A. Thejll, Danish Meteorological Institute, Solar-Terrestrial Physics Division, Lyngbyvej 100, DK-2100 Copenhagen Ø, Denmark, e-mail thejll@dmi.dk
J. MacDonald, Department of Physics and Astronomy, University of Delaware, Newark DE 19716, USA, e-mail jim@physics.udel.edu
L. Girardi, Instituto de Fisica, UFRGS, Caixa Postal 15051, 91501–970 Porto Alegre, RS, Brazil, e-mail lgirardi@astrpd.pd.astro.it

EXTRAGALACTIC ABUNDANCES OF HYDROGEN, DEUTERIUM AND HELIUM

New Steps, Missteps and Next Steps

CRAIG J. HOGAN

Departments of Physics and Astronomy, University of Washington, Box 351580, Seattle, WA 98195, USA

Abstract. Estimates of the deuterium abundance in quasar absorbers are reviewed, including a brief account of incorrect claims published by the author and a brief review of the problem of hydrogen contamination. It is concluded that the primordial abundance may be universal with a value $(D/H)_P \approx 10^{-4}$, within about a factor of two, corresponding to $\Omega_B h_{0.7}^2 \approx 0.02$ or $\eta_{10} \approx 2.7$ in the Standard Big Bang. This agrees with current limits on primordial helium, $Y_P \leq 0.243$, which are shown to be surprisingly insensitive to models of stellar enrichment. It also agrees with a tabulated sum of the total density of baryons in observed components. Much lower primordial deuterium ($\approx 2 \times 10^{-5}$) is also possible but disagrees with currently estimated helium abundances; the larger baryon density in this case fits better with current models of the Lyman-α forest but requires the bulk of the baryons to be in some currently uncounted form.

1. Introduction

Material in the Earth and other planets, the solar neighborhood, and indeed our entire Galaxy has experienced significant chemical evolution that has modified the traces of light elements from the Big Bang. We are now beginning to sample abundances in more distant environments with a variety of different chemical histories. Some are nearly pristine and serve as fossil beds preserving the original chemistry, particularly those at at high redshift before much of the primordial gas first formed into galaxies and stars. In addition to insights about chemical evolution under different circumstances the new measurements give us unique information about the history of fine-grained structure over an enormous spacetime volume; for example they test the idea that primordial chemistry is the same everywhere and that the primordial gas was precisely uniform on small scales. New techniques also now allow us to tabulate a more reliable direct estimate of the mean baryon density of the Universe in various forms, allowing a further test of Standard Big Bang Nucleosynthesis for which mean total baryon density is the principal parameter.

The theory and its concordance have been extensively reviewed in the literature from many points of view (e.g. Walker *et al.*, 1991; Smith *et al.*, 1993; Copi *et al.*, 1995; Sarkar, 1996; Fields *et al.*, 1996; Hogan, 1997; Schramm, 1998), and many of the topics covered here are reviewed in more detail elsewhere in this volume. This review explores a selection of interesting contradictions among new extragalactic datasets but also highlights the general concordance with the Standard Big Bang Model.

Space Science Reviews **84**: 127–136, 1998.

2. The Cosmic Deuterium Abundance

The measurement of deuterium abundances by analyzing Lyman series absorption lines in stellar spectra from foreground diffuse gas (Rogerson and York, 1973) has yielded a reliable abundance for the Galactic interstellar medium, accurate to about $\pm 50\%$. Thanks to new technology the same technique applied to quasar spectra now yields the abundance of much more distant gas (see Tytler, 1998; Vidal-Madjar, 1998). High resolution, high signal-to-noise spectra allow the study of resolved absorption features of low optical depth which accurately count atomic column densities and thus in principle give reliable abundances. The subject is still very young however; effects of finite resolution and saturation still contribute a subtle source of error (see Levshakov et al., 1998), and the systematic errors are not yet well calibrated since the physical model of the absorbing material is not mature. (It is certainly not, as assumed in the spectral fits, in discrete isothermal slabs). At the moment there is an apparent polarization between "high" and "low" values which seems likely to disappear with time.

2.1. EVIDENCE FOR A HIGH ABUNDANCE

An early Keck spectrum of QSO 0014+813 allowed the first estimate of an extra-galactic deuterium abundance, yielding a remarkably high value $D/H \approx 2 \times 10^{-4}$ (Songaila et al., 1994). Multiple lines of the Lyman series gave a good agreement of hydrogen and deuterium redshift, and a reliable estimate of both column densities, so the most significant uncertainty in this measurement was the amount of contamination of the deuterium absorption feature by interloping hydrogen.

Subsequent analysis of this same data (Rugers and Hogan, 1996a) showed that at full resolution the Lyman-α feature of deuterium was split into two components, each of which was too narrow for hydrogen. The authors used this fact to argue that the hydrogen contamination did not dominate the deuterium feature and therefore that the high abundance could be trusted. The split of the feature was traced back to an electron excess in the raw echellogram data so was not a reduction artifact.

However, in a classic illustration of systematic error it now appears that the feature was not a real spectral feature of the quasar. Subsequent spectra by both the Hawaii group and by Tytler et al. (1997), which have better signal-to-noise than the original spectrum, do not confirm it. Apparently the ≈ 60 electrons contributing to the earlier signal, whatever their origin, were neither simply noise nor photoelectrons from the quasar light. (Similar problems also appear to plague another candidate feature in the same spectrum, proposed by Rugers and Hogan, 1996b). The deuterium feature in the real spectrum is instead smooth and is well fit by a single thermally broadened component, so the linewidth is consistent with significant hydrogen contamination. On the other hand the width is the same as the associated hydrogen (24 km/sec) and is therefore also consistent with being caused by deuterium if the broadening is mostly turbulent.

Indeed, the better data allow the redshift of the deuterium and hydrogen features to be compared more precisely and they differ by 10 km/sec, indicating that there is some hydrogen contaminating the deuterium (Tytler *et al.*, 1997). However the total column density of contaminating hydrogen required to move the centroid is small, so the known presence of some contamination does not imply that the best estimate of the abundance changes appreciably from that originally given by Songaila *et al.*

Because of the good constraints on both hydrogen and deuterium column densities, and in spite of confusing claims made by this author, the new data on the Q0014+813 absorber still displays good evidence of a high primordial abundance, about as convincing as the original claim by Songaila *et al.* (1994). In another quasar (BR 1202-0725) a "detection or upper limit" at $D/H = 1.5 \times 10^{-4}$ was found by Wampler *et al.* (1996). A high abundance (2×10^{-4}) may also be detected in Q0402-388, although it is required only if OI/HI is assumed to be constant in the fitted components (Carswell *et al.*, 1996). The same high abundance (2×10^{-4}) was also found by Webb *et al.* (1997) in Q1718+4807, although this is also not yet conclusive; the present analysis relies on a SiIII line to fix the redshift of the hydrogen, the D column is based only on a Lyman-α fit and the H column on a low resolution spectrum of the Lyman limit. The agreement between these estimates is at least suggestive of a high universal abundance, although none of the evidence is yet conclusive.

2.2. EVIDENCE FOR A LOW ABUNDANCE

Burles and Tytler (1996) and Tytler *et al.* (1996) have presented evidence for a low D/H in two quasar absorbers. Of these the stronger case at present is in Q1937-1009 since high quality data are available up to the Lyman limit. The estimated abundance is $2.3 \pm 0.3 \times 10^{-5}$, nearly an order of magnitude less than the high values discussed above.

Unfortunately the total column in this case is high so the HI absorption is optically thick even past the Lyman limit and the column density must be estimated from saturated features. This has led to a debate in the literature on the allowed range for the HI column and for D/H (e.g. Songaila *et al.*, 1997; Burles and Tytler, 1997; Songaila, 1997), which is likely to end up somewhere in the middle: although the total error in the abundance is probably larger than originally quoted by Tytler *et al.* (1996), the HI column is probably well enough constrained to exclude very high values like those in Q0014+813.

Therefore the dispersion in abundance between the best high and low estimates appears on the surface to be real. What is going on? There have been many suggestions (e.g., Jedamzik and Fuller, 1996). Perhaps the primordial abundance is not uniform; perhaps the low-D systems have experienced stellar destruction of their deuterium; perhaps the high-D systems have found some exotic source of nonprimordial deuterium. The most prosaic explanation however is that the high-D features are all dominated by contaminating hydrogen.

2.3. STATISTICAL APPROACHES TO THE CONTAMINATION PROBLEM

The latter possibility deserves serious attention since the effect is known to be there and known to bias abundance estimates upwards. However, a quantitative estimate of the effect shows that it is unlikely to be important most of the time. The mean number of hydrogen lines in a velocity interval δv per $\ln(N[HI])$ interval can be estimated from the line counts of Kim $et\ al.$ (1997) to be about

$$\delta P(N[\text{HI}]) \approx 5 \times 10^{-3} \left(\frac{\delta v}{10\text{km sec}^{-1}} \right) \left(\frac{N[\text{HI}]}{10^{13}\text{cm}^{-2}} \right)^{-0.4}.$$

If the contaminating hydrogen lines are distributed at random, we can use Poisson statistics to estimate the probability of contamination. For example, in the case of Q0014+813 the column density of the DI feature is about $10^{13.2}\text{cm}^{-2}$; the probability of an HI line close to this column density appearing in the right redshift range to mimic deuterium (within an interval of about $\delta v \approx 20\text{km sec}^{-1}$) is only about $P \approx 1\%$.

Of course, smaller amounts of contamination are more likely. They tend to bias the deuterium abundance estimates upwards and create a nongaussian (power-law) error distribution allowing low D/H with nonnegligible probability. However the magnitude of the bias is still small in this range of column density; for example, the chance of a $\approx 10\%$ contamination (for which there is indeed some evidence in Q0014+813 in the line profiles) is greater than the probability of 100% contamination by a factor of about $10^{0.4}$; but this is still only a few percent.

Furthermore this calculation does not yet allow for the additional coincidence required in the Doppler parameters. Real deuterium cannot be wider than its corresponding hydrogen. Contaminating hydrogen has a linewidth drawn at random from the parent population, which is not compatible with a deuterium identification if it happens to exceed the width of the corresponding hydrogen feature. In the case of Q0014+813, since the D feature linewidth of 24 km sec^{-1} is typical of HI Lyman-α forest lines we should multiply the probabilities by about half to account for this effect. This factor is smaller in situations where the features are unusually narrow.

It is important to verify that the lines are uncorrelated as we have assumed. Although there are disagreements over the amplitude of line correlations the best data on the Lyman-α forest shows correlations smaller than unity at velocity separations of $\simeq 100\text{km sec}^{-1}$; for example in the line lists of Kim $et\ al.$ (1997) the amplitude is less than 10% for all lines. We have checked this also specifically for correlations with high HI column density features similar to those studied in the D candidates; although the sample is smaller and the result noisier, the amplitude of the correlation is still less than a few tenths. In this study we used one-sided correlations (counting line companions only to the red side of each Lyman-α line) in order to exclude first-order effects of deuterium contamination in the hydrogen sample. (Similar conclusions were recently reached by Songaila, 1997).

Thus even without a thorough physical understanding of the absorbing material, these empirical studies indicate that correlations among the lines are too weak to alter significantly the simple estimates made on the basis of Poisson statistics. Therefore the best guess is that the deuterium abundance in cases where it appears to be high really is high, at least where the DI column is greater than about 10^{13}cm^{-2} (which is in any case required for a reasonably precise column estimate at realistic signal-to-noise). Noting that the full error in the fitting technique has not yet been calibrated on realistic physical models of the clouds, it still seems reasonable to guess that the current data are consistent with a universal primordial abundance with a factor of two of 10^{-4}. A much more convincing measurement will be possible with a few more good targets.

2.4. NEXT STEPS

The contamination problem is less for lower redshift, where the Lyman-α forest thins out. Although low redshift quasar spectroscopy at high resolution is costly as it requires a large investment of time with the Hubble Space Telescope, the new STIS two-dimensional spectrograph can observe the entire Lyman series at the same time; the greater efficiency will give more solid results on cases such as Q1718+4807 which are already known to be interesting (e.g. Webb *et al.*, 1997) .

Contamination can also be reduced by studying high column density systems. Especially interesting are damped HI absorbers where reliable columns can be obtained for both HI and DI (Jenkins, 1996), although targets are also rarer and tend to be in evolved galaxies. The best target of this type so far identified is Q2206-199, which has a low metal abundance and a very low velocity dispersion (Pettini and Hunstead, 1990).

In the long term this problem will be solved by a larger sample of absorbers in bright quasars observed from the ground. Although the current number of suitable target quasars is now quite small, the target sample in coming years will grow by about two orders of magnitude as a result of the Sloan Digital Sky Survey, so progress in this field will be limited by observing time rather than by the availability of targets.

3. The Cosmic Helium Abundance

3.1. A BAYESIAN APPROACH TO HELIUM ENRICHMENT

Helium abundances in extragalactic HII regions have for many years been the principal observational constraint on Big Bang Nucleosynthesis. Flourescent nebular emission lines of hydrogen and helium reveal quite precisely the number of electrons recombining into each species and thereby the abundances of each (e.g. Peimbert and Torres-Peimbert, 1976; Pagel et al., 1992; Skillman and Kennicutt, 1993). The techniques and estimates of systematic errors are discussed by Høg et al. (1998), Skillman et al. (1998), and Steigman et al. (1998). Here I make one simple point: the dominant source of systematic error in the primordial abundance Y_P, especially in the upper limit, probably lies not in the model used to extrapolate to zero metallicity but in the physical models used to estimate the present-day abundances Y from observations of nebulae.

Even though the bulk of the helium of the Universe originates in the Big Bang, the additional helium enrichment by stars cannot be ignored in estimating the primordial abundance from observations of present-day helium. The most widely used approach to estimate the nonprimordial enrichment is to correlate Y with metallicity Z. However, one of the limitations of this method is the need to assume a linear relation between Y and Z, which is not well motivated. Moreover most of the information on the primordial abundance is contained in the lowest metallicity points, where the correlation is not very reliably established; this information is not being efficiently used in regression fits dominated by highly enriched regions. Another approach has been to simply take the lowest, best measured points and use them as estimates of (or at least limits on) the primordial abundance. Clearly however some correction has to be made for the obvious Malmquist-like bias introduced, and some statistical way needs to be found to combine more than one region.

Hogan, Olive and Scully (1997) introduced a new statistical method to estimate the primordial helium abundance Y_p directly from helium observations, without using metal abundances. They constructed a likelihood function using a Bayesian prior, encapsulating the key assumption that the true helium abundance must always exceed the primordial value, by an amount which may be as large some maximum enrichment w. They computed the likelihood as a function of the two parameters Y_p and w using samples of measurements compiled from the literature.

Some results from published samples are shown in figure 1. Estimates of Y_p vary between 0.221 and 0.236, depending on the specific subsample and prior adopted, consistent with previous estimates using different techniques. Evidence for stellar enrichment ($w \neq 0$) appears even in the lowest metallicity subsamples, but in all samples the most conservative upper bound on Y_p occurs for $w = 0$, yielding a nearly model-independent bound $Y_p < 0.243$ at 95% confidence. The main uncertainty in the Y_p bound is not the model of stellar enrichment but possible common systematic biases in the estimate of Y in each individual HII region.

3.2. HELIUM AT HIGH REDSHIFT

In highly ionized environments singly ionized helium (HeII) is typically orders of magnitude more common than HI, making it the most cosmically abundant absorber, detectable even in the most rarefied regions between the Lyman-α forest

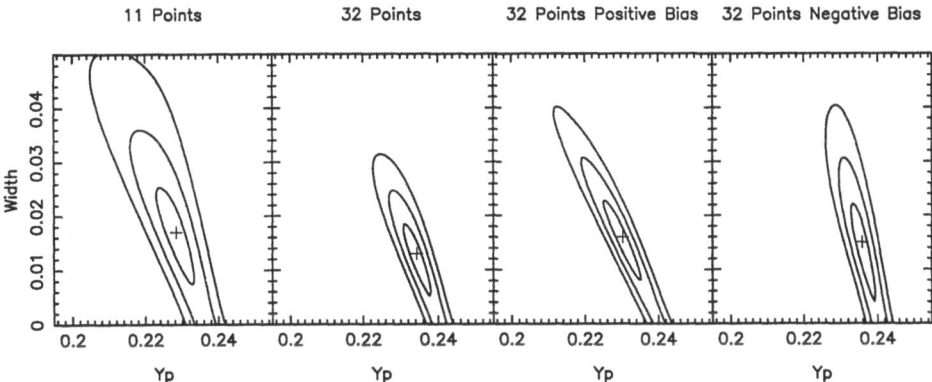

Figure 1. Likelihood function showing 1σ, 2σ and 3σ contours in the (Y_p, w) plane, from Hogan, Olive and Scully (1997). The +'s indicate the peaks of the likelihood functions. The two left panels show results using a top-hat prior and two different subsamples of 11 and 32 lowest metal points. The two right panels show the results for positive and negative bias priors, for the 32 point sample. In all cases the conservative 2σ limit occurs at $w = 0$ and yields a limit $Y_P \leq 0.243$.

clouds. Lyman-α absorption by HeII nearly continuously fills redshift space at optical depth of the order of unity as recently verified in at least three quasars (Jakobsen *et al.*, 1994; Davidsen *et al.*, 1996; Hogan *et al.*, 1997; Reimers *et al.*, 1997).

The need to model ionization states precludes a truly precise measurement of helium abundance. An absolute helium abundance can be estimated from HI and HeII Lyman-α absorption if: (1) the helium abundance is uniform; (2) HeII and HI are in ionization equilibrium dominated by photoionization; (3) we know the shape of the ionizing spectrum; (4) absorption is unsaturated, so column densities of absorbing species can measured. The total redshift integrated columns of HeII and HI are then both proportional to the same line integral $\int d\ell\, n_e^2$ with coefficients depending on the abundance and the ionizing spectrum.

These conditions are certainly never met in detail, but in some situations data can still be used to set constraints on the abundance and its variations. In some regions, the HeII fraction may be close to unity (Reimers *et al.*, 1997); in others, the strength and/or spectrum of the ionizing field may be known, for example for gas close to the quasar (Hogan *et al.*, 1997); and in some circumstances the spectrum may be approximately constant over a large pathlength along the line of sight, allowing the universality of the abundance to be constrained even if its absolute value is uncertain.

The new results are however most important not as abundance determinations but as probes of structure formation and ionization history. If we assume that the helium abundance is close to the Big Bang prediction, helium observations put stringent constraints on the density of diffuse gas at high redshift which limit the range of possible baryonic histories. For example, it is not possible to place the bulk of the baryons in a diffuse medium at $z = 3$ without exceeding helium absorption limits unless the gas is hot enough to be thermally ionized; on the other hand a substantial fraction of the baryons appear to be necessary in gaseous form (in clouds) to account for the Lyman-α forest (Rauch et al., 1997; Weinberg et al., 1997; Zhang et al., 1997).

4. The Cosmic Baryon Abundance and the Concordance of Standard Big Bang Nucleosynthesis

In a recent survey, Fukugita, Hogan and Peebles (1997) estimate the mean density of the Universe observed in various components of baryons, summarized in Table 1. Not included in the table are two "uncounted" components, MACHOs and hot plasma ($T \approx 2 \times 10^6 \text{K}$), which are known to exist locally. For both of these almost no meaningful upper or lower bounds can be set on the global density. In the case of MACHOs, a new population of dark compact objects, probably baryonic, has been detected by microlensing in the direction of the LMC; depending on assumptions used to extrapolate, the global density of this population could either be negligible or could dominate all other forms of baryons combined. Similarly, a thermal background from hot gas is detected, but could either be from a globally insignificant portion of the Galactic corona or from a globally distributed plasma containing the bulk of the baryons. These two components are therefore possible repositories of additional baryons.

The point here of course is to compare the observed number of baryons with the number expected on the basis of light element abundances. For the nucleosynthesis entries we adopt upper and lower limits for the primordial deuterium $2 \times 10^{-5} \leq (D/H)_P \leq 2 \times 10^{-4}$. For primordial helium we adopt a central value of $Y_P = 0.23$ and a 2σ limit as described above, $Y_P \leq 0.243$. Note that this limit is not compatible with the low deuterium values. The lithium abundance allowing for some depletion is taken to be $\text{Li/H} \leq 4 \times 10^{-10}$ from Galactic stars (see Cayrel, 1998); even though it does not offer a principal constraint on baryon density in the Standard Model, it is important as a constraint on departures from the Standard Model, such as small scale inhomogeneities (e.g. Kurki-Suonio et al., 1997).

It is clearly instructive to contemplate these numbers at length. Most of the baryons today are still in the form of ionized gas, which contribute a mean density uncertain by a factor of about four (due to uncertainties in extrapolating from observed x-ray emission). For the best-guess plasma density, stars are a relatively minor component – all stars and their remnants comprise only about 17% of the baryons, while populations contributing most of the blue light comprise less than 5%. The formation of galaxies and of stars within them appears to be a globally inefficient process – an effect not fully understood in models of galaxy formation. The sum over the budget, expressed as a fraction of the critical Einstein-de Sitter density, is in the range $0.01 \leq \Omega_B \leq 0.04$, with a best guess $\Omega_B \sim 0.02$ (at Hubble parameter 70 km s^{-1} Mpc^{-1}). This is close to the prediction from the Standard

Table I

The Baryon Budget (Fukugita et al., 1997)

	Component	Optimum	Maximum	Minimum	Grade
	observed at $z \approx 0$:				
1	stars in spheroids	0.0026 $h_{0.7}^{-1}$	0.0043 $h_{0.7}^{-1}$	0.0016 $h_{0.7}^{-1}$	A
2	stars in disks	0.00086 $h_{0.7}^{-1}$	0.00129 $h_{0.7}^{-1}$	0.00051 $h_{0.7}^{-1}$	A−
3	stars in irregulars	0.000069 $h_{0.7}^{-1}$	0.000116 $h_{0.7}^{-1}$	0.000033 $h_{0.7}^{-1}$	B
4	neutral atomic gas	0.00033 $h_{0.7}^{-1}$	0.00041 $h_{0.7}^{-1}$	0.00025 $h_{0.7}^{-1}$	A
5	molecular gas	0.00022 $h_{0.7}^{-1}$	0.00029 $h_{0.7}^{-1}$	0.00014 $h_{0.7}^{-1}$	A−
6	plasma in clusters	0.0026 $h_{0.7}^{-1.5}$	0.0044 $h_{0.7}^{-1.5}$	0.0014 $h_{0.7}^{-1.5}$	A
7	plasma in groups	0.014 $h_{0.7}^{-1}$	0.03 $h_{0.7}^{-1}$	0.007 $h_{0.7}^{-1}$	B
	sum (at $h = 0.7$)	0.02	0.04	0.01	
	observed at $z \approx 3$:				
10	in damped absorbers	0.001– 0.002 $h_{0.7}^{-1}$	0.0027 $h_{0.7}^{-1}$	0.007 $h_{0.7}^{-1}$	A−
11	in the Lyman-α forest	0.04 $h_{0.7}^{-2}$		0.025 $h_{0.7}^{-2}$	A−
12	between the forest clouds		0.01 $h_{0.7}^{-1.5}$	0.0001 $h_{0.7}^{-1}$	B
	nucleosynthesis:				
9a	deuterium		0.051 $h_{0.7}^{-2}$	0.013 $h_{0.7}^{-2}$	A
9b	helium	0.010 $h_{0.7}^{-2}$	0.027 $h_{0.7}^{-2}$		A−
9c	lithium		0.06 $h_{0.7}^{-2}$	0.007 $h_{0.7}^{-2}$	B

Big Bang for moderately high D/H $\approx 10^{-4}$. If the deuterium abundance is high, this suggests we may be close to a complete survey of the major states of the baryons as well as a concordance with nucleosynthesis. On the other hand if D is low (and Y_P has been underestimated) the baryon budget is likely to be dominated by currently uncounted components. Although galaxy formation models prefer the higher baryon density (e.g. Rauch et al., 1997; Zhang et al., 1997) they do not yet predict which of the hidden forms dominates today (see e.g. Fields and Schramm, 1998) – very diffuse gas or very compact cold bodies.

Acknowledgements

I am grateful to my collaborators, especially S. F. Anderson, M. Fukugita, K. A. Olive, M. H. Rugers, P. J. E. Peebles, and S. T. Scully, for many insights, and A. Songaila and L. L. Cowie for generous sharing of their superb data. This work was supported by NASA and NSF at the University of Washington.

References

Burles, S., and Tytler, D.: 1996, astro-ph/9603070.

Burles, S., and Tytler, D.: 1997, astro-ph/9707176.
Carswell, R.F., *et al.*: 1994, *M.N.R.A.S.* **268**, L1.
Carswell, R.F., *et al.*: 1996, *M.N.R.A.S.* **278**, 506.
Cayrel, R.: 1998, this volume.
Copi, C. J., Schramm, D. N. and Turner, M. S.: 1995, *Science* **267**, 192.
Davidsen, A., Kriss, G. A., and Zheng, W.: 1996, *Nature* **380**, 47.
Fields, B.D., Kainulainen, K., Olive, K.A., and Thomas, D.: 1996 *New Astronomy* **1**, 77.
Fields, B., and Schramm, D. N.: 1998, this volume.
Fukugita, M., Hogan, C. J. and Peebles, P. J. E.: 1996, *Nature* **381**, 489.
Fukugita, M., Hogan, C. J. and Peebles, P. J. E.: 1997, in preparation.
Høg, E. *et al.*: 1998, this volume.
Hogan, C. J.: 1997, in *Critical Dialogues in Cosmology*, ed. N. Turok (Princeton: Princeton University Press; astro-ph/9609138).
Hogan, C. J., Anderson, S. F. and Rugers, M. H.: 1997, *Astron. J.* **113**, 1495.
Hogan, C. J., Olive, K. A. and Scully, S. T.: 1998, *Ap. J. Lett.*, in press (astro-ph/9705107).
Jakobsen, P. *et al.*: 1995, *Nature* **370**, 35.
Jedamzik, K., and Fuller, G.: 1996, *ApJ* **483**, 560, astro-ph/9609103.
Jenkins, E.B.: 1996, *ASP conference ser.* **99**, 90.
Kim, T-S., Hu, E., Cowie, L. L., and Songaila, A.: 1997, *Astron. J.* **114**, 1.
Levshakov, S. A., Kegel, W. H. and Takahara, F.: 1998, this volume; see also astro-ph/9708064.
Pagel, B., *et al.*: 1992, *M.N.R.A.S.* **255**, 325.
Peimbert, M. and Torres-Peimbert, S.: 1976, *et al.* **203**, 581.
Pettini, M. and Hunstead, R. W.: 1990, *Australian J. Phys.* **43**, 227.
Rauch, M. *et al.*: 1997, in press, astro-ph/9612245.
Reimers, D. *et al.*: 1998, *Astron. Astrophys.*, in press, astro-ph/9707173.
Rogerson, J., and York, D.: 1973, *Astrophys. J. Letters* **186**, L95.
Rugers M., and Hogan, C.J.: 1996a, *Astrophys. J. Letters* **549**, L1.
Rugers M., and Hogan, C.J.: 1996b, *Astron. J.* **111**, 2135.
Sarkar, S., 1996, *Rep. Prog. Phys* **59**, 1493.
Schramm, D.N.: 1998, this volume.
Skillman, E. and Kennicutt, R. C.: 1993, *ApJ* **411**, 655.
Skillman, E., Terlevich, E. and Terlevich, R.: 1998, this volume.
Smith, M. S., Kawano, L. H., and Malaney, R. A.: 1993, *ApJ* **85**, 219.
Songaila, A., Cowie, L.L., Hogan, C.J., and Rugers, M.: 1994, *Nature* **368**, 599.
Songaila, A., Wampler, E.J., and Cowie, L.L.: 1997, *Nature* **385**, 137.
Songaila, A.: 1997, astro-ph/9709293.
Steigman, G., Viegas, S. M., and Gruenwald, R.: 1998, this volume.
Thuan, T. X.: 1998, this volume.
Tytler, D., Fan, X.-M., and Burles, S.: 1996, *Nature* **381**, 207.
Tytler, D., Burles, S. and Kirkman, D.: 1997, *ApJ*, in press, astro-ph/9612121.
Tytler, D.: 1998, this volume.
Vidal-Madjar, A.: 1998, this volume.
Walker, T. P., Steigman, G., Schramm, D. N., Olive, K. A., and Kang, H.: 1991, *ApJ* **376**, 51.
Wampler, *et al.*: 1996, *Astron. Astrophys.* **316**, 33.
Webb, J.K., Carswell, R.F., Lanzetta, K.M., Ferlet, R., Lemoine, M., Vidal-Madjar, A., and Bowen, D.V.: 1997, *Nature* **388**, 250.
Zhang, Y. *et al.*: 1997, *ApJ*, in press, astro-ph/9706087.

III: LOW-Z STARS

THE LOW METALLICITY TAIL OF THE HALO METALLICITY DISTRIBUTION FUNCTION

T.C. BEERS and S. ROSSI*
Department of Physics & Astronomy, Michigan State University, USA

J.E. NORRIS
Mt. Stromlo & Siding Springs Observatory, Australian National University, AUSTRALIA

S.G. RYAN
Royal Greenwich Observatory, Cambridge, UK

P. MOLARO
Usservatorio Astronomico di Trieste, ITALY

R. REBOLO
Instituto de Astrofísica de Canarias, Tenerife, CANARY ISLANDS

Abstract. Ongoing spectroscopy and photometry of stars selected in the HK objective-prism/interference-filter survey of Beers and colleagues has resulted in the identification of many hundreds of additional stars in the halo (and possibly the thick disk) of the Galaxy with abundances [Fe/H] ≤ -2.0. A new calibration of the technique for estimation of metal abundance based on a CaII K index as a function of broadband $B - V$ color is applied to obtain metallicities for stars observed with the SSO 2.3m and INT 2.5m telescopes. This new data is combined with other samples of extremely metal-deficient stars (Ryan and Norris, 1991a; Beers *et al.*, 1992; Carney *et al.*, 1994) to form a large database of objects of low metallicity. The combined sample is examined and compared with expectations derived from a Simple Model of Galactic chemical evolution. There appears to be a statistically-significant *deficit* of stars more metal-weak than [Fe/H] $= -3.0$. An abundance of [Fe/H] ≈ -4.0 can be taken as the low-metallicity limit for presently-observable stars in the Galaxy.

1. Introduction

The halo metallicity distribution function (MDF) provides direct information about the initial stages of galaxy formation, in that it is sensitive to the bulk chemical properties of the interstellar gas from which the earliest generations of stars were born. A compilation of the relative numbers of low metallicity stars in the halo enables comparisons with alternative models for Galactic chemical evolution, and can be used to place constraints on the primordial rate of Type II supernovae, the star formation rate, and the time scale for the redistribution of elements in the early Galaxy.

Previous attempts to address this question have been hampered by rather small samples of stars at the lowest abundances; as a result a variety of disparate conclusions about the early Galaxy have been obtained. Even the most naive models could not be adequately tested. With the addition of a new large sample of stars with abundances less than 1% of the solar composition, the picture of the early epochs of our Galaxy can now be drawn with more clarity.

* On leave from Department of Astronomy, Instituto Astronômico E Geofísico, Universidade de São Paulo

Space Science Reviews **84**: 139–144, 1998.
© 1998 Kluwer Academic Publishers

2. The HK Survey – An Update

Beers *et al.* (1992) describe an objective-prism/interference-filter technique for the
identification of large numbers of candidate low-metallicity stars in the Galactic
halo. The HK survey, as it has come to be known, is based on some 300 prism
plates covering roughly 1/6th of the sky in the northern and southern hemisphere.
Visual inspection of the survey plates (with a 10 X microscope) has resulted in the
selection of some 10,000 candidates in the magnitude range $10.5 \leq B \leq 15.5$.
For the past 15 years, follow-up medium-resolution (1–2 Å) spectroscopy and
broadband UBV photometry of these stars has been conducted with 1m- to 2m-
class telescopes from Las Campanas, Kitt Peak National Observatory, the European
Southern Observatory, Siding Springs Observatory, and the Canary Islands. Spectra
for roughly half of the low-metallicity candidates have been obtained to date.
Photometry has been obtained for some 3500 candidates. The observational efforts
from Siding Springs and the Canary Islands are essentially finished; observations
from Kitt Peak and ESO will be completed within the next year.

3. Metal Abundances from Medium-Resolution Spectra; A New Calibration

In spectra of limited resolution, lines of FeI and FeII are too weak to be reliably
detected for stars with metallicities [Fe/H]< -1.0. Thus, in order to extract a
reasonable estimate of stellar abundance, other methods must be employed. Beers
et al. (1990) describe a method based on measurement of the equivalent width of the
CaII K line at 3933 Å, used in combination with a broadband $B - V$ color. Clearly,
if abundances are to be measured on the [Fe/H] scale, this technique relies on the
existence of a monotonic, and ideally, scatter-free relationship between [Ca/Fe]
and [Fe/H] for the abundance range under consideration. Fortunately, it appears
that these conditions hold, to a rather high degree, for the great majority of halo
stars (Edvardsson *et al.*, 1993; Ryan *et al.*, 1996). Beers *et al.* (1990) show that
metal abundances in the range $-4.5 \leq$ [Fe/H] ≤ -1.5 could be assigned, using
the CaII K technique, with an external accuracy on the order of 0.2 dex. Above
[Fe/H]$= -1.5$, especially for cooler stars, the CaII K line begins to saturate, and
the method suffers from increased scatter as well as a zero-point offset.

In order to address the known deficiencies of the Beers *et al.* (1990) calibration,
we have undertaken a new approach to the problem. A first-pass estimate of the
metal abundance of a given star is derived by comparison with a grid of some
700 synthetic spectra constructed from the recent Kurucz (1993) models (over
the ranges of physical parameters $-5.0 \leq$ [Fe/H] ≤ 0.0, $3500 \leq T_{\text{eff}} \leq 7500$ K,
$0.0 \leq \log g \leq 5.0$), using colors obtained from the Revised Yale Isochrones (Green
et al., 1987). A resistant locally-weighted surface fit was used to capture the run of
CaII K equivalent width with $B - V$ color over the grid. Such a fit is not expected
to be perfect, owing to problems in the model atmospheres, synthetic colors, or

equivalent-width measurements; hence it must be calibrated externally with stars of known [Fe/H].

In the course of recent spectroscopic follow-up, observations of some 400 stars with available fine-analysis abundance estimates were obtained (a five-fold increase over the number used in the Beers *et al.* 1990 calibration). Residuals between the first-pass estimates of abundance and the standard stars were then fit with a two-dimensional surface in a similar manner as before, and corrections between the model estimates and the calibration-star abundances were applied. A detailed analysis of the final residuals indicates that for stars with CaII equivalent width index $KP < 6.0$, [Fe/H] can be estimated via this technique with an accuracy between 0.15 and 0.30 dex; the higher accuracy applying to the more metal-weak stars. For stars with $KP > 6$ (the more metal-rich stars), the precision of the method decreases to between 0.4 and 0.6 dex.

In an effort to decrease the scatter in estimates of metallicity for the stars with [Fe/H] > -1.5, we have used the same grid of synthetic spectra to explore abundance estimation via an auto-correlation technique (Ratnatunga and Freeman, 1989). The auto-correlation method appears to be quite satisfactory for stars with abundances [Fe/H] ≥ -2.0, but loses power at lower metallicities. In the final step of our calibration procedure, the two methods are combined using a weighted average in order to obtain abundance estimates which are accurate to 0.20–0.25 dex over the entire range $-4.5 \leq$ [Fe/H] ≤ 0.0. For more details see Beers *et al.* (1997) .

4. Sample Selection

The HK survey is not the only source for stars of extremely low abundance. Other large survey efforts have contributed stars with [Fe/H] < -1.5, primarily based on proper-motion selection criteria. In Table I we summarize the available stars from three programs. The first set are the kinematically-selected samples of Ryan and Norris (1991a) and Carney *et al.* (1994). The second is the published list of stars from the HK survey (Beers *et al.*, 1992), labelled here as BPS II (note that abundances for these stars have been estimated anew using the methodology described above). The third subsample is the combined set of HK follow-up stars from the Siding Springs and INT efforts.

Note that for the stars in Table I with [Fe/H] ≤ -3.0 and ≤ -4.0 we have listed parenthetically the numbers of stars with [Fe/H] ≤ -2.8 and ≤ -3.8, respectively, in order to emphasize the effect which small changes in metallicity estimates might have on the cumulative numbers of stars below a given abundance. The factor of two to three discrepancy seen in the numbers of stars with [Fe/H] ≤ -2.8 and [Fe/H] ≤ -3.0 underscores the need for a uniform set of high-resolution determinations of the abundances of ALL stars in our sample with [Fe/H] ≤ -2.5 in order to

Table I
Summary of the Metal-Poor Star Sample

Sample	N ≤ −1.5	N ≤ −2.0	N ≤ −2.5	N ≤ −3.0	N ≤ −3.5	N ≤ −4.0
LCL/RN	578	256	79	11 (29)	4	1 (1)
BPSII	726	540	197	29 (73)	5	1 (1)
SSO/INT	547	356	155	26 (59)	4	0 (2)
TOTAL	1851	1152	431	66 (161)	13	2 (4)

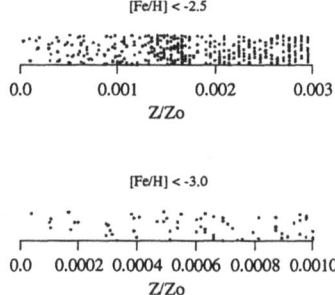

Figure 1. Density Plots of observed abundances for stars in the combined sample. The position of each star has been randomized along the vertical direction.

be absolutely confident of the derived MDF. For now, we proceed with a method which explicitly takes the expected errors in abundances into account.

5. The Metallicity Distribution Function

Selection effects in the HK survey candidate list (at the metal-rich end) limit the abundance range over which our derived MDF can be considered as representative of the Galactic halo. Comparison with the kinematically-selected samples (which should not suffer from the same bias) indicates that the populations are consistent with being drawn from an identical parent population for [Fe/H] ≤ −2.2. For the purposes at hand we will only test stars with [Fe/H] ≤ −2.5, in order to ensure that bias does not play a large role.

In the limit of extremely low metal abundance (well below the effective-yield parameter) the Simple Model of Galactic chemical evolution (Hartwick, 1976) predicts that the MDF, expressed as a function of the linear variable $z/z_\odot$, goes over to a constant, in other words to a uniform distribution. Figure 1 shows the distribution of abundances for the combined sample of stars with [Fe/H] ≤ −2.5 and [Fe/H] ≤ −3.0, respectively. The apparent "overdensity" of stars with [Fe/H] ≈ −2.75 is certainly of potential interest, but beyond the scope of the present discussion.

Table II
Tests of the Uniform Distribution Hypothesis

[Fe/H]	N ≤ [Fe/H]		$\overline{U}$	p-value
		Full Sample		
−2.5	431		0.56	0.025
−3.0	66		0.61	0.005
		$|Z| \leq 1$ kpc		
−2.5	203		0.57	0.010
−3.0	26		0.62	0.005
		$|Z| > 1$ kpc		
−2.5	228		0.55	0.050
−3.0	40		0.59	0.010

A simple test may be employed by comparing these samples with the hypothesis of selection from a uniform distribution. Transformation of the stellar abundances to uniform deviates on the interval $0 \leq U_i \leq 1$ permits calculation of the statistic $\overline{U}$, the mean deviate. If stars are selected from a uniform parent population, then $\overline{U} = 0.5$, within sampling fluctuations. Table II indicates that *neither* sample of stars satisfies this test; both fail in the sense of having too high a value of the mean deviate, indicating a lack of the lowest metallicity stars.

Beers and Sommer-Larsen (1995) have argued (following Morrison *et al.*, 1990) for the existence of a metal-weak thick-disk population. If such a population exists, it could certainly dominate our sample of low-metallicity stars close to the Galactic plane. Hence, as a further test, we split our samples on distance from the plane. Still, *both* subsamples in *both* metallicity regimes allow rejection of the uniform hypothesis. Figure 2 presents the MDFs of the two subsamples selected on distance from the plane of the Galaxy, as compared to the prediction of a Simple Model. Visual inspection confirms the detailed statistical analysis – there is a clear lack of stars at the lowest metallicities as compared to the numbers predicted from a Simple Model.

6. Interpretation and Future Work

Audouze and Silk (1995) used the scatter in abundances among individual elements for stars with [Fe/H] ≤ -2.5 to predict that the expected minimum abundance for stars in the Galactic halo should be on the order of [Fe/H]= -4.0. The data seem to bear this out, though of course a detailed model has yet to be put forward. Now that the set of stars with very low abundance is substantially larger, it would be of interest to compare the observed tail of the halo MDF to the stochastic model discussed most recently by Ryan and Norris (1991b).

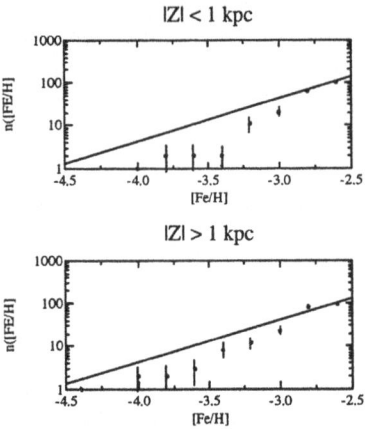

Figure 2. MDFs for the combined samples with [Fe/H] ≤ -2.5. Bins are 0.2 dex in width, and error bars indicate the $\sqrt{N}$ noise associated with each bin. The lines are predicted curves for a Simple Model with $y_{\text{eff}} = -1.6$.

References

Audouze, J., and Silk, J.: 1995, 'The First Generation of Stars: First Steps Toward Chemical Evolution of Galaxies', *ApJ* **451**, L49–L52.

Beers, T.C., and Sommer-Larsen, J.: 1995, 'Kinematics of Metal-Poor Stars in the Galaxy', *ApJS* **96**, 175–221.

Beers, T.C., Preston, G.W., and Shectman, S.A.: 1992, 'A Search for Stars of Very Low Metal Abundance. II. ', *AJ* **103**, 1987–2034.

Beers, T.C., Preston, G.W., Shectman, S.A., and Kage, J.A.: 1990, 'Estimation of Stellar Metal Abundance. I. Calibration of the CaII K Index', *AJ* **100**, 849–883.

Beers, T.C., Rossi, S., Norris, J.E., Ryan, S.G., and Shefler, T.: 1997, 'Estimation of Stellar Metal Abundance. II. Refinements of the CaII K Technique and the Auto-correlation Technique', in preparation.

Carney, B.W., Latham, D.W., Laird, J.B., and Aguilar, L.A.: 1994, 'A Survey of Proper Motion Stars. XII. An Expanded Sample', *AJ* **107**, 2240–2289.

Edvardsson, B., Andersen, J., Gustafsson, B., Lambert, D.L., Nissen, P.E., and Tomkin, J.: 1993, 'The Chemical Evolution of the Galactic Disk. I. Analysis and Results', *A&A* **275**, 101–152.

Green, E.M., Demarque, P., and King, C.R.: 1987, 'The Revised Yale Isochrones and Luminosity Functions', Yale University Observatory, New Haven.

Hartwick, F.D.A.: 1976, 'The Chemical Evolution of the Galactic Halo', *ApJ* **209**, 418–423.

Kurucz, R.L.: 1993, 'CD-ROM 13: ATLAS9 Stellar Atmospheres Programs and 2 km/s Grid', Smithsonian Astrophysical Observatory, Cambridge.

Morrison, H.L., Flynn, C., and Freeman, K.C.: 1990, 'Where Does the Disk Stop and the Halo Begin? Kinematics in a Rotation Field', *AJ* **100**, 1191–1222.

Ratnatunga, K.U. and Freeman, K.C.: 1987, 'Field K Giants in the Galactic Halo. II. Improved Abundance and Kinematic Parameters', *ApJ* **339**, 126–148.

Ryan, S.G., and Norris, J.E.: 1991a, 'Subdwarf Studies. II. Abundances and Kinematics from Medium Resolution Spectra', *AJ* **101**, 1835–1864.

Ryan, S.G., and Norris, J.E.: 1991b, 'Subdwarf Studies. III. The Halo Metallicity Distribution', *AJ* **101**, 1865–1878.

Ryan, S.G., Norris, J.E., and Beers, T.C.: 1996, 'Extremely Metal-Poor Stars. II. Elemental Abundances and the Early Chemical Enrichment of the Galaxy', *ApJ* **471**, 254–278.

LITHIUM ABUNDANCES IN LOW-Z STARS

R. CAYREL

Observatoire de Paris, 61, avenue de l'Observatoire F-75014 Paris, France

Abstract. An historical view of the discovery and subsequent studies of lithium in low-Z stars is presented. The determination of the lithium abundance in extremely low-Z stars, found in the vast Beers, Preston and Shectman survey, is reviewed. The problem of the exact connection between the lithium abundance found in the photospheres of low-Z stars and the true cosmological abundance of lithium is discussed, and identified as the most important problem to be solved in the coming years.

Key words: lithium, subdwarfs, mixing

1. Introduction

For many years (see for example Herbig, 1965; Müller *et al.*, 1975) it has been known that lithium has a high abundance in young solar type stars (close to its solar system meteoritic value) whereas it is strongly depleted in the Sun itself. It is usually thought that the depletion in old G stars is due to nuclear burning of lithium, by slow mixing with regions just below the convective zone, where the temperature is high enough ($\simeq 2.6 \times 10^6$ K) to destroy lithium.

At the beginning, lithium has not been searched in subdwarfs, probably because it was thought that, if it was destroyed in a 4.6 Gyr star, it was *a fortiori* destroyed in a 15 Gyr old star. Spite and Spite (1982) found that lithium was, contrary to expectation, present in most subdwarfs, provided that their effective temperature be above 5500 K. Below this effective temperature there is evidence of depletion, with practically no detectable lithium below $T_{\text{eff}} = 5000$ K. The logarithmic abundance of lithium found in subdwarfs having an effective temperature between 5500 K and the turnoff temperature ($\simeq 6400$ K) was about 2.05, in the scale $\log(n(\text{H})) = 12.0$. After 15 years and almost 50 papers on the subject, this number is still valid, with an accuracy of 0.1 dex, and this abundance is totally independent of the abundance of the other elements, behaviour absolutely unique among all elements measurable in subdwarfs. This is readily explained if the lithium found in subdwarfs is of cosmological origin.

The story could be closed here if there was not the following (theoretical) question: if lithium is depleted by a factor of 100 in 4.6 Gyr in the Sun, are we really sure that it is not at all depleted in 14 Gyr old subdwarfs? The next section will explain why lithium may have survived 14 Gyr in subdwarfs, whereas section 3 introduces the various theories of lithium depletion applied to pop. I and pop. II stars. Section 4 addresses the hot question of the lithium depletion on the Spites' plateau itself. Section 5 discusses the special case of lithium depletion in tidally-locked binaries. Section 6 gives recent results on lithium abundance in extremely metal-poor stars. Section 7 concludes this talk.

Space Science Reviews **84:** 145–154, 1998.

R. CAYREL

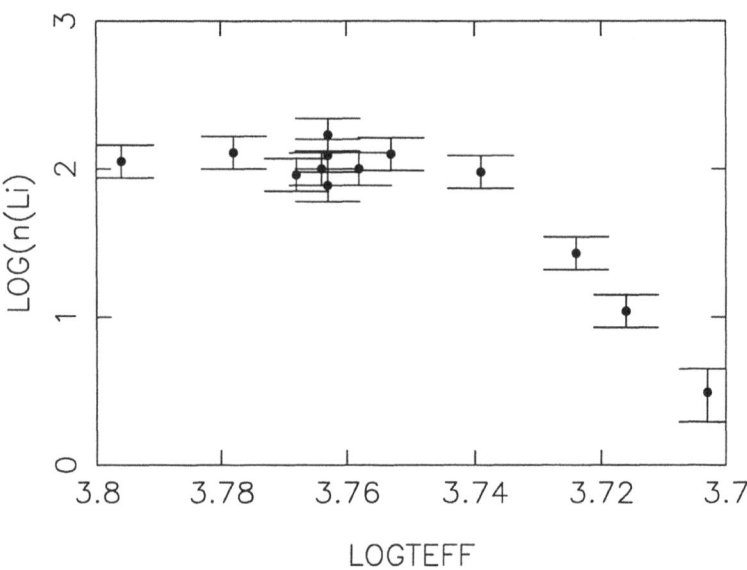

Figure 1. The so called Spites' plateau according to the original paper. The upper limit 0.5 given by the Spites for HD 103095 has been replaced by the recent value 0.49 (King, 1997)

2. Lithium depletion and metallicity

The obvious question which arises, when the survival of ^{7}Li in subdwarfs is known, is: can this be possibly due to the change in internal structure induced by the low metallicity of these objects? Fig. 2a and 2b show the internal structures of the unevolved Sun and of an unevolved subdwarf, with the boundary of the convection zone and the depth at which ^{7}Li is locally burned in a time very short with respect to the lifetime of the star. It is clear from Fig. 2 that, just below the convective zone (CZ), lithium is more exposed to burning in the Sun than in a subdwarf of metallicity [Fe/H] $= -2.0$ and of mass $0.8M_\odot$, corresponding to an effective temperature of 5950 K, right on the Spites' plateau. So, there is at least a reasonable structural effect, opening the possibility of a much better protection of ^{7}Li in a subdwarf than in a normal metallicity star. The effect is still more striking when the computation is made at the same mass: in a solar metallicity star of mass $0.8M_\odot$, ^{7}Li is burning fast within the convective zone itself! But because there is evidence of lithium depletion even when the burning zone does not touch the bottom of the convective zone, several papers have investigated the effect of "slow mixing" between the convective zone and deeper layers, under various potential hydrodynamic effects. The seminal paper is by Schatzman (1977). He proposed to explain the mixing of the CZ with the subjacent burning zone by turbulent diffusion in the intermediate region. A diffusion coefficient of about 1000 cm^{2}s^{-1} produces the right abundance of lithium in the Sun, but a too low abundance of beryllium if the same value is kept in deeper layers (personal computations of the author of this

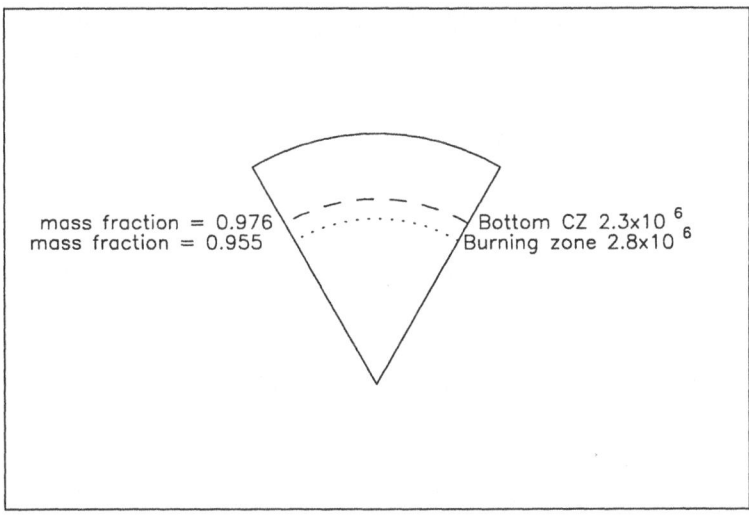

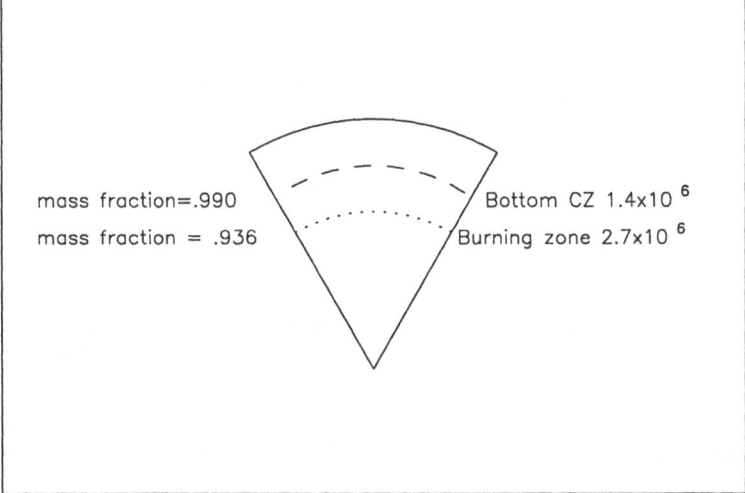

Figure 2. This figure compares the respective positions of the bottom of the convective zone and the top of the layers in which ^{7}Li burning becomes effective, in two cases: the young Sun (upper panel), and a typical subdwarf sitting on the Spites' plateau (lower panel). Note that there is a thick layer (5 times the mass of the convective zone itself!) to cross, to bridge the two regions in the case of the subdwarf, whereas this layer is quite thinner in mass and size for the unevolved Sun. This makes clear that lithium may burn more easily in the case of the Sun than in the case of the subdwarf.

paper). Also the use of the same diffusion coefficient would completely destroy lithium in subdwarfs. Various attempts have been made to identify the source of the turbulence, and to insert the turbulent diffusion process in a more comprehensive hydrodynamical scheme. We discuss these approaches in more detail in the next section.

3. Theories of depletion

Because, in the Sun, lithium is depleted even though the temperature at the bottom of the CZ is not quite high enough to burn Li by the reaction $^7Li(p,\alpha)^4He$, some ways of bringing the lithium of the CZ deeper have been considered. The simplest assumption made was to add some overshooting to the convection (Straus *et al.*, 1976). The result was not satisfactory because if the overshooting was adjusted to solve the solar depletion, then the depletion curve of the Hyades was not reproduced at all (Cayrel *et al.*, 1984).

The second attempt was to find a source for the turbulent diffusion advocated by Schatzman (1977). Rotation has been taken as the culprit. But here a branching occurs: Pinsonneault *et al.* (1992) assume that the turbulence is driven by the radial gradient of the rotational velocity, whereas Charbonnel *et al.* (1992) assume that the turbulence is driven by the differential rotation in latitude. Both theories contain a free parameter (the efficiency of the generation of turbulence in the radial direction) adjusted either to reproduce the lithium depletion in the Sun or in open clusters. Keeping the same value for this efficiency coefficient, the model is then applied to metal-poor stars. Generally speaking these models predict a non-zero depletion of lithium in subdwarfs, and this is the source of a debate between those who claim, on the ground of their theoretical results that the abundance of lithium observed in subdwarfs is not its cosmological abundance, but lower, and those who take the subdwarf abundance at face value, on the ground that it is constant over a wide range of stellar parameters, unlikely circumstance if there is a significant depletion.

Other mechanisms have been considered, not involving rotation. Reeves (1974), Guzick *et al.* (1987), have proposed that mass-loss has moved the depleted region into the CZ. This requires stellar winds considerably stronger than the present mass-loss rate of the Sun, and this mechanism is, in recent work (Vauclair and Charbonnel, 1995), invoked to explain the absence of signature of microscopic diffusion of lithium, rather than for explaining the depletion itself.

Microscopic diffusion (Michaud, 1970) has been considered also as mechanism of lithium depletion, with the difference that lithium may be just stored below the convective zone and not destroyed. However, this depletion mechanism is considerably more effective in the hottest subdwarfs than in those in the middle of the plateau, and one should observe this feature in the depletion curve. It is not the case, and this is why Vauclair and Charbonnel (1995) have proposed a stellar wind of the order of $10^{-12}M_\odot$ per year to avoid the unseen feature.

Mixing not induced by rotation, but by gravity waves, has been studied by Montalban (1994), and Montalban and Schatzman (1996). Although the computation of mixing by this mechanism is fairly difficult (non linear effects caused mostly by the radiative damping of an otherwise oscillatory motion), this mechanism is now a very serious competitor to rotationally induced mixing. We shall see why later on. Finally let us mention that the problem of the solar spin-down, thought to be strongly linked to the lithium depletion (Zahn, 1992), has been studied by

Table I

Summary of arguments for and against depletion on the plateau

Arguments for	Arguments against
depletion found from Sun-calibrated models	too large difference in structure between the Sun and the subdwarfs
predicted dispersion on Li at the same T_{eff}	the dispersion does not really exist in homogeneous data
a few low-Z stars have a lithium abundance above the plateau	these stars are very rare and present sometimes other anomalies
the ^{6}Li argument is still marginally established	^{6}Li detection in a few plateau stars show that ^{7}Li is not depleted
the curvature is below clear detection under noise condition	when depletion is strong the plateau is not a true plateau but is curved

Charbonneau and MacGregor (1993), with a detailed modelisation of the magnetic coupling between the deep layers and the outer Sun (corona) where the angular momentum is actually lost. All these works are of considerable interest, and should eventually give us the true physical picture of lithium depletion and when lithium is depleted. But it must be kept in mind that there is not one, but several possibilities, and that the present state of theoretical computations does not allow to make predictions certain by more than a factor of 2 to 3.

4. Is there any depletion on the plateau itself?

This question is crucial for cosmology. If there is no depletion, the abundance of lithium measured in the photosphere of subdwarfs is the primordial cosmological abundance of lithium. If there is some depletion, one has to know the amount of depletion, to recover the cosmological abundance of lithium, left by the Big Bang. There is no consensus on the matter, so I shall give the arguments of both sides. The first argument of the pros is that most models predict some amount of depletion, once the efficiency of mixing has been calibrated with the Sun. This amount is currently of the order of 0.5 to 1 dex (Pinsonneault *et al.*, 1992) for the radial shear models, or 0.2 to 0.5 dex (Charbonnel *et al.*, 1992) for the differential rotation shear models.

However, as we have seen in Fig. 2, there is such a difference in structure between the Sun and a very-metal poor star, that there is some doubt that the efficiency coefficient calibrated with the Sun is valid for a subdwarf. The result crucially depends upon the mixing over a large gap below the convective zone, not existing in the Sun, and in which the depth dependence is different according to the physical cause of the mixing (rotationally induced or gravity wave driven).

The spread of lithium abundances on the Spites' plateau has been put forward as an argument of the existence of rotationally induced depletion (Ryan *et al.*, 1996). The spread in initial rotation is then taken as the source of a variable amount of depletion. The problem with this argument is that, if a homogeneous set of observation is considered, and a uniform method of analysis is followed, then there is no more evidence for an intrinsic spread, in addition to the unavoidable spread caused by the observational uncertainties (Spite *et al.*, 1996; Molaro *et al.*, 1995).

A stronger argument for possible depletion on the plateau (King *et al.*, 1996; Deliyannis *et al.*, 1995) is the discovery of a few stars with a lithium abundance *above* that of the plateau. Apparently these mavericks are always pop. II subgiants and not subdwarfs. The real problem is to resolve if it is easier to avoid depletion in just a few objects, or to enrich just a few object by the process which has brought the lithium from its abundance on the plateau to its present abundance in the young population I stars.

An argument which has been used to claim that ^{7}Li has suffered very little depletion on the plateau is the fact that at least two stars of the plateau, HD 84937 (Smith *et al.*, 1993) and BD+26 3578 (Smith, 1996) have been found to contain some ^{6}Li. Because ^{6}Li is much more fragile than ^{7}Li and burns at a lower temperature, it means that ^{7}Li has been only slightly depleted. The presence of ^{6}Li in HD 84937 (at a level of about 5 to 6 per cent of the amount of ^{7}Li) has now been confirmed by several authors. Supporters of a depleted plateau cope with this argument claiming that the ^{6}Li detection is still marginal, below the 95 per cent of confidence limit. It would be more convincing if ^{6}Li was found in more stars. But it is likely that only metal-poor stars having a minimal convective zone (at the turnoff) can save the small amount of ^{6}Li they have got from the "reverse spallation" process (Vangioni-Flam *et al.*, 1997). If the existence of ^{6}Li is confirmed in more stars, it will very probably bring support to the internal gravity wave mixing mechanism, because of its sharp decline below the bottom of the convective zone, a feature needed to protect ^{6}Li. An attractive test would be to observe with very high S/N ratio, the 4 abundances ^{6}Li, ^{7}Li, Be, and B in a star. Thanks to the remarkable discovery (Duncan *et al.*, 1995) that the abundances of Be and B vary linearly with the abundance of oxygen in very metal-poor stars, bringing support to the reverse spallation process (the broken C,N,O nuclei are those ejected by SNe, not those of the surrounding ISM), it is plausible that the abundances of the other spallation light elements (^{6}Li and ^{7}Li) vary also in the same proportion. The production ratios of the four isotopes are predicted in Vangioni-Flam *et al.* (1997). It is then possible to infer the initial amount of ^{6}Li in the star from the abundances of the more robust elements Be and B, as well as the small amount of non-cosmological ^{7}Li. The connection between the inferred initial ^{6}Li amount and the observed one, puts a limit to the depletion of ^{6}Li, and consequently on the related amount of depletion of ^{7}Li (Lemoine, 1997). This is probably the strongest constraint which can be brought to the actual level of depletion of ^{7}Li on the plateau, and the most powerful test of validity for depletion theories, *in metal-poor stars* (Lemoine, 1997).

Another important test for discriminating between rotationally induced mixing and other mechanisms, not involving rotation, is the study of lithium in tidally locked binaries, that we consider now in the following section.

5. Lithium in tidally-locked binaries

For rotationally induced mixing it is believed that lithium depletion is closely linked to loss of angular momentum (Zahn, 1992). A short period binary has its rotational period equal to its orbital period, and therefore experiences an extremely slow decline of its rotation, contrary to single stars as the Sun, which exhibit a considerable spin-down. It is thus a prediction of rotationally induced lithium depletion that tidally-locked binaries should more or less keep their original lithium. Several authors (Ryan and Deliyannis, 1995; Spite *et al.*, 1994; Spite *et al.*, 1995) have studied tidally-locked binaries. The conclusions to draw are not obvious. A positive effect is seen in open clusters, as in the case of vB 22 in the Hyades (Thorburn *et al.*, 1993). In metal-poor stars no effect is visible on the plateau (expected if the plateau is *not* depleted), and contradicting results appear in cooler stars, for which depletion is certain, with a trend to show the expected effect for objects below $T_{\text{eff}} = 5100$ K.

6. Lithium in very metal-poor subdwarfs

Thanks to the major effort developed by Beers *et al.* (1985; 1992) a great number of very metal-poor stars have been discovered in the last 15 years. A number of them have now been studied at high spectral and S/N resolution. Lithium abundances have been obtained for many stars in the metallicity range [Fe/H] = -4 to -2.5 extending the Spites' plateau to the lowest metallicities. We show the result in Fig. 3.

The total independence of the lithium abundance from that of the other elements, over almost three orders of magnitude of the iron/hydrogen ratio is a phenomenon absolutely unique among all elements, with the exception of ^{4}He, which unfortunately cannot be spectroscopically determined in these cool stars, but only indirectly in globular clusters. The evidence of this independence is an extraordinary tribute to the standard Big Bang Nucleosynthesis. At the sight of Fig. 3 one may perhaps still argue if the cosmological abundance of lithium is the value of the plateau or slightly above (depletion!) *but the cosmological origin of the element is beyond possible doubt.*

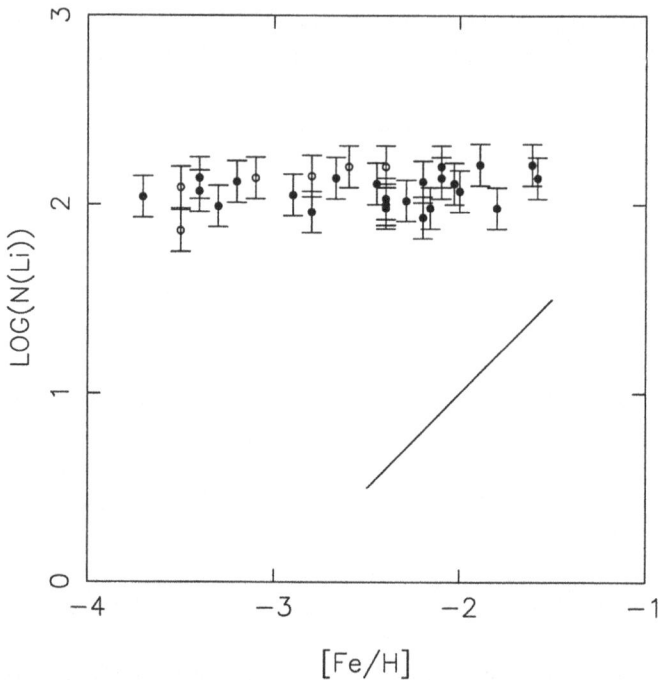

Figure 3. Lithium abundance in very metal-poor stars. In abscissa [Fe/H], in ordinate logarithmic abundance of Li, in the conventional scale $\log(n(H)) = 12.0$. The source is the compilation in Cayrel (1996) complemented by new results (open circles) from observations by M. Spite and R. Cayrel at the NTT of ESO. A few "dropouts" (a few per cent of objects with no measurable lithium lines) are not plotted, being below the detection line. Only stars with T_{eff} larger than 5500 K have been kept for avoiding the depletion obvious in cooler stars. The vertical and terminating horizontal bars represent the expected errors. The straight line of slope 1 represents the behaviour of the other elements.

7. Conclusions

• The major point is the evidence of an abundance of lithium in low-Z stars, independent of the abundance of elements produced by stellar nucleosynthesis, down to the lowest observable Z (at present), i.e. $Z = 3 \times 10^{-6}$. Whatever is a possible depletion-correction to apply to this abundance, (2.1 in the usual scale $\log(n(H)) = 12$), this makes sense only with a cosmological origin of lithium, as predicted by the Big Bang theory.

• Because of the importance of this abundance for Cosmology, it is urgent to establish on firm grounds if the metal-poor star abundance can be taken at face-value, or if some depletion has occurred during the life of the star, as in the more metal-rich stars. For this, the major mechanism of depletion should be identified, among those presently studied.

• It is proposed as a very powerful constraining check, to study more stars in the area of the HR diagram around HD 84937, in which cosmological ^{7}Li coexists with a small amount of ^{6}Li, ^{7}Li, Be and B, produced by reverse spallation. The depletion in ^{6}Li can be inferred from the production ratio of ^{6}Li with Be and B in the reverse spallation process, the abundances of B and Be, and the observed ^{6}Li abundance. The depletion mechanism depleting ^{6}Li by the proper amount has all chances to correctly predict the depletion of ^{7}Li.

References

Beers, T.C., Preston, G.W. and Shectman, S.A.: 1985,'A Search for Stars of very low Metal Abundance. I', *Astron.J.* **90**, 2089.

Beers, T.C., Preston, G.W. and Shectman, S.A.: 1992, 'A Search for Stars of very low Metal Abundance', *Astron.J.* **103**, 1987.

Cayrel, R., Cayrel de Strobel, G., Campbell and B., Däppen, W.: 1984, 'The Lithium Abundance of Hyades main-sequence Stars', *Astrophys.J.* **283**, 205.

Charbonneau, P. and MacGregor, K.B.: 1993, 'Angular Momentum Transport in Magnetized Stellar Radiative Zone. II. The Solar Spin-Down', *Astrophys.J.* **417**, 762.

Charbonnel, C., Vauclair and S., Zahn, J.-P.: 1992, 'Rotation-Induced Mixing and Lithium Depletion in Galactic Clusters', *Astron.Astrophys.* **255**, 191.

Duncan, D.K., Primas, F., Coble, K.A., Rebull, L.M., Boesgaard, A.M., Delyannis, C.P., Hobbs, L.M., King, J.R. and Ryan, S.G.: 1995, 'Some Surprises Concerning the Origin of the Light Elements' *AAS* **186**, 5805.

Deliyannis, C.P., Boesgaard, A.M. and King, J.R.: 1995, 'Evidence of Higher Primordial Lithium from Keck Observations of M92', *Astrophys.J.* **452**, 13.

Guzick, J.A., Willson, L.A. and Brunish, W.M.: 1987 'A comparison between mass-losing and standard solar model', *Astrophys.J.* **319**, 957.

Herbig, G.H.: 1965, 'Lithium Abundances in F5-G8 Dwarfs', *Astrophys.J.* **141**, 588.

King, J.R., Deliyannis, C.P. and Boesgaard, A.M.: 1996, 'Constraints of the Origin of the Remarkable Lithium Abundance in the Halo Star BD+23 3912' , *Astron.J.* **112**, 2839.

Lemoine, M., Schramm, D.N., Truran, J.W. and Copi, C.J.: 1997, 'On the Significance of Population II ^{6}Li Abundances', *Astrophys.J.* **478**, 554.

Michaud, G.: 1970, 'Diffusion Processes in Peculiar A Stars', *Astrophys.J.* **160**, 641.

Molaro, P., Bonifacio, P. and Primas, F.: 1995, 'About the lithium abundance of halo stars', *Mem.Soc.Astron.It.* **66**, 323.

Montalban, J.: 1994, 'Mixing by internal waves: Lithium depletion in the Sun', *Astron.Astrophys.* **281**, 421.

Montalban, J. and Schatzman, E.: 1996, 'Mixing by internal waves.II. Li and Be depletion rate in low-mass main sequence', *Astron.Astrophys.* bf 305, 513.

Müller, E.A., Peytremann, E., de la Reza, R.: 1975, 'The Solar Lithium Abundance. II.', *Solar Phys.* **41**, 53.

Pinsonneault, M.H., Deliyannis, C.P. and Demarque, P.: 1992, 'Evolutionary models of halo stars with rotation. II - Effects of metallicity on lithium depletion, and possible implications for the primordial lithium abundance', *Astrophys.J.S.S.* **78**, 179.

Reeves, H.: 1974, 'On the origin of the light elements', *Ann.Rev.Astron.Astrophys.* **12**, 437.

Ryan, S.G. and Deliyannis, C.P.: 1995, 'Lithium in Short-Period Tidally Locked Binaries: A Test of Rotationally Induced Mixing', *Astrophys.J.* **453**, 819.

Ryan, S.G., Norris, J.E. and Beers, T.C.: 1996, 'Extremely Metal-poor Stars. II. Elemental Abundances and the Early Chemical Enrichment of the Galaxy', *Astrophys.J.* **471**, 254.

Schatzman, E.: 1977, 'Turbulent transport and lithium destruction in main sequence stars', *Astron.Astrophys.* **56**, 211.

Smith, V.V.: 1996, in *Stellar abundances* eds. B. Barbuy, W.J. Maciel, J.C. Gregorio-Hetem, Instituto Astronomico e geofisico da USP, Sao Paulo, Brazil, p. 13.

Smith, V.V., Lambert, D.L. and Nissen, P.E.: 1993, 'The ^{6}Li/^{7}Li ratio in the metal-poor halo dwarfs HD 19445 and HD 84937', *Astrophys.J.* **458**, 543.

Spite, F. and Spite, M.: 1982, 'Abundance of Lithium in Unevolved Halo Stars and Old Disk Stars: Interpretation and consequences', *Astron.Astrophys.* **115**, 357.

Spite, M., Pasquini, L. and Spite, F.: 1994, 'Lithium in old binary stars', *Astron.Astrophys.* **290**, 217.

Spite, M., Fleming, T., Cayrel, R., Pasquini, L. and Spite, F.: 1995, 'Lithium in metal-deficient binaries', *Mem.Soc.Astron.It.* **66**, 337.

Spite, M., François, P., Nissen, P.E. and Spite, F.: 1996, 'Spread of the lithium abundance in halo stars', *Astron.Astrophys.* **307**, 172.

Straus, J.M., Blake, J.B. and Schramm, D.N.: 1976, 'Effects of convective overshoot on lithium depletion in main-sequence stars', *Astrophys.J.* **204**, 481.

Thorburn, J.A., Hobbs, L.M., Deliyannis, C.P. and Pinsonneault, M.H.: 1993, 'Lithium in the Hyades. I - New observations', *Astrophys.J.* **415**, 150.

Vangioni-Flam, E., Cassé, M. and Ramaty, R., 1997 'Light element production by low-energy nuclei from massive stars', in *'The transparent Universe'*, proceedings of the 2nd INTEGRAL workshop, (Saint-Malo, France), ESA, SP-382, p. 123.

Vauclair, S. and Charbonnel, S.: 1995, 'Influence of a stellar wind on the lithium depletion in halo stars: a new step towards the lithium primordial abundance', *Astron.Astrophys.* **295**, 715.

Zahn, J.-P.: 1992, 'Circulation and turbulence in rotating stars', *Astron.Astrophys.* **265**, 115.

LITHIUM ABUNDANCE IN POP. II STARS
A *post-HIPPARCOS discussion*

F. SPITE, M. SPITE and V. HILL
Observatoire de Paris, DASGAL et URA 335 du CNRS
Place Janssen, 92195 Meudon Cedex, France

Abstract. The relation between the lithium abundance observed in Population II stars and the primordial abundance, is still an open question (see Cayrel and Duncan, this meeting). A few recent results are discussed. HIPPARCOS data show that the standard model of stellar evolution can explain the ^{6}Li detection in HD 84937, suggesting a negligible depletion of ^{7}Li. A slope in the Li/T_{eff} relation for Pop II dwarfs and a spread of their Li abundance have been advocated, and both used as arguments in favor of Li depletion. The slope is not confirmed when two other independent temperature scales are used. The Li scatter around the plateau is hardly larger than the scatter predicted from determination errors. Hints from a scatter of Li in subgiants of the globular cluster M92 are not completely conclusive. The determination of more accurate Li abundances in the Pop II stars is an urgent but difficult task, requiring better model atmosphere (better convection treatment) and the help of observational data about Pop II stars (such as long base interferometry).

Key words: Stars: abundances, lithium, nucleosynthesis: primordial

1. Introduction

In spite of progresses in observational facilities (detectors sensitivity and size of telescopes) lithium abundances in stars are not yet as accurate as wished. HIPPARCOS data bring a useful contribution. Many points deserve discussion, we select here only a few ones. Definite conclusions require progresses: more accurate data and better ways for interpreting them.

2. The ^{6}Li isotope problem

The ^{6}Li isotope is more fragile than the ^{7}Li isotope, so that its detection in a star is an indication of a necessarily moderate (or negligible) depletion of ^{7}Li. The ^{6}Li isotope has been detected in the star HD 84937 (Smith *et al.*, 1993), and confirmed by Hobbs and Thorburn (1994). The metallicity of the star is low: $[Fe/H] = -2.2$ dex, i. e. this star is clearly included among the stars which define a plateau (Spite and Spite, 1982) for the lithium abundance (relative both to the metallicity and to the effective temperature T_{eff}).

The classical parallax indicates that the star is a dwarf. Chaboyer (1994), using the standard model of stellar evolution with improved opacity, computed that, in such a dwarf, the ^{6}Li isotope is strongly destroyed. Then, owing to its presumably small initial abundance, the ^{6}Li would not be detectable. Chaboyer deduced that in

Space Science Reviews **84**: 155–160, 1998.

order to fit the observations, the star should in fact be a subgiant (a star of higher mass and therefore with a better preservation of lithium), and predicted its parallax: 0.011 arcsec. Deliyannis and Malaney (1995) also admitted that in such a dwarf the ^{6}Li should be strongly destroyed, and also suggested a parallax around 0.011 arcsec. HIPPARCOS' parallax is 0.0124 ± 0.0011 arcsec (ESA 1997), showing that the star is a subgiant (Crifo et al., 1997), reconciling observations and standard model computations suggesting a negligible depletion of ^{7}Li. However, models different from the standard model, such as rotational induced mixing (Pinsonneault et al., 1992), or such as models with segregation and wind (Vauclair and Charbonnel, 1995; see also Vauclair, this meeting) can explain the detection of ^{6}Li with a significant depletion of ^{7}Li (for example factors of 2 to 3), so that the detection of ^{6}Li in this star, is not yet a definite proof that the ^{7}Li abundance in this star is near the primordial abundance.

A detection of ^{6}Li has recently been made by Smith et al. (1996) in the star BD 26 3578. The star is a subgiant, following its HIPPARCOS parallax, and this result, is a nice confirmation of the ^{6}Li detection made in the very similar star HD 84937. A few Pop II stars have also been observed for ^{6}Li detection without reaching a clear detection (Smith et al., 1993; Hobbs and Thorburn, 1994; 1997). The measurements are compatible with the predictions of the standard model (Chaboyer, 1994).

However, the ^{6}Li isotope is detected as a small alteration in the profile of the lithium line (dominated by ^{7}Li) which could possibly be due to slightly unadequately compensated velocity fields. Large telescopes should enable to ascertain the ^{6}Li detection, by providing measurements of higher accuracy, and by extending the detections to a larger sample of stars.

3. The slope of the plateau

In some published works, the lithium abundance seems to be slightly increasing with the effective temperature T_{eff} and with the metallicity [Fe/H]. The slope with T_{eff} has sometimes been interpreted as the signature of a strong lithium depletion, stronger at effective temperatures around 5700 K (smaller mass stars) than at temperatures around 6200 K (higher mass stars).

In a recent work, Ryan et al. (1996) found a significant slope, using a temperature scale similar to the temperature scale of Carney (1983) and therefore significantly different (Carney et al., 1994) from the IRFM scale (Blackwell et al., 1991). Changing for the IRFM scale in the analysis is enough for nearly cancelling the slope. Other independent temperature scales also provide a flat plateau (Spite et al., 1996; Molaro et al., 1995a; 1995b). When Bonifacio and Molaro (1997) adopt the scale of Alonso et al. (1996) who use the IRFM method, the slope is again very small, similar to the prediction of the standard model. We are back at a difficulty already mentioned: the discrimination between several theories of Li depletion require very accurate lithium abundances. The currently available stellar model

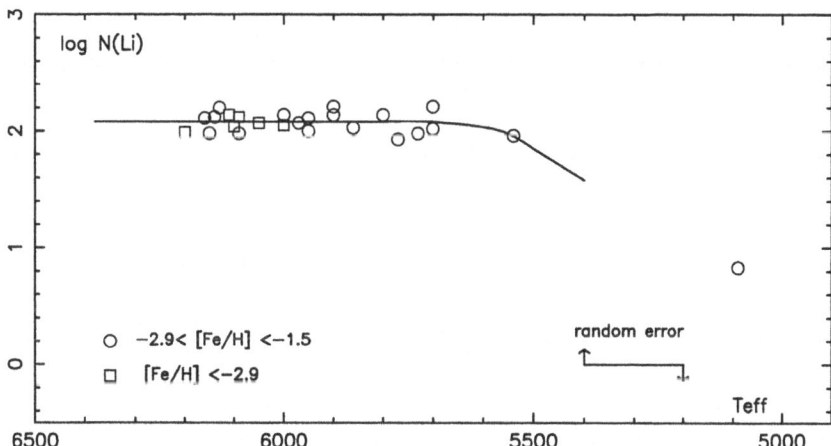

Figure 1. The data of Spite *et al.* (1996), in the temperature scale of Nissen *et al.* (1994) indicate a negligible slope of the plateau and a small spread (this figure). Other independent temperature scales also provide a negligible slope.

atmospheres are not realistic enough (not taking into account properly convection, inhomogeneity etc.) for predicting very accurately the colors and therefore the effective temperatures T_{eff}, required for accurate Li abundances (see also Kurucz, 1997).

This is why the astronomers have been trying, with contrasted success, to develop and use semi-empirical relations, relating colors to T_{eff} with the adequate corrections (for metallicity and gravity effects). Progresses in convection theory, in the treatment of inhomogeneities and in very long base stellar interferometry (constraining the temperature gradient of the models) should solve the problem in the future.

4. The spread around the plateau

All theories predicting lithium depletion on the plateau also predict a spread of lithium abundance, because the depletion processes, acting differently (acting on stars of different masses, metallicities, ages, evolution phase, histories etc.) necessarily produce different depletions from the uniform initial (primordial) lithium abundance, leading to different observable abundances, and therefore to a lithium spread around the plateau.

In analyzing this spread, the systematic errors in lithium abundance determinations are unimportant for similar stars: the spread is found from the differential comparison of similar stars. Only the random errors are important.

The two main sources of errors are the measurement of the equivalent width of the lithium line, and the error on the temperature T_{eff}. In the discussion of her sample, Thorburn (1994) analysed the temperature error, but her values should be underestimated, since for example unrecognized binarity and dereddening errors are not taken into account. The equivalent width measurement errors, in principle rather small, should not be forgotten. For example, the use of a direct integration routine, for the measurement of equivalent widths of absorption lines, may, in some cases (low S/N ratio per pixel), introduce a systematic error, (Spite, 1997). Taking into account the temperature and measurement errors, the combined error is about 0.08 dex (one sigma for one star): it is similar to the observed spread, when a few obvious binaries have been discarded from the observed sample (Spite *et al.*, 1996). Rebolo (this meeting) indicates an even smaller observed spread when excluding two discrepant stars. Therefore the intrinsic Li spread must be small, if existing at all. Admittedly, a few stars are considerably below the plateau (Li-poor). Norris *et al.* (1997) analyze three of these stars and confirm the binarity of one of them (they are all suspected binaries). No clear signature of mass transfer is found, but the Li line may be altered by the companion, as well as the color, resulting in a poor determination of the Li abundance, the real amount of Li depletion being uncertain. Norris *et al.* (1997) propose that *all* the Pop II stars have suffered a more or less identical Li depletion, the Li-poor stars suffering, for some reason, an enhanced depletion, whatever the physical process (wind, diffusion, rotation, internal waves...). Another possible interpretation is that the stars of the plateau are not depleted, the Li-poor stars beeing examples of the normal depletion predicted by the various physical processes noted hereabove. Let us note that G 139-8 could be a subgiant following HIPPARCOS.

The small value of the observed Li spread, found in several independent analyses, for the stars of the plateau, is a severe constraint for the theories predicting a significant lithium depletion.

5. The lithium abundance of the plateau

Since the stars of the plateau could provide a lithium abundance close to the primordial abundance, it is important to find the correct temperature scale, since the Li abundance depends strongly on the effective temperature T_{eff}. There are however conflicting arguments about such a choice.

There are good arguments for using the IRFM scale, and in practice the temperatures provided by Alonso *et al.* (1996) are a good choice. It happens that the scale of Alonso is relatively similar to the scales derived from the wings of the Balmer lines (Fuhrmann *et al.*, 1995 and references therein; Spite *et al.*, 1996): this criterion is relative to rather deep layers of the star. These scales provide "high" temperatures (and high Li abundances), relative to the scale of Edvardsson *et al.* (1993) based on the b-y color of the MARCS model, and checked by Nissen *et al.* (1994) as providing the excitation equilibrium.

Coherent abundance determinations require the excitation equilibrium, suggesting the choice of the T_{eff} scale of Nissen *et al.* (1994). This choice is correct only if the gf values are free of systematic errors and if the possible NLTE effects are small or independent of the excitation potential).

The solution: by some indirect arguments (which cannot be exposed here due to lack of space), it may be suggested that the correct temperature for Li abundances should be somewhere between the Nissen's scale and the IRFM scale, providing therefore an "intermediate" value of the Li abundance on the plateau. The disagreement between the T_{eff} obtained by excitation equilibrium and by Balmer lines fitting remains to be explained, whether it would be an incorrect temperature gradient in the models or some other reason. The real value of the Li abundance of the plateau will be reached through progresses of laboratory measurements (or computations) of atomic structures (gf values), and through progresses in stellar atmosphere models (convection, inhomogeneity) with the help of progresses in stellar photometry and interferometry.

6. Conclusion

The data about ^{6}Li are limited, very delicate to interpret, and need some extension before leading to a definitive conclusion. The current data are compatible with the standard model.

The slope of the plateau, sometimes considered as the signature of a significant Li depletion, disappears when other temperature scales, different from Carney's scale, are chosen. More precisely, using the IRFM scale, Bonifacio and Molaro (1997) find that the plateau follows nearly exactly the shape predicted by the standard model.

The observed spread of Li abundances around the plateau is found to be small, hardly larger than the value predicted by determination errors. An even smaller value is proposed by Rebolo *et al.* (this workshop). This point is a severe constraint on the theories predicting a significant Li depletion on the plateau.

The value of the Li abundance on the plateau depends on the choice of the temperature scale. In principle, the excitation equilibrium should be realized. There are reasons to think that the correct temperature could bc intcrmcdiate between the value provided by the Nissen's scale and the IRFM scale, the correct lithium abundance of the plateau beeing therefore intermediate between the value of Spite *et al.* (1996) and Bonifacio and Molaro (1997).

Some progresses remain to be accomplished, both in observational and theoretical fields for achieving the extreme accuracy in abundance determination required by the problem of lithium in Pop II stars.

References

Alonso, A., Arribas, S., and Martínez-Roger, C.: 1996, *A&AS* **117**, 227.
Blackwell, D. E., Lynas-Gray, A. E., and Petford, A. D.: 1991, *A&A* **245**, 567.

Boesgaard, A. M., Deliyannis, C. P., Stephens, A., and King, J. R.: 1997 (preprint).
Bonifacio, P. and Molaro, P.: 1997, *MNRAS* **285**, 847.
Carney, B. W.: 1983, *AJ* **88**, 623.
Carney, B. W., Laird, J. B., Latham, D. W., and Aguilar, L. A.: 1994, *AJ* **107**, 2240.
Chaboyer, B.: 1994, *ApJ* **432**, L47.
Crifo, F., Spite, F., and Spite, M.: 1997, in "Proc. HIPPARCOS Venice 1997 Symposium", ESA SP-402 (in press).
Deliyannis, C. P. and Malaney, R. A.: 1995, *ApJ* **453**, 810.
Edvardsson, B., Andersen, J., Gustafsson, B., Lambert, D. L., Nissen, P. E., and Tomkin, J.: 1993, *A&A* **275**, 101.
ESA 1997, The HIPPARCOS Catalogue, ESA SP-1200.
Fuhrmann. K., Axer. M., and Gehren, T.: 1995, *A&A* **301**, 492.
Hobbs, L. M. and Thorburn, J. A.: 1994, *ApJ* **428**, L25.
Hobbs, L. M. and Thorburn, J. A.: 1997, *ApJ* (in press).
Kurucz, R. L.: 1997, in "Fundamental Stellar Properties", IAU Sympos 189, eds. T. Bedding *et al.*, Kluwer, Dordrecht, p. 217.
Molaro, P., Primas, F., and Bonifacio, P.: 1995a, in " Lithium and primordial nucleosynthesis", Proc. IAU JD 11, eds. F. Spite and R. Pallavicini, *Mem. Soc. Astr. It.* **66**, 323,
Molaro, P., Primas, F., and Bonifacio, P.: 1995b, *A&A* **295**, L47.
Nissen, P. E., Gustafsson, B., Edvardsson, B., and Gilmore, G.: 1994, *A&A* **285**, 440.
Nissen. P. E., Høg. E., and Schuster. W. J. B.: 1997, in "Proc. HIPPARCOS Venice 1997 Symposium", ESA SP-402.
Norris, J. E., Ryan, S. G., Beers, T. C., and Deliyannis, C. P.: 1997, *ApJ*, (in press).
Pinsonneault, M. H., Deliyannis, C.P., and Demarque, P.: 1992, *ApJ Sup. Ser.* **78**, 179.
Ryan, S. G., Beers, T. C., Deliyannis, C., and Thorburn, J. A.: 1996, *ApJ* **458**, 543.
Smith, V. V., Lambert, D. L., and Nissen, P. E.: 1993, *ApJ* **408**, 262.
Smith, V. V., Lambert, D. L., and Nissen, P. E.: 1996, in "Stellar abundances", eds. B. Barbuy, W. J. Maciel, Gregòrio-Hetem, IAGUSP, São Paulo, p. 13.
Spite, M.: 1997 in "Fundamental Stellar Properties", IAU Sympos 189, eds. T. Bedding *et al.*, Kluwer, Dordrecht, p. 185.
Spite, F. and Spite, M.: 1982, *A&A* **115**, 357.
Spite, M., François, P., Nissen, P. E., and Spite, F.: 1996, *A&A* **307**, 172.
Thorburn, J. A.: 1994, *ApJ* **421**, 318.
Vauclair, S. and Charbonnel, C.: 1995, *A&A* **295**, 715.

Address for correspondence: F. Spite, Observatoire, Batiment 11, F-92195 Meudon Cedex.

GALACTIC EVOLUTION OF THE LIGHT ELEMENTS: A NEW SET OF B OBSERVATIONS

F. PRIMAS *

Dept. of Astronomy and Astrophysics, The University of Chicago, 5640, S. Ellis Avenue, Chicago, IL 60637

Abstract. The boron 2500 Å spectral region has been observed with the Goddard High Resolution Spectrograph (GHRS) of the *Hubble Space Telescope* (HST) in a new set of metal-poor stars and analyzed by spectrum synthesis technique, adopting the most recent model atmospheres. By taking into account the Li and Be abundances available from the literature for this same set of objects, the resulting patterns of their light elements abundances cannot be easily justified with the currently known stellar structure scenarios. The finding of real differences in the B content between stars with very similar stellar characteristics suggest that also production effects, rather than depletion and/or mixing only, should be taken into account as a possible and valuable explanation.

Key words: abundances, light elements, stellar structure

1. Introduction

Abundances of the light elements Li, Be, and B play a critical role in understanding Big-Bang nucleosynthesis (BBN), Galactic chemical evolution, and stellar mixing. The cosmological interests are related to the fact that standard Big Bang nucleosynthesis predicts primordial production of ^{7}Li, but not of ^{6}Li, ^{9}Be, and ^{10,11}B, which arise from GCR spallation reactions. The most recent analyses of Be and B (e.g. Boesgaard, 1996; Primas, 1996; Duncan *et al.*, 1997) have pointed out that this GCR scenario needs to be revised in order to reproduce the direct proportionality observed in a plot Be, B vs. [Fe/H]. On the other hand, the stellar structure interest stems from the fact that Li, Be, and B are destroyed by (p,α) reactions in the outer layers of stellar atmospheres at progressively higher effective temperatures (2.5, 3.5 and 5.0×10^5 K). Only by combining information about all these three elements can an overall understanding of light element production over Galactic history or stellar mixing be achieved.

2. Data Reduction and Analysis

The Goddard High Resolution Spectrograph (GHRS) of the Hubble Space Telescope (HST) was used with the G270M grating to obtain spectra of resolution 26,000 and typical S/N of 35 per pixel in the BI region (2500 Å).

* Current address: European Southern Observatory, Karl-Schwarzschild str. 2, D-85748 Garching b. München

Space Science Reviews **84**: 161–166, 1998.
© 1998 *Kluwer Academic Publishers*

B abundances were determined via spectrum synthesis using the latest Kurucz model atmospheres. For this purpose, the version of SYNTHE distributed by Kurucz (1993) on CD-ROM # 18, but modified by Steve Allen (University of California at Santa Cruz) to run on UNIX SPARC stations, was used. The model grid, released on CD-ROM # 13 (Kurucz, 1993) incorporates the additional blanketing of nearly 60 million atomic and molecular features. The final adopted line list is the one adopted by Duncan et al. (1997) in their study of B abundances in the Hyades giants, although slight adjustments were made in order to achieve a good overall match to the absorption features near the BI lines (cf. Primas et al., 1997). The adopted solar B abundance is the canonical value $\log n(B) = 2.60$ (Anders and Grevesse, 1989). Stellar parameters were selected from the literature, but further checked by running several syntheses.

Considerable care was spent in determining the error bars associated to each B determination, taking into account different sources of uncertainties. Previous literature determinations of the stellar parameters for this sample of stars suggest ± 75 K, ± 0.25, and ± 0.10 dex as the typical uncertainties to be associated with T_{eff}, $\log g$, and [Fe/H] respectively. In particular, the uncertainty in stellar metallicity accounts for the likely existence of metal lines blending with the B feature. It is well known that the Co I 2496.716 Å feature blends with the B I 2496.772 Å, but our spectra do not resolve this blend. Assuming a higher metallicity would attribute more of the blend to Co, decreasing the derived B abundance. But it is important to notice that the findings that Co remains scaled-solar down to [Fe/H] $= -2.5$ (McWilliam et al., 1995) and that our previous B determination in HD 140283 (Duncan et al., 1997) is in very good agreement with the value found by Edvardsson et al. (1993), who partially resolve the Co-B blend, argues against a strong Co overabundance at the metallicities concerned by this sample. Placement of the continuum is a significant source of random errors, which arise from choosing the proper normalization. We estimated it by looking at the effects caused by adopting different regions for normalization: the range of vertical shift we found could still (barely) fit the B feature and continuum within the errors was $\simeq 1.5 - 2\%$. After combining these different contributions, we decided to assign a conservative average error bar of 0.22 dex to each object analyzed, since differences from star to star were found to be negligible (ranging from 0.20 dex to 0.22 dex). This mean value does not include the possible systematic error that could arise from NLTE effects. The analysis of such an error source is very delicate because NLTE calculations are still uncertain, being limited by our incomplete understanding of the atomic physics involved and by the uncertainties in the uv flux values at precise wavelengths. Calculations performed by Kiselman (1994) and Kiselman and Carlsson (1996) find that NLTE effects should be more significant in the most metal-poor stars only. A more detailed description of our error analysis (including NLTE effects, which are under investigation) will be reported in Primas et al. (1997).

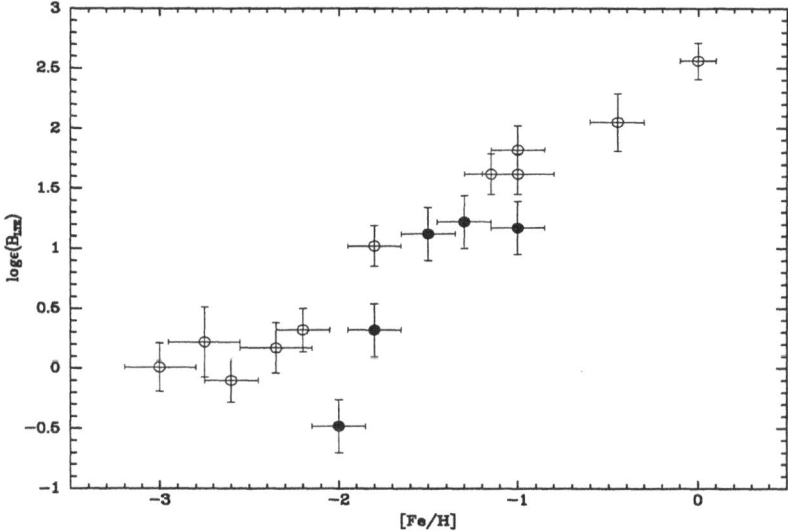

Figure 1. log ε(B) vs. [Fe/H], combining the stars analyzed by Duncan *et al.* (1997) and this work (filled circles).

3. Results

Because Li, Be, and B burn at relatively low temperatures, circulation and destruction of the light elements can result in observable abundance changes, which can provide an invaluable probe of stellar structure and mixing. Since Li is burnt by H most easily and B least easily, the surface layer in which their main-sequence abundances remain unchanged is thinnest for Li and thickest for B. This also means that Be and B, which burn at slightly higher T_{eff} than Li are preserved to a deeper mass fraction. Therefore in a standard scenario we do not expect to detect any Li, if Be has already undergone some depletion.

The main purpose of this analysis was to determine the B abundance in halo stars characterized by peculiar Li and Be abundances. Therefore we chose to target some of the Be-weak stars known from the literature, with the aim of testing the possible different mechanisms responsible for the observed LiBeB patterns. The striking feature emerging from this new set of data (cf. Fig. 1) is the existence of real differences in the B abundances among stars at almost the same metallicity.

Although in some cases the combined LiBeB information for a single star can be easily accommodated within known stellar structure scenarios (e.g. HD 221377, in which the Li, Be, and B depletion factors follow the pattern predicted by a standard stellar structure model), it has been generally more difficult to justify the different B absorption strengths observed between stars characterized by very similar stellar parameters, i.e. stars which have supposedly shared a common evolutionary history

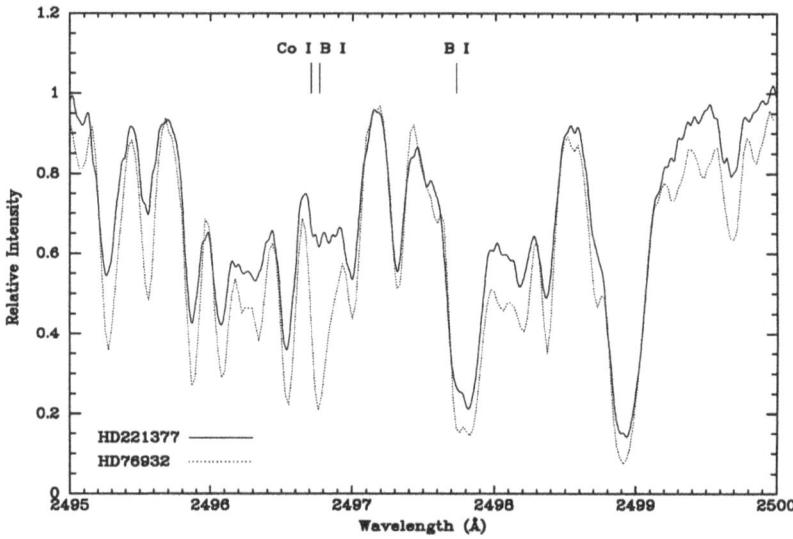

Figure 2. HD 221377 overplotted to HD 76932, a star with similar stellar parameters (cf. Duncan *et al.*, 1997).

(cf. Fig. 2, where the B spectra of HD 221377 and HD 76932 - Duncan *et al.*, 1997 - are directly compared).

For one case in particular (HD 160617) no stellar structure model is able to account for the observed LiBeB pattern. HD 160617 is a metal-poor ([Fe/H] $\approx$ -1.80) subgiant star, with an effective temperature of about 6000 K. Its Li abundance ($\log N(\text{Li}) = 2.2$, where $\log N(\text{Li}) = \log(N(\text{Li})/N(\text{H})) + 12$) is very close to the Spite plateau value, but it has a lower Be abundance with respect to stars with similar stellar parameters. Its subgiant status cannot account for a B depletion of approximately a factor of 3 (cf. Fig. 3), primarily because a 6000 K star is not expected to have experienced severe dilution yet, if any (Deliyannis *et al.*, 1997).

The low B content observed in HD 160617 is at odds not only with the predictions of standard stellar evolutionary theory, but also with all the other possible additional mechanisms, as diffusion, meridional circulation, rotational mixing, mass loss, ... Because all the above scenarios do not account for a depleted B (and Be) abundance, while Li is "unchanged" from what is currently considered its primordial value (the Spite plateau), we are then faced with at least two possible scenarios: we can make the hypothesis that the primordial Li abundance was originally much higher, and assume that the measured Li abundance at a level of 2.2 has already experienced some depletion of a factor larger than 1 dex; or HD 160617 might have formed with unusually low B (and possibly Be), which should be quite easy to test in stars with similar stellar parameters.

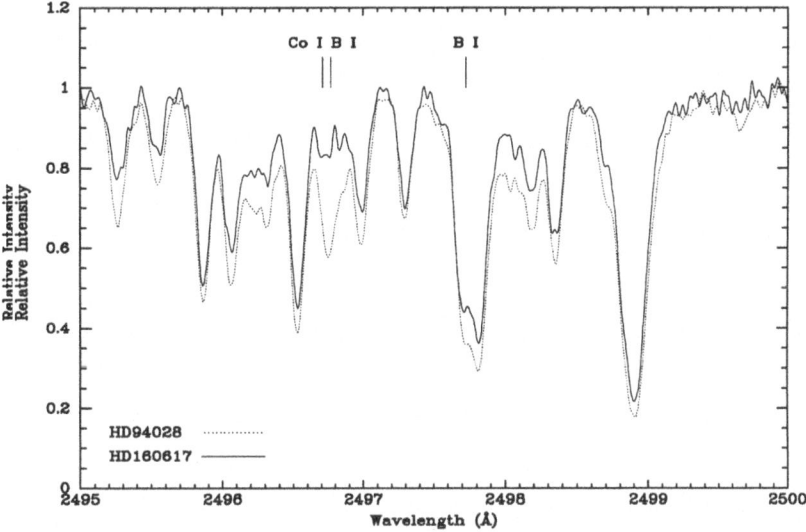

Figure 3. HD 160617 overplotted to HD 94028, a star with similar stellar parameters, although slightly more metal-rich.

4. Concluding Remarks

Although at a preliminary stage, this analysis has pointed out for the first time that also B differentiates among stars with very similar stellar characteristics, as already observed for Li and Be for which many more determinations are available. The difficulties encountered in trying to account for the observed B abundances in our new set of stars, when combined with previously determined Li and Be abundances, might suggest that we have been looking for a solution in the wrong direction: these differences might be evidence of spatial variation of element production in the early Galaxy, rather than destruction. The main achievement is that a set of 15 halo and disk stars for which Li, Be, and B are known has now finally been gathered, offering us one of the most powerful tools for investigating and testing different predictions of stellar structure models and chemical evolution scenarios.

Acknowledgements

The author wishes to thank the ISSI Institute and the local organizers, Dr. J. Geiss and Dr. R. von Steiger, for having offered the opportunity to attend such an interesting workshop.

References

Anders, E. and Grevesse, N.: 1989, 'Abundances of the Elements – Meteoritic and Solar', *Geochim. Cosmochim. Acta* **53**, 197–214.

Boesgaard, A.M.: 1996, 'Light Element Abundances in the Halo', *A.S.P. Conf. Series* **92**, 327–336.

Deliyannis, C.P., Boesgaard, A.M., King, J.R. and Duncan, D.K.: 1997, 'New Observations of Beryllium in the Galactic Halo', *ApJ*, submitted.

Duncan, D.K., Primas, F., Rebull, L.M., Boesgaard, A.M., Deliyannis, C.P., Hobbs, L.M., King, J.R. and Ryan, S.G.: 1997a, 'The Evolution of Galactic Boron and the Production Site of the Light Elements', *ApJ* **488**, 338–349.

Duncan, D.K., Peterson, R.C., Thorburn, J.A. and Pinsonneault, M.H.: 1997, 'Boron Abundances and Internal Mixing in Stars I: The Hyades Giants', *ApJ*, submitted.

Kiselman, D.: 1994, 'A NLTE Study of Neutral Boron in Solar-Type Stars', *A&A* **286**, 169–180.

Kiselman, D. and Carlsson, M.: 1996, 'The NLTE Formation of Neutral-Boron Lines in Cool Stars', *A&A* **311**, 680–689.

Kurucz, R.L.: 1993a, 'ATLAS9 Stellar Atmospheres Programs and 2 km/s Grid', *Smithsonian Astroph. Obs.* **CD-ROM no. 13**.

Kurucz, R.L.: 1993b, 'SYNTHE Spectrum Synthesis Programs and Line Data', *Smithsonian Astroph. Obs.* **CD-ROM no. 18**.

Primas, F.: 1996, 'The Challenge of Be Observations', *PhD Thesis – University of Trieste*.

Primas, F., Duncan, D.K. and Thorburn, J.A.: 1997, 'A New Set of B Observations: Implications for Stellar Mixing and Cosmology', *in preparation*.

Address for correspondence: ESO – Karl-Schwarzschild str. 2, D-85748 Garching b. München

KEY QUESTIONS FOR LOW METALLICITY STARS

Rapporteur Summary of the Working Group on Low-Z Stars

D. DUNCAN

Dept. of Astronomy and Astrophysics, University of Chicago,
5640 S. Ellis Ave., Chicago, IL 60637, USA

Abstract. An overview of the discussions of the working group on Low-Z stars is presented. Key questions addressed include how the abundances of lithium observed in these stars should be compared to that produced in the Big Bang. Evidence for and against a small star-to-star variation in Li abundances is reviewed, and whether such a variation, if real, necessarily indicates that stellar depletion has occurred, necessitating correction to the value compared to primordial nucleosynthesis calculations. A second key question concerns how and where the light elements are produced. Taken together, their abundance ratios strongly suggest that in low-Z stars the light elements other than ^{7}Li are produced by cosmic ray spallation. The most recent evidence suggests that a minority of this spallation happens in the general interstellar medium, and that a larger fraction might happen in the immediate vicinity of Supernovae, possibly producing observable star-to-star variation. Finally, the question of the overall metallicity of the Galaxy is discussed. How homogeneous in space and time is its evolution? Can we identify subsystems or individual stars which indicate a pregalactic contribution to the galactic metallicity?

Key words: subdwarfs – abundances – lithium – beryllium – boron

1. Introduction

The system of lowest metallicity galactic stars presents many questions and, if correctly interpreted, carries many answers about the history of the Galaxy and, indeed, the early history of the universe. From a great number of such questions posed by working group members, we decided to focus attention on a few key ones which draw together multiple lines of evidence.

2. Lithium

The experimental "cosmological" study of the light elements began with the discovery by Spite and Spite (1982) that lithium was detectable in most subdwarfs of effective temperature above about 5500 K. Since theories of primordial nucleosynthesis (e.g. Yang *et al.*, 1984; Schramm, 1998) predict that ^{7}Li is formed in the Big Bang and can serve to constrain the baryonic density of the universe, considerable work has focussed on determining whether the abundances measured in the oldest stars can be taken as the *unaltered* primordial abundance. The fact that over more than two orders of magnitude in metallicity, [Fe/H] ≈ -1.3 to less than [Fe/H] ≈ -3.5, and over the range in temperature of 5600 K to 6300 K more

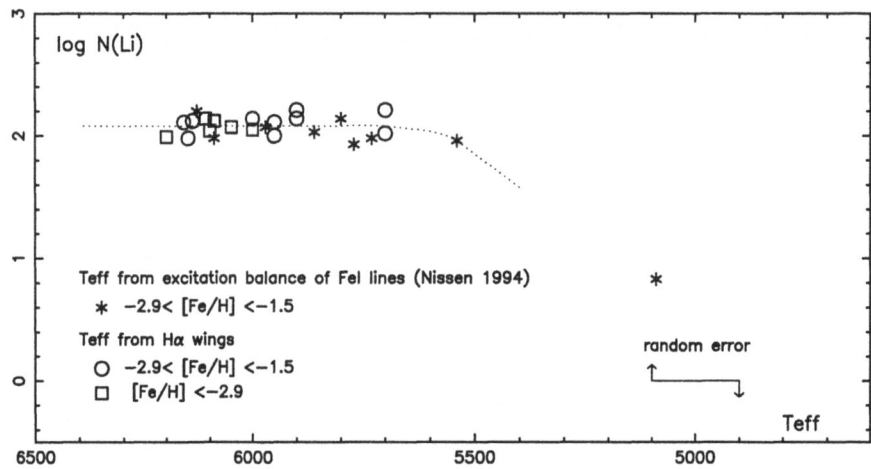

Figure 1. Li in stars of various metallicities (plotted with different T scales). Presented by F. Spite (1998).

than 100 surveyed halo stars show a similar Li abundance, N(Li) $\approx$ 2.1 (a trend now referred to as the Spite Plateau), shows that we are seeing a very uniform abundance, as would be expected from a primordial source.

However, as observational techniques have become better, higher measurement accuracy has become possible, and claims of star-to-star variation among the plateau stars have been presented. We therefore extensively discussed the following questions: Is there real variation in halo star Li abundances, and, if so, does this indicate depletion? If depletion has occurred, can we determine the amount?

2.1. ARE THERE REAL LI ABUNDANCE DIFFERENCES AMONG THE LOWEST Z STARS?

By the time of the thesis of Thorburn (1994), about 100 low-Z stars had been measured for Li. All but a small number of these showed very similar Li abundances. At the present workshop, both F. Spite and R. Rebolo presented further very accurate abundances in very metal poor stars. Rebolo concluded that there is no evidence for any real dispersion, with a limitation $2\sigma \leq 0.15$ dex. Data presented by F. Spite is shown in Figure 1. RMS scatter about the mean is 0.08 dex, which he concludes is consistent with the errors of measurement.

Deliyannis, Pinsonneault, and Duncan (1993) intercompared the Li observations of three entirely independent investigations (different astronomers, telescopes, and spectrographs). This external comparison indicated a scatter of about ±20% (0.08 dex) in measured equivalent width at a fixed color. However, the interpretation of their paper was that the scatter *cannot* be due to equivalent width measurement errors, since the external comparison showed that these were determined to higher accuracy. The scatter was attributed to a small, real scatter in Li abundances.

Thorburn (1994) also concluded that there is scatter about the Spite Plateau in excess of errors of measurement and analysis. She used typical values for such errors of 3 mÅ in equivalent width and 100 K in temperature, not unreasonable for modern, accurate measurements. Only by increasing both of these values by ≈55% could she reproduce the observed scatter. Furthermore, she found a slight slope with temperature to the plateau at fixed metallicity, or a slight increase with metallicity at a fixed temperature.

It is important to note that it possible for there to be scatter in excess of measurement error which does not necessarily reflect scatter in the primordial production of Li. As the metallicity of the Galaxy increases, Li from other sources, notably cosmic ray (CR) spallation, will start to be incorporated into stars. The strongest evidence that the origin of Be and B is CR spallation is that the abundances follow each other in a ratio very close to that of the spallation cross sections as the metallicity of the Galaxy increases. This may be seen in Figure 2. Since the cross section ratios are known and constant (except near threshold energies), the same ratios can be used to predict how much Li (^{7}Li and ^{6}Li) would be formed if the B or Be abundances are known. The amount increases as the Be and B abundances do, so that it is smallest at the lowest metallicities and larger for the more metal-rich halo stars. Such CR "contamination" might be thought to account for some of the variation noted by Thorburn, especially the trend with metallicity. However, the effect is too small to provide such an explanation. For metallicities of [Fe/H] $= -3.0, -2.5$, and -2.0, the CR-produced Li increases the primordial production only by approximately $0.01, 0.02$, and 0.03 dex respectively. This would raise the "plateau" insignificantly in Figure 2.

As pointed out by Spite, Rebolo, and others, other physical differences between stars might cause differences in equivalent width even for uniform abundances. Uncorrected reddening or any other source of mis-estimated temperature could do this. Undetected binarity could also have an effect. Adoption of different temperature scales for pop. II stars can introduce or remove a slight slope with temperature from the Spite Plateau.

It is worth noting that a small number of halo stars have no Li apparent in their spectra. Approximately 5% of the large sample of stars studied by Thorburn are in that category. It is generally assumed that some unknown sporadic event has affected those few stars and not the others. Figure 3 presents the data of Duncan *et al.* (1997), augmented by new data presented by Primas (1998) and Duncan at the present workshop. It shows that a similar effect is now detected in B abundances – a small number of stars below the mean trend. The work of Primas (1996) and of Molaro *et al.* (1997) shows the same to be true of Be abundances. Although the sample size is small, the percentage of depleted stars appears to be higher than 5% for the B and Be samples.

In conclusion, we can now be sure that among most low-Z stars the variation in Li abundance is small – less than 0.08 dex RMS. We cannot be sure if a smaller variation is present.

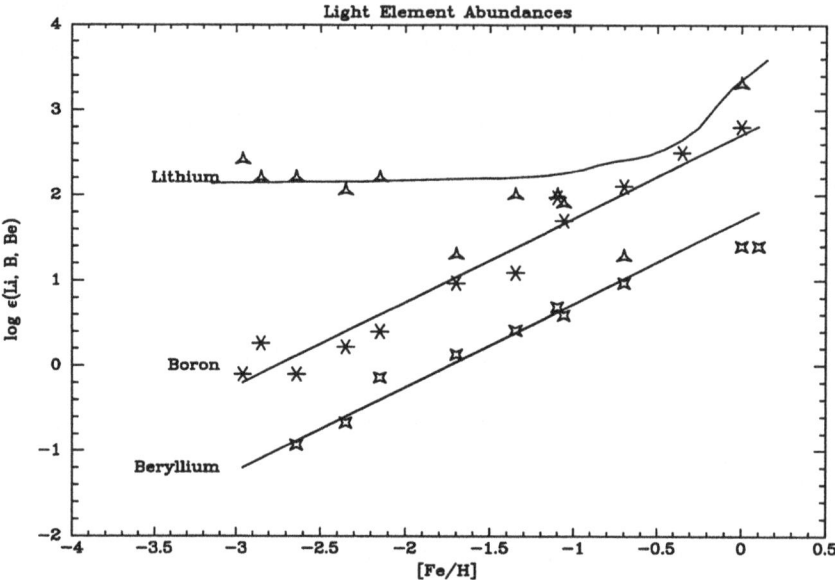

Figure 2. B and Be abundances follow close to the spallation cross-section ratio of 10 to 1; Li abundances reach a plateau at low metallicities.

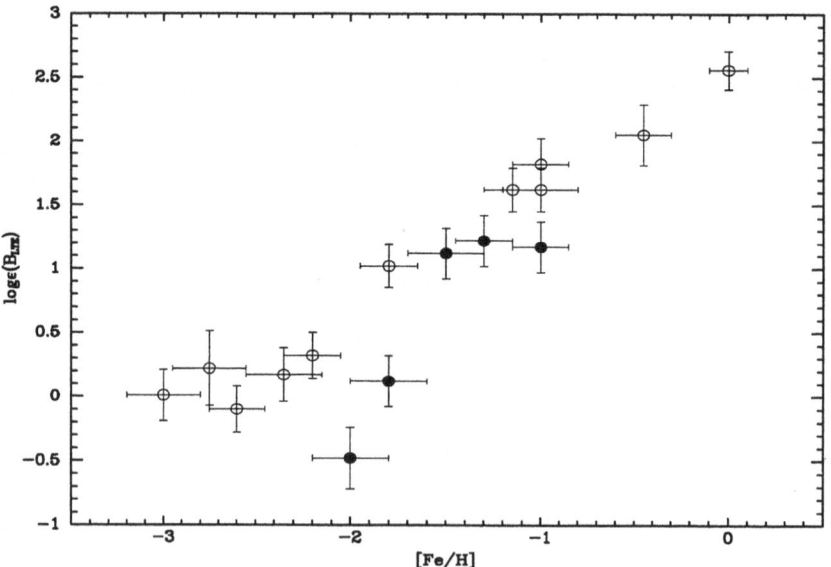

Figure 3. B abundances from Duncan (1997) and those reported by Primas (1998) and Duncan (this workshop).

2.2. DOES A POSSIBLE SCATTER IN Li ABUNDANCES INDICATE DEPLETION?

The light elements Li, Be, and B undergo nuclear reactions at the relatively low temperatures of approximately 2.5, 3.5, and 5×10^6 K at densities similar to those in the Sun. The temperature at the base of the solar convection zone is very close to 2.0×10^6 K, and the fact that solar Li is depleted by approximately a factor of 100 has led to many theoretical investigations attempting to determine what causes "extra mixing" below the convection zone.

Different possible mechanisms are nicely summarized by Cayrel (1998) in this Volume, so only a brief mention will be made here. Microscopic diffusion was suggested by Michaud (e.g. Michaud and Charbonneau, 1991). The Yale group (e.g. Pinsonneault, Deliyannis, and Demarque, 1992) have presented a series of models in which turbulence driven by rotation causes mixing. Charbonnel *et al.* (1992) present a different model of rotationally driven turbulence. A potentially important newer suggestion is mixing by gravity waves (e.g. Montalban and Schatzman, 1996). The amount of depletion suggested by these models ranges from a few tenths dex in models of microscopic diffusion, through about half a dex for the models of Charbonnel *et al.*, to as much as 1 dex for some of the models of the Yale group.

However, the models mentioned in the previous paragraph are usually calibrated to reproduce the depletion observed in the sun and other population I stars. One of the most important contributions to the working group discussion was a figure by R. Cayrel depicting the structural differences between the young sun and a young subdwarf, highlighting the convection zone boundaries and the depths at which Li is destroyed. This is reproduced as Fig. 2 in his contribution to the present Volume. These differences are so substantial that it seems exceedingly dangerous to depend in a quantitative way on such scaling between pop. I and pop. II stars. Furthermore, 3-D modelling of convection as well as high-resolution observations of the sun show that convection actually consists of cells in which narrow descending fingers of gas coexist with rising elements of larger filling factor. Modelling this with scaled one dimensional approximation is a dangerous if necessary simplification. It is probably best to take the models more as predictors of possible depletion rather than quantitative measures of such depletion.

Fortunately, other observational constraints can be placed on the amount of Li depletion. Smith *et al.* (1993) have found ^{6}Li in the star HD 84937, and this difficult observation has been confirmed by Hobbs and Thorburn (1994). ^{6}Li burns at a lower temperature than ^{7}Li and its detection means that ^{7}Li has been at most slightly depleted.

R. Cayrel (1998) has made the important point that since ^{6}Li is made by the same CR spallation process which produces Be and B, the abundances of these elements should be used to exactly predict the initial amount of ^{6}Li in a given halo star. The difference between the predicted initial ^{6}Li abundance and the observed

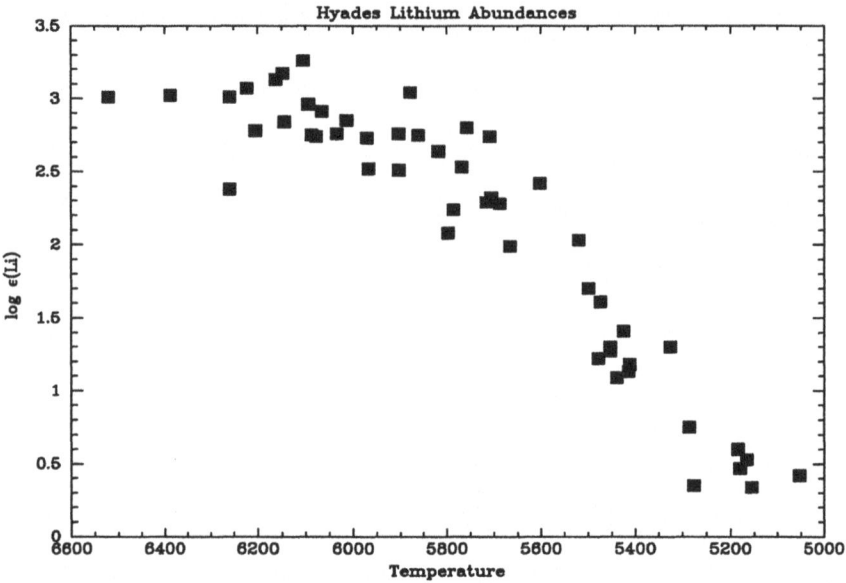

Figure 4. Li abundances in the Hyades (Thorburn, 1997).

one gives a measure (not just a limit) of the depletion of ^{6}Li, and consequently the related amount of depletion of ^{7}Li (Lemoine *et al.*, 1997).

A general point is sometimes made that with Li destruction so sensitive to temperature and the mixing caused by rotation it is unlikely that large amounts of Li destruction could take place without causing large star-to-star variations. However, C. Deliyannis pointed out that there are counterexamples among the pop. I stars. In the Hyades, for instance, Li depletion of more than an order of magnitude has taken place for cooler solar-type stars. Nevertheless, for stars of a given temperature, the scatter is small. Figure 4 presents Hyades Li abundances from a large, homogeneous sample provided by J. Thorburn (1997). The Hyades also shows remarkably homogeneous rotation for stars like the ones plotted in Figure 4 (Raddick *et al.*, 1987). It is not certain whether this has an influence on the small scatter in Figure 4 since such complete rotation information is not available for stars outside the Hyades.

Deliyannis also presented several spectra of globular cluster stars which seem to show Li abundances significantly different from those of the Spite plateau. These are difficult observations, of very faint stars, but the possibility of very interesting results was evident to all and should clearly be pursued with the largest telescopes.

We conclude that a modest amount of Li depletion in the halo stars is not ruled out. More quantitative limits on the amount may be possible if B, Be, ^{6}Li, and ^{7}Li can each be measured in the same stars.

3. Production of Lithium, Beryllium, and Boron

As was pointed out in Figure 2 above, the ratio of B to Be seen in low-Z stars is a strong indication of an origin due to CR spallation. Such an origin was suggested long ago by Reeves, Fowler, and Hoyle (1970), and developed in more detail by Meneguzzi, Audouze, and Reeves (1971) and Reeves, Audouze, Fowler, and Schramm (1973). Light element formation was attributed to the (p, α) reaction of galactic cosmic rays (GCR; primarily protons and α particles) impinging on CNO nuclei in the interstellar medium (ISM).

The data of Figs. 2 and 3 indicate that the original theories must be seriously incomplete. That is because even in more complex, recent versions of the theory (e.g. Prantzos, Cassé, and Vangioni-Flam, 1993), the slope of Be or B with [Fe/H] or [O/H] is predicted to be close to 2, whereas the figures show the data has a slope very close to unity. [The theoretical slope of 2 arises because CR production is dependent on the product of CR flux × target cross section. The cross section depends on the metallicity of the ISM, which is proportional to the integral of the number of supernovae (SN) up to a given time in the evolution of the Galaxy. The CR flux itself is usually thought to be proportional to the SN rate. The two factors combine to make a quadratic dependence on metallicity.]

Duncan *et al.* (1997) suggests a substantially different solution to the problem. The data indicate that B as well as Be follow Fe in direct proportion from the earliest times to the present. Plots vs. O are similar, though with more scatter, presumably due to the lower accuracy of the O abundances. A straightforward interpretation of this is that the rate of production of B and Be does *not* depend on the CNO abundances in the ISM, which are much lower at early times. This can be achieved if spallation does not primarily result from protons and α particles colliding with CNO nuclei in the interstellar medium. Instead, it results from the "reverse" process of C and O nuclei in the regions of massive star formation colliding with ambient protons and αs. This decouples light element production from the metallicity of the ISM and results in the linear relationship observed. The originally suggested spallation in the general ISM would still occur, but especially for the low-Z stars it would be less important than the "reverse" spallation.

An intriguing possible piece of evidence in favor of spallation near SN was presented by F. Primas. Such a process might be expected to produce more star-to-star variation than spallation in the general ISM, and the data of Figure 3 show two stars with significantly less B than others of the same metallicity. In the case of one of them, HD 160617, the lower abundance is almost certainly due to lower production and not due to depletion. That is because both Be and B are lower in the star, but the Li abundance is the normal plateau value. Since Li is more easily destroyed than Be and B, it is difficult to think of a destruction process which could reduce Be and B but not Li. On the other hand, the lower B and Be abundances appear to still be in the CR spallation ratio of 10 to 1, indicating that lower overall production could be a viable explanation. More details are presented by Primas (1998).

4. Evolution of the Overall Galactic Metallicity

Discussion of spatial variation of the light elements led to discussion of the homogeneity in space and time of the overall galactic metallicity. Does the Galaxy begin near zero metallicity? At a given epoch, how significant is the range in metallicity? T. Beers presented data from approximately 400 of the very lowest metallicity stars (to [Fe/H] $\approx$ -4.0). It does appear that the Galaxy did not start with zero metallicity – that the stars of lowest Z give evidence for pregalactic formation of some metals. Examination of the kinematics of the stars of lowest metallicity may give us clues as to whether this might have arisen from the accretion of a pre-galactic subsystem. The possibility of such accretion was also suggested in later discussion by M. Rees.

References

Cayrel, R., 1998, *Space Sci. Rev.*, this volume.
Charbonnel, C., Vauclair, S., and Zahn, J.-P.: 1992, *A&A* **255**, 191.
Deliyannis, C.P., Pinsonneault, M.H., and Duncan, D.K.: 1993, *ApJ* **414**, 740.
Duncan, D.K., Primas, F., Rebull, L., Boesgaard, A.M., Deliyannis, C.P., Hobbs, L.M., King, J.R., and Ryan, S.G.: 1997, *ApJ* **488**, 338.
Hobbs, L.M., and Thorburn, J.A.: 1994, *ApJ* **428**, L25.
Lemoine, M., Schramm, D.N., Truran, J.W., and Copi, C.J.: 1997, *ApJ* **478**, 554.
Meneguzzi, M., Audouze, J., and Reeves, H.: 1971, *A&A* **40**, 110.
Michaud, G., and Charbonneau, P.: 1991, *Space Sci. Rev.* **57**, 1.
Molaro, P., Bonifacio, P., Castelli, F., and Pasquini, L.: 1997, *A&A* **319**, 593.
Montalban, J., and Schatzman, E.: 1996, *A&A* **305**, 513.
Pinsonneault, M.H., Deliyannis, C.P., and Demarque, P.: 1992, *ApJS* **78**, 181.
Prantzos, N., Cassé, M., and Vangioni-Flam, E.: 1993, *ApJ* **403**, 630.
Primas, F.: 1996, Ph. D. Thesis, Univ. of Trieste.
Primas, F.: 1998, *Space Sci. Rev.*, this volume.
Raddick, R. R., Thompson, D.T., Lockwood, G.W., Duncan, D.K., and Baggett, W.E.: 1987, *ApJ* **321**, 459.
Reeves, H., Audouze, J., Fowler, W.A., and Schramm, D.N.: 1973, *ApJ* **179**, 909.
Reeves, H., Fowler, W.A., and Hoyle, F.: 1970, *Nature* **226**, 727.
Schramm, D.N.: 1998, *Space Sci. Rev.*, this volume.
Smith, V.V., Lambert, D.L., and Nissen, P.E.: 1993, *ApJ* **458**, 543.
Spite, F. and Spite, M.: 1982, *A&A* **115**, 357.
Spite, F., Spite, M., and Hill, V.: 1998, *Space Sci. Rev.*, this volume.
Thorburn, J.A.: 1994, *ApJ* **421**, 318.
Thorburn, J.A.: 1997, private communication.
Yang, J., Turner, M.S., Steigman, G., Schramm, D.N., and Olive, K.A.: 1994, *ApJ* **281**, 493.

IV: GALACTIC DISK, GALACTIC EVOLUTION

MEASUREMENTS OF THE $^{12}C/^{13}C$ RATIO IN PLANETARY NEBULAE AND IMPLICATIONS FOR STELLAR EVOLUTION

F. PALLA and D. GALLI
Osservatorio Astrofisico di Arcetri
L.go E. Fermi, 5 - Firenze (Italy)

R. BACHILLER and M.PÉREZ GUTIÉRREZ
Observatorio Astronómico Nacional
Alcala de Henares (Spain)

Abstract. We present the results of a study aimed at determining the $^{12}C/^{13}C$ ratio in two samples of planetary nebulae (PNe) by means of mm-wave observations of ^{12}CO and ^{13}CO. The first group includes six PNe which have been observed in the $^{3}He^{+}$ hyperfine transition; the other group consists of 23 nebulae with rich molecular envelopes. We have determined the isotopic ratio in 14 objects and the results indicate a range of values between 9 and 23. In particular, three PNe have ratios well below the value predicted by standard evolutionary models ($\gtrsim 20$), indicating that some extra-mixing process has occurred in these stars. We briefly discuss the implications of our results for standard and nonstandard stellar nucleosynthesis.

Key words: stars: abundances - planetary nebulae: general - radio lines: stars, ISM

1. Introduction

In the PN phase, stars more massive than solar return to the ISM material that has been processed in the stellar interior. This matter mixes with the surrounding medium and modifies the original abundances of elements. The contribution of PNe to the galactic chemical evolution is particularly important for ^{3}He which, together with deuterium (D), plays a fundamental role in testing the standard Big Bang nucleosynthesis model. While the evolution of D is well understood, that of ^{3}He still encounters serious problems which cast doubts on the usefulness of this isotope as a test of Big Bang nucleosynthesis models (e.g. Galli *et al.*, 1995). In fact, observations of ^{3}He toward PNe and HII regions give values of the abundance that differ by almost two orders of magnitude: $[^{3}He/H] \sim 10^{-3}$ in PNe and $\sim 10^{-5}$ in HII regions and in the solar system. However, the abundance in PNe is exactly that predicted by standard stellar evolution models for stars of mass 1–1.5 $M_{\odot}$. The main question is then: if low mass stars produce a lot of ^{3}He and return it to the ISM during the PN-phase, why don't we see it at a level much higher than observed in HII regions and the solar system as all standard galactic evolutionary models predict?

Possible solutions to this question have been extensively discussed in this Workshop (cf. the reviews by Tosi, 1998, and by Charbonnel, 1998). The most interesting suggestions invoke the existence of nonstandard mixing mechanisms which operate during the red giant phase of stars with $M_{*} \lesssim 2\ M_{\odot}$. If such mechanisms are

Space Science Reviews **84**: 177–183, 1998.

indeed at work, an unavoidable consequence is that the ratio of $^{12}C/^{13}C$ in PN ejecta should be much *lower* than in the standard case. For a 1 $M_\odot$ star, the predicted ratio is about 5 against the standard value of 25–30 (Charbonnel, 1995; Sackmann and Boothroyd, 1997). Therefore, it is very important to have a precise measure of the isotopic ratios in those PNe where the ^3He abundance has been determined. Should these stars show a $^{12}C/^{13}C$ ratio close to 25–30, then no modifications to the standard stellar models would be required. Otherwise, one has to invoke another selective process (mixing, diffusion etc.) that operates on some isotopes but not on ^3He. However, the number of PNe with ^3He measurements is small (see Rood *et al.*, 1998), whereas the suggested physical processes should be quite general and should affect the nucleosynthetic yields of all stars of mass less than $\sim 2\ M_\odot$. Thus, the interest of measuring the carbon isotopic ratio in a sample of PNe as large as possible. In this contribution, we present the initial results of such a study.

2. Measuring the isotopic ratio from mm-wave observations

Molecular line observations at mm-wavelengths provide the most powerful method to estimate the $^{12}C/^{13}C$ ratio in PNe. In particular, the $^{12}CO/^{13}CO$ ratio should faithfully reflect the atomic $^{12}C/^{13}C$ ratio, since the mechanisms which could alter the $^{12}CO/^{13}CO$ ratio are not expected to be at work in PNe. Namely, (i) the kinetic temperatures in PN envelopes (25–50 K) is high enough that the isotopic fractionation should not operate, and (ii) selective photodissociation is expected to be compensated by the isotope exchange reaction $^{12}CO+^{13}C^+ \rightarrow ^{13}CO+C^+$ which is faster than the ^{13}CO photodestruction in PN envelopes (e.g. Likkel *et al.*, 1988). However, although the J=1–0 and J=2–1 lines of ^{12}CO have been extensively observed in PNe (e.g. Huggins *et al.*, 1996), very few observations of the ^{13}CO lines are available and the value of the isotopic ratio is presently unknown.

Our project consists of two steps. In the first one, we have carried out high quality observations of ^{12}CO and ^{13}CO in six PNe where the ^3He abundance is known from the observations of Rood and collaborators. In the second step, a larger sample of nebulae with strong ^{12}CO line emission has been observed in ^{13}CO lines in order to determine the isotopic ratio in PNe *without* ^3He measurements. Galli *et al.* (1997) have argued that extra-mixing processes need not to be at work in *all* low-mass stars in order to reconcile the predictions of the galactic evolution of ^3He with the observational constraints: acceptable results are obtained if about 70%–80% of stars with mass lower than 2 $M_\odot$ undergo extra mixing. This suggestion can be tested by determining the isotopic ratio in a statistically significant sample of PNe.

2.1. RESULTS: PNE WITH ^{3}HE MEASUREMENTS

We have observed the six PNe studied by Balser *et al.* (1997) with the IRAM 30-m telescope in an observing run in November 1996. The observations were made in the J=2–1 and J=1–0 lines of ^{12}CO and ^{13}CO, simultaneously. Bachiller *et al.* (1993) have shown that PNe shells are characterized by an intrinsic CO (2–1)/(1–0) line ratio in the range 2–5, indicating that the J=2–1 line is more effective for CO searches. The results of the observations are given in Table I with the PN name, the distance, mass, the ^{3}He number abundance (given by Balser *et al.*, 1997), the intensities I_{21} of the J=2–1 lines and the carbon isotopic ratio. With the exception of NGC 6720, the values of I_{21} represent upper limits to the intensity and have been estimated from the line widths deduced from the expansion velocities listed in the catalog of Acker *et al.* (1992).

Disappointingly enough, this sample of PNe shows little emission in CO: the main and isotopic lines have been firmly detected only in NGC 6720. Bachiller *et al.* (1989) had already detected carbon monoxide emission in this object and the emission showed a kind of clumpy ring, resembling the optical appearance of the nebula. We have observed in ^{13}CO the most prominent clumps (three positions) and detected emission in all cases. Thus, the present data allow to produce a map of the isotopic ratio across the nebula. The resulting value of ^{12}C/^{13}C=22 is in agreement with a previous estimate of Bachiller *et al.* (1997). Unfortunately, NGC 6720 has an estimated progenitor mass of $\sim$2 $M_{\odot}$, at the borderline of the mass range where the nonstandard mixing mechanism is expected to significantly decrease the isotopic ratio. Thus, the derived ratio is consistent with both standard and nonstandard evolutionary models.

We also detected a line around the ^{12}CO J=1–0 frequency in the central position of NGC 6543, but we failed to detect the J=2–1 line at relatively low levels. This indicates that the line near the J=1–0 frequency is probably not due to CO. It is interesting to recall that the H38α recombination line is only separated by 3 MHz (7.8 km/s) from the ^{12}CO J=1–0 line. As discussed in Bachiller *et al.* (1992), the H38α line can dominate the emission around the ^{12}CO J=1–0 frequency in some nebulae with little or no molecular gas. We believe that this is the case in the central position of NGC 6543.

Following the discussion of Rood *et al.* (1998), the best object for testing the mixing hypothesis is NGC 3242 where the ^{3}He measurements are the most reliable ones *and* the progenitor mass is sufficiently small ($\sim$1.2 $M_{\odot}$) that the predicted isotopic ratio should be about 10 (instead of $\sim$30). However, the excellent signal to noise ratio of our observations and the fact that we searched for CO emission across the whole circumstellar shell imply that CO is really absent in the nebula. Therefore, we conclude that the isotopic ratio cannot be determined in this important object with millimeter line observations.

2.2. RESULTS: PNE WITHOUT ^{3}HE MEASUREMENTS

We have searched for ^{13}CO emission in 22 PNe using the IRAM 30-m telescope in May 1997. The objects were selected on the basis of their strong ^{12}CO emission from the list of Huggins *et al.* (1996). The results are: 13 detections, 6 tentative

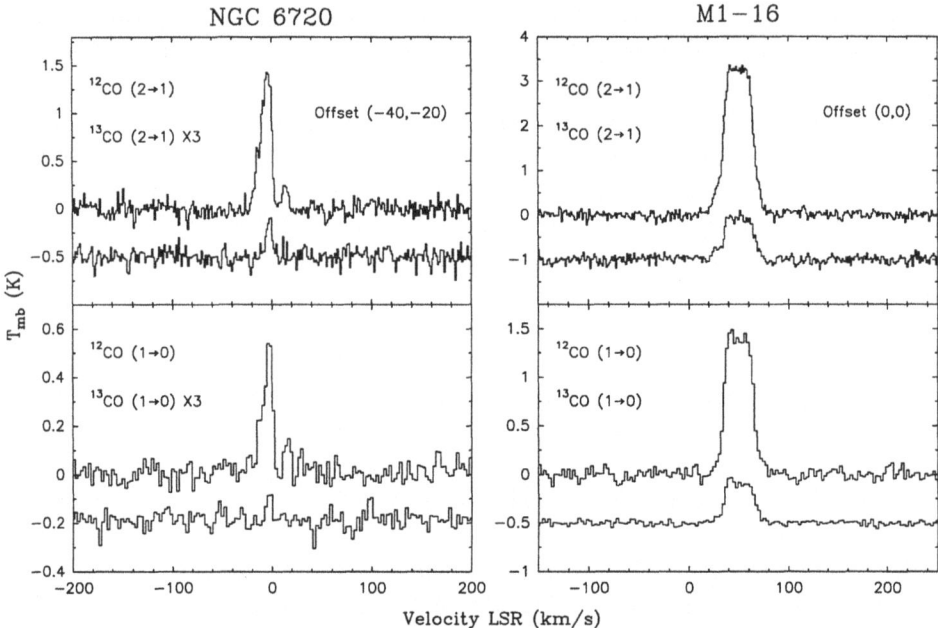

Figure 1. CO emission in PNe. Two examples are shown from the sample of PNe with (*left*) and without (*right*) ^{3}He measurements. The offset in arcsec from the central PN is also indicated. The stronger J=2–1 transitions are shown in the upper panels, while the J=1–0 transitions in the bottom panels. The isotopic lines are detected in all cases.

Table I

Distances, Progenitor Masses and Abundances

PN name	D (kpc)	M ($M_\odot$)	^{3}He/H ($\times 10^{-3}$)	$I_{21}(^{12}CO)$ (K km s^{-1})	$I_{21}(^{13}CO)$ (K km s^{-1})	$^{12}C/^{13}C$
IC 289	1.43	<1.6	<6.7±2.6	0.90		
NGC 3242	0.88	1.2 ± 0.2	0.92±0.18	0.16		
NGC 6543	0.98	1.6 ± 0.2	≤0.49±0.27	0.19		
NGC 6720	0.87	2.2 ± 0.6	<0.54±0.29	20.0	.9	22
NGC 7009	1.2	1.4 ± 0.2	≤0.80±0.31	0.12		
NGC 7662	1.16	1.2	<0.22	0.12		

detections, and 3 upper limits. The parameters of the detected sources together with the derived isotopic ratios are listed in Table II.

In order to estimate the $^{12}CO/^{13}CO$ isotopic ratio, one needs to make a number of approximations. First, we assume that the emitting regions fill the antenna beams in the lines of both molecules, or that the filling factor is the same (in the case of an extended clumpy medium). Second, we assume that the rotational levels are thermalized at a representative uniform temperature of 25 K (see e.g. Bachiller *et al.*, 1997). Thermalization is indeed a reasonable assumption for ^{12}CO and ^{13}CO, since the dipole moment is very low (about 0.1 Debye). Third, if we assume that the emission is optically thin for both the ^{12}CO and ^{13}CO lines, then the ^{12}CO/^{13}CO column density ratio is given by the ratio of the integrated intensities.

The last assumption is likely to be accurate in the case of evolved nebulae like the Ring, the Helix, NGC 2346, etc. In fact, Large-Velocity-Gradient (LVG) models confirm that the ^{12}CO and ^{13}CO line emission is optically thin in these cases. On the other hand, the same assumption is not appropriate for the CO lines in young objects such as CRL 2688, CRL 618, NGC 7027, and M 1-16. In such cases, the derived CO column densities and ^{12}CO/^{13}CO column density ratios represent only crude lower limits.

Finally, one could have weak emission arising from small optically thick clumps very diluted within the ^{12}CO and ^{13}CO beams. This could be the case in some compact CO envelopes such as the Butterfly nebula, M 2-9. The CO in M 2-9 is concentrated in an expanding clumpy ring which has a mean diameter of 6 arcsec (Zweigle et al., 1997). Individual clumps in this ring have sizes < 4 arcsec. The weakness of the ^{12}CO and ^{13}CO lines we observe could be due to the important dilution of such small clumps within the 30-m beam. The clumps could be optically thick in CO, and the reported ^{12}CO/^{13}CO intensity ratio would just represent a lower limit to the abundance ratio.

In summary, we believe that the isotopic ratios reported here are reasonably robust estimates for the extended evolved nebulae, namely NGC 6720, NGC 2346, NGC 7293, NGC 6781, M 4-9, M 2-51, and very likely for M 1-17 and IC 5117. The values of the isotopic ratio in these nebulae are in the range 9 to 23 (within a factor of less than 2), significantly lower than the Solar System value of 89. The reported values are however nearly within the range 12 to 36 found in the envelopes of massive AGB stars (Greaves and Holland, 1997), which could indicate that the progenitors of such nebulae were relatively massive. Unlike the PNe discussed in the previous subsection, we have not yet obtained estimates of the progenitor mass for these stars and therefore we cannot say if the three PNe with ratios definitely below 20 originated from low-mass stars. These are the most promising objects for testing the hypothesis of nonstandard burning processes in the AGB phase and work is in progress to obtain mass estimates (this part is done in collaboration with L. Stanghellini and M. Tosi). Similarly, they are prime targets for future measurements of ^{3}He abundances.

3. Discussion

The method of using mm-wave transitions is not the only one adequate for measuring the isotopic abundance in PNe. Clegg (1985) first pointed out the possibility of using the CIII multiplet near 1908 Å to obtain a direct estimate of the isotopic ratio in the *ionized* gas of PNe. Very recently, Clegg et al. (1997) have successfully detected the extremely weak isotopic line ^{13}C $^3_{1/2}$P$^o_0$-$^1_{1/2}$S$_0$ in two PNe (plus a tentative detection in a third one), using the HST Goddard High Resolution Spectrograph. The ^{12}C/^{13}C ratio has been measured to be 15 ± 3 and 21 ± 11, respectively. In either case, no measurements at mm-wavelengths have been made to independently check the derived values. However, as Clegg *et al.* (1997) point out, the two types of measurements are complementary since they cannot be performed on the same objects: the UV transitions require high excitation conditions, while the opposite is true for the mm-wave lines studied by us.

What are the implications of our results for the understanding of stellar nucleosynthesis? The fact that the majority of the PNe shows an isotopic ratio of ~20 implies a moderate degree of depletion from the constant value achieved during the first dredge-up in the red giant branch. Such constant value depends on both mass and metallicity of the stars and varies between 30 for 1 $M_\odot$ and 23 for 2 $M_\odot$ at Z=0.02. Thus, our results indicate that the observed PNe had rather massive progenitors. On the other hand, we also find some PNe with an isotopic ratio below this constant value, approaching the equilibrium value in the CN cycle, namely 4–9. Similar low values have been measured in field population II stars and in globular cluster giants and have provided the motivation to introduce some extra-mixing process in the standard evolution (e.g. Charbonnel, 1995). It will be interesting to try to estimate the progenitor mass of these PNe since nonstandard mechanisms are effective only for stars less massive than ~ 2 $M_\odot$. These stars should also have very low ^{3}He abundances, and future observations should address this issue. Finally, we remind that in order to reconcile the ^{3}He abundances with galactic chemical evolution models, the majority ($\gtrsim70\%$) of the PNe should show an isotopic ratio well below the standard value. It is interesting to note that 3 out of the 7 PNe with reliable ratios have values well below 20. However, the small number statistics does not allow us to draw any conclusion from this preliminary study, but clearly calls for more radio observations on a larger sample of PNe.

As emphasized by Rood, the key object to test theories is NGC 3242. The lack of a molecular envelope around NGC 3242 can be a consequence of the low-mass of the progenitor. It is well known that the transition from the AGB phase to the PN stage is slow for low-mass stars and therefore the gas remains exposed to ionizing and dissociating radiation for a longer time than in the case of more massive stars. This is also consistent with the fact that NGC 6720 has a rich molecular envelope and is the most massive objects of our sample. Therefore, the lesson is clear: there is a trade-off between the need to go to low-mass PNe in order to discriminate between theoretical models and the higher detection rate of molecular envelopes around more massive objects. Thus, the determination of isotopic ratios may be a tricky business, but future observations should take up such a challenge.

Table II
Distance, CO Intensity and Isotopic Ratio

PN name	D (kpc)	$I_{21}(^{12}CO)$ (K km s^{-1})	$I_{21}(^{13}CO)$ (K km s^{-1})	$^{12}C/^{13}C$
NGC 7027	0.70	448.7	17.3	>25
NGC 2346	0.80	14.2	0.58	23
NGC 7293	0.16	17.5	1.8	9.3
NGC 6781	0.70	28.4	1.72	20
M 1-16	5.45	109.0	32.3	>3
M 1-17	7.36	66.2	3.2	22
M 2 9	1.0	3.3	2.0	>2
M 2-51	1.92	33.1	2.3	15
M 4-9	1.8	32.2	1.8	18
CRL 2688		312.5	112	>3
CRL 618	1.80	182.2	41.2	>4
OH 09+1.3	8.0	5.0	2.5	> 2
IC 5117	2.10	19.6	1.4	14

Acknowledgements

It is a pleasure to thank R. Rood, T. Bania and D. Balser for numerous and fruitful discussions on the ^{3}He measurements. This work is done in collaboration with M. Tosi and L. Stanghellini.

References

Acker, A., Ochsenbein, F., Stenholm, B., Tylenda, R., Marcout, J., and Schohn, C.: 1992, *Strasbourg-ESO Catalogue of Galactic Planetary Nebulae.*
Bachiller, R., Bujarrabal, V., Martín-Pintado, J., and Gómez-González, J.: 1989, *A&A* **218**, 252.
Bachiller, R., Huggins, P.J., Martín-Pintado, J., and Cox, P.: 1992, *A&A* **256**, 231.
Bachiller, R., Huggins, P.J., Cox, P., and Forveille, T.: 1993, *A&A* **267**, 177.
Bachiller, R., Forveille, T., Huggins, P.J., and Cox, P.: 1997, *A&A*, in press.
Balser, D.S., Bania, T.M., Rood, R.T., and Wilson, T.L.: 1997, *ApJ* **483**, 320.
Charbonnel, C.: 1995, *ApJ* **453**, L41.
Charbonnel, C.: 1998, *Space Sci. Rev.*, this volume.
Clegg, R.E.S.: 1985, in *Production and Distribution of C, N, O Elements*, eds. J. Danziger *et al.*, Munich: ESO, p.261.
Clegg, R.E.S., Storey, P.J., Walsh, J.R., and Neale, L.: 1997, *MNRAS* **284**, 348.
Galli D., Palla, F., Ferrini, F., and Penco, U.: 1995, *ApJ* **443**, 536.
Galli, D., Stanghellini, L., Tosi, M., and Palla, F.: 1997, *ApJ* **477**, 218.
Greaves, J.S., and Holland, W.S.: 1997, *A&A*, in press.
Huggins, P.J., Bachiller, R., Cox, P., and Forveille, T.: 1996, *A&A* **315**, 284.
Likkel, L., Forveille, T., Omont, A., and Morris, M.: 1988, *A&A* **198**, L1.
Rood, R.T., Bania, T.M., Balser, D.S., and Wilson, T.L.: *Space Sci. Rev.*, this volume.
Sackmann, I.-J., and Boothroyd, A.I.: 1997, *ApJ*, in press.
Tosi, M.: 1998, *Space Sci. Rev.*, this volume.
Zweigle, J., Neri, R., Bachiller, R., Bujarrabal, V., and Grewing, M.: 1997, *A&A*, in press.

HELIUM-3: STATUS AND PROSPECTS

R. T. ROOD

Dept. of Astronomy, University of Virginia, Charlottesville VA 22903

T. M. BANIA

Dept. of Astronomy, Boston University, Boston MA 02215

D. S. BALSER

National Radio Astronomy Observatory, Green Bank WV 24944

T. L. WILSON

Submillimeter Telescope Observatory, University of Arizona, Tucson AZ 85721

Abstract. We report on our continuing efforts to determine ^{3}He abundances in H II regions and planetary nebulae. Our detections of ^{3}He in some PNe show that some stars produce large amounts of ^{3}He. However the H II region abundances show no evidence for this production. From our sample of > 40 H II regions, the subsample which should yield the most reliable abundances has ^{3}He/H abundances which scatter between 1–2×10^{-5}. There is no trend with either galactocentric distance or metallicity. Even if we do not understand the underlying mechanisms, we see empirically that stars neither produce nor destroy ^{3}He in a major way. We thus suggest that the level of the "^{3}He Plateau" (^{3}He/H $= 1.5^{+1.0}_{-0.5} \times 10^{-5}$) is a reasonable estimate for the primordial ^{3}He.

Key words: 3-Helium, cosmic abundance, planetary nebula, chemical evolution

1. Introduction

Determining the abundance of any astrophysical species is a difficult multistep procedure, and ^{3}He is no exception. Its abundance beyond the local interstellar medium (LISM) can only be derived from measurements of the hyperfine line of ^{3}He$^+$ which has a rest wavelength of 3.46 cm. Obtaining high signal-to-noise spectra for the extremely weak (~ 1 mK) and wide (~ 1 MHz) spectral lines of ^{3}He$^+$ is a non-trivial feat. Going from the measured line parameters to an abundance (^{3}He/H) is also a formidable task. The strength of the collisionally excited hyperfine line is proportional to the source density, whereas the hydrogen abundance which is derived from the thermal free-free bremsstrahlung radio continuum emission depends on the square of the density (Rolfs and Wilson, 1996). Thus the source density structure must be modeled. Further, since we only observe ^{3}He$^+$, it is obvious that we must be concerned with ionization as well. Since 1978 we have measured (or attempted to measure) ^{3}He in planetary nebulae (PNe) and H II regions distributed throughout the Milky Way (Rood, Wilson, and Steigman, 1978; Rood, Bania, and Wilson, 1984 [RBW84]; Bania, Rood, and Wilson, 1987 [BRW87]; Rood, Bania, and Wilson, 1992 [RBW92]; Balser *et al.*, 1994 [BBBRW94]; Rood *et al.*, 1995 [RBWB95]).

^{3}He is one of the isotopes produced in cosmological nucleosynthesis. In addition, the classic papers by Iben (1967), Truran and Cameron (1971), and Iben and Truran

Space Science Reviews **84**: 185–198, 1998.

(1978) showed that low-mass stars could produce ^{3}He. Standard stellar evolution models predict that red giant branch (RGB) and asymptotic giant branch (AGB) stellar winds, and planetary nebulae of stars with masses $M < 2\,M_\odot$ should be substantially enriched in ^{3}He. Because the PNe and their preceding winds are a major source of mass input to the interstellar medium (ISM), one might expect substantial evolution of ^{3}He (Rood, Steigman and Tinsley, 1976 [RST]). RST argued that ^{3}He/H should grow with time and be higher in those parts of the Galaxy where there has been substantial stellar processing (RST). Specifically: (1) the protosolar value should be less than that found in the present ISM; and (2) there should be a ^{3}He/H abundance gradient with the highest abundances occurring in the highly-processed inner Galaxy.

Primordial D is also a source of ^{3}He. The $d\,(p, \gamma)\,^3$He cross-section is so large, that the D in any primordial gas which has passed through a star has been converted into heavier isotopes including some ^{3}He. For many years chemical evolution models included only this source of ^{3}He despite the simplicity of the case for the synthesis of ^{3}He in low mass stars. Recently, however, many models including stellar production of ^{3}He have appeared (Dearborn, Steigman, and Tosi, 1996; Vangioni-Flam and Cassé, 1995; Olive et al., 1995; Galli et al., 1995; Fields, 1996; Prantzos, 1996). These give results consistent with standard stellar evolution and RST.

Our earlier measurements of ^{3}He in H II regions (RBWB95, BBBRW94), in protosolar material (Geiss, 1993), and in the LISM (Gloeckler and Geiss, 1996) all indicate a value of ^{3}He/H $\sim 2 \times 10^{-5}$ (by number). There was no evidence for a gradient in the Galaxy. Nor did the H II regions, which are zero-age objects compared to the age of the solar system, show any evidence for stellar ^{3}He enrichment during the last 4.5 Gyr. On the other hand, RBW92 detected ^{3}He$^+$ in the planetary nebula NGC 3242, showing that at least one star produced ^{3}He as predicted by standard stellar models. The fact that our observed abundances of ^{3}He throughout the Galaxy disagreed so strongly with chemical evolution models coupled with our observation that some stars do indeed produce ^{3}He led to what is called "The ^{3}He Problem" (e.g., Galli et al., 1997). The existence of this problem has prompted investigations of many non-standard models of both stellar and galactic chemical evolution. Since 1994/95 we have made significant progress in a number of areas and take this opportunity to provide a status report on our experiment updated to include results obtained since the conference.

2. ^{3}He: September 1997

2.1. Do Stars Actually Make ^{3}He?

To see if stars actually do produce ^{3}He, we have made a survey for ^{3}He$^+$ in PNe (Balser et al., 1997). Our sample includes 6 PNe with a solid detection in

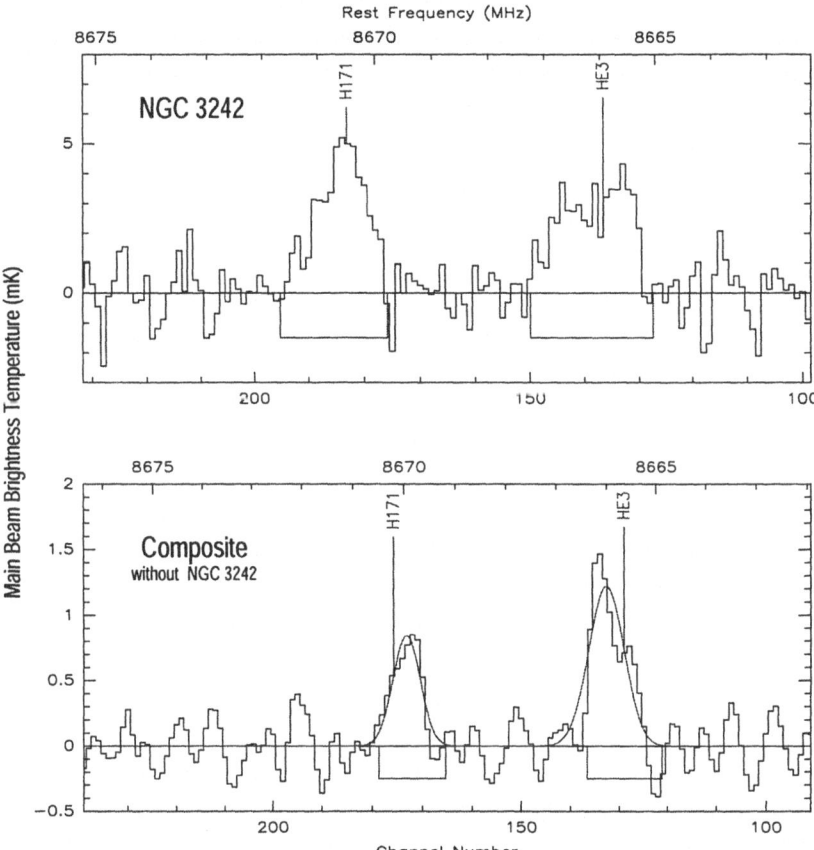

Figure 1. Top panel: Shown is the ^{3}He$^+$ spectrum for NGC 3242. Bottom panel: a composite spectrum which is the average of the five other PNe for which we have significant observations. The H171 and HE3 flags show the expected centers of the H171η and ^{3}He$^+$ lines.

NGC 3242 (Fig. 1 top). In this survey, we are working at the limit of the MPIfR 100 m telescope and do not have a definitive detection in any other single PN. However, there are possible detections in 2 additional PNe, and hints in 2 others. To make a more definitive statement about ^{3}He in PNe other than NGC 3242, we have made a composite PN spectrum which is shown in bottom panel of Fig. 1. This is an average of 5 of the PNe in our sample not including NGC 3242. We do this to build-up signal-to-noise and to increase the number of effective observing epochs (which reduces baseline problems). We consider this composite spectrum as being representative of a "generic" planetary nebula. Although the observed PNe have somewhat different physical properties, they are similar in most respects.

Note that there is clearly a ^{3}He$^+$ line in the composite spectrum. As in NGC 3242 the wings of the line are square, and there appears to be a double-peak to the line shape*. From models which include the ionization structure of each individual source we estimate a ^{3}He/H abundance of 2–3×10^{-4} for this composite spectrum. We do not wish to imply that this is the average abundance of the sample. So many factors would enter in the average that it is difficult to know how the individual spectra should be weighted to obtain an average. *We do take the presence of the ^{3}He$^+$ line in the composite spectrum as an indication that the lines in the individual sources which we consider to be probably real are indeed real.* We thus conclude that *some* (i.e., > 1) PNe produce ^{3}He as expected by standard stellar evolution theory. Because our sample was purposefully biased (see BBRW97 for details) to maximize the likelihood of finding ^{3}He, whether this is true for all PNe remains to be shown.

2.2. SOURCE MODELING

While we made some primitive attempts at modeling density structure, our early papers gave abundances based on the assumption that our sources were uniform density spheres (UρS). The first really significant effort in modeling was Balser (1994), and some of those results were reported in RBWB95. Since then we have made considerable progress toward constructing even more accurate source models. These are described in detail in Balser (1994), Balser *et al.* (1995, 1998) [BBRW95, BBRW98], and Bania *et al.* (1997) [BBRWW97]. It is important to bear in mind that ^{3}He is not ^{4}He. Our results are significant for cosmology and chemical evolution if we achieve an accuracy of ± 1 in the **first** significant figure. If we could definitively say whether ^{3}He/H was 1×10^{-5} rather than 2×10^{-5} this would be an important result. By comparison ^{4}He requires accuracy of ± 2 or 3 in the **third** significant figure. For ^{4}He, whether its mass fraction Y is 0.235 or 0.240 makes for a heated debate. To approach the required accuracy our models include:

High resolution continuum modeling: We have made higher resolution 8.7 GHz continuum maps with the MPIfR 100 m telescope and with the VLA. These have been used along with our Green Bank continuum observations to produce nested multiple core/halo models for each source. The latest source models use continuum maps made by combining the VLA and 100 m data such that all the zero-spacing flux is recovered.

Recombination lines: We observe many radio recombination lines (RRL) of H and ^{4}He. Earlier we used these solely as a monitor of system performance. We now also use them to determine: (1) the extent of non-LTE effects; and (2) the level

---

* Using our modeling as described in §2.2 we can understand how in NGC 3242 (Fig. 1) the ^{3}He$^+$ line shape is "Bactrian," while at the same time the recombination lines are distinctly "dromedarian." The squarish line shape arises because the Doppler broadening in the PNe is due to the expansion of the nebula rather turbulence as in H II regions. The observed PNe all have a dense core surrounded by a diffuse halo. The recombination lines originate primarily from the core. The hyperfine ^{3}He$^+$ has contributions both from the core and low density halo. Since the halo is typically larger than the telescope beam, the resulting ^{3}He$^+$ line can be double peaked.

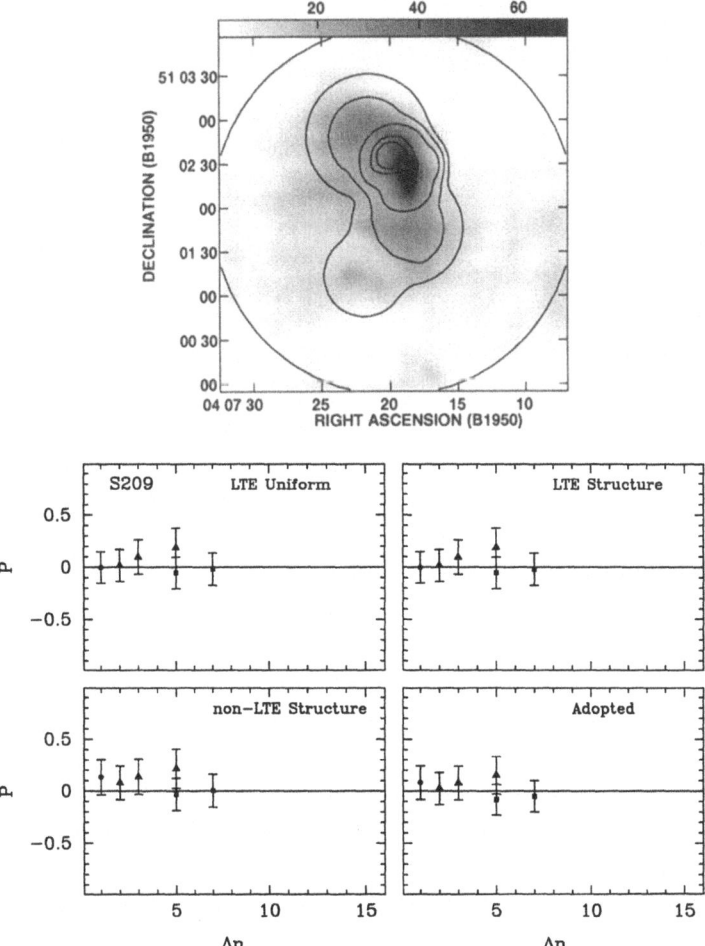

Figure 2. Comparison of model results for the "simple" H II region S209 with the observations. The greyscale images are the 8.7 GHz continuum maps based on VLA D-array and 100 m data. The contours are the model results. The fours panels at the bottom of each figure compare the observed and calculated line-to-continuum ratios for several transitions all at approximately the same frequency (P = observed/calculated − 1). *LTE Uniform:* a single-component UρS under the conditions of LTE with no pressure broadening. *LTE Structure:* the structure measured by the high resolution data is included by modeling various components as UρS's in LTE with no pressure broadening. *non-LTE Structure:* the same as "LTE Structure" except that LTE is not assumed and pressure broadening by electron impacts is included. *Adopted:* The adopted model. Starting with the "non-LTE Structure" model, we vary the electron temperature and the filling factor to obtain the best fit. The greyscale flux density ranges from 0.7 to 69.2 mJy beam^{-1}. The contour levels are 5, 10, 30, 50, 70, 80, 90, and 95 K.

of micro-clumping not resolved by the VLA. We model any micro-clumping via a volume filling factor in the core components.

Pressure broadening: In the densest environments pressure broadening can lead to very wide wings for recombination lines originating from levels with large

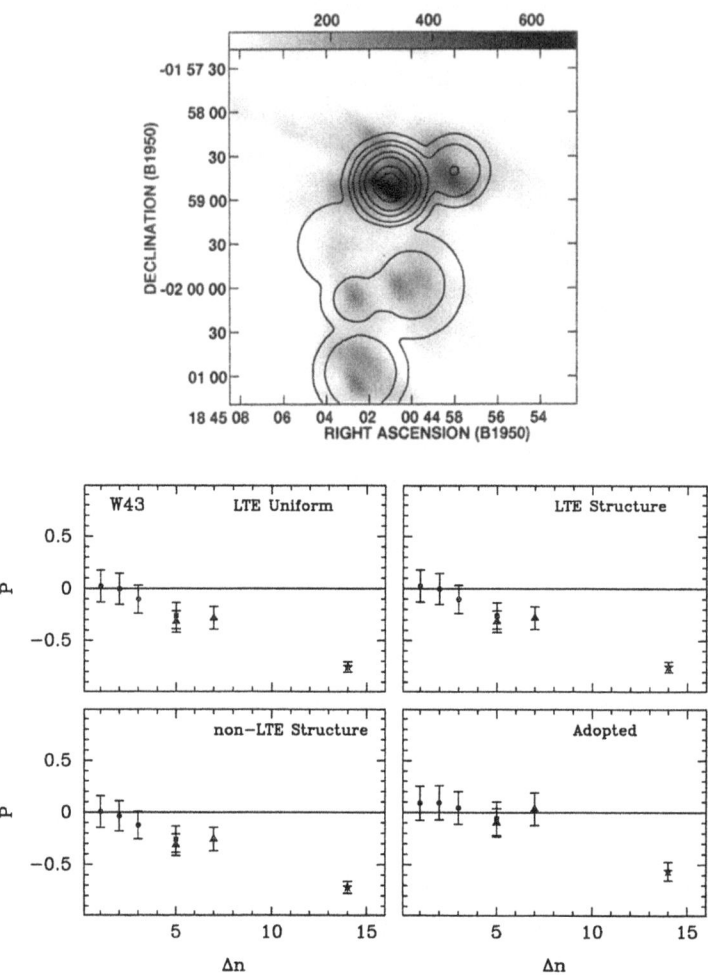

Figure 3. Comparison of model results for the "complex" H II region W43 with the observations. The greyscale flux density ranges from 6.8 to 676.5 mJy beam^{-1}. The contour levels are 2, 10, 30, 50, 70, 80, 90, and 95 K.

principle quantum number. In particular, the wings of the H213ξ line can extend into the region of the ^{3}He$^+$ line. This can lead to ^{3}He/H abundances higher than the true value, although we suspect the wide line wings are ordinarily removed along with our spectral baseline.

Ionization effects: We use the ^{4}He recombination lines as constraints for the ionization modeling. Sources with significantly low ^{4}He$^+$ are eliminated from our sample.

Our models show that the "structure correction" we derive will always increase the ^{3}He/H abundance as compared with the UρS value. To see this consider the following simple example: Suppose we observe 2 slabs along the line of sight,

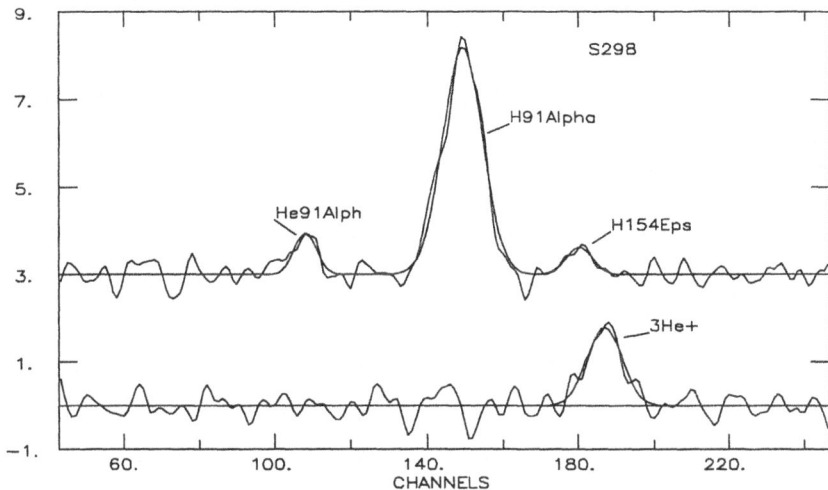

Figure 4. The lower spectrum shows the ^{3}He$^+$ from the H II region S298. The intensity scale is in mK of antenna temperature. The upper line shows the H91α spectrum. The intensity scale is the same as for the ^{3}He$^+$ line but has been offset for clarity. On the same scale, the W43 ^{3}He$^+$ line would be about twice as high as that of S298, but showing the W43 H91α would require a plot 5 m high (and would not fit the Kluwer format).

each with thickness ℓ and filling the telescope beam. If the densities of the slabs are respectively n and fn where $0 < f < 1$, and the ^{3}He$^+$/H$^+$ abundance is A_3, then the ^{3}He$^+$ line temperature is $T_L \propto A_3 n \ell (1 + f)$. The continuum temperature is $T_c \propto n^2 \ell (1 + f^2)$. This yields $A_3 \propto T_L \sqrt{1 + f^2} / \sqrt{T_c \ell}(1 + f)$. If instead one "modeled" two slabs as one uniform slab with thickness 2ℓ, the abundance is $A_3^{\text{uniform}} \propto T_L / \sqrt{2 T_c \ell}$. The ratio of the true abundance to the "uniform" slab abundance is $A_3 / A_3^{\text{uniform}} \propto \sqrt{2(1 + f^2)}/(1 + f) \geq 1$.

One very interesting consequence of our modeling results is that the H II regions can be divided into two categories: "Simple sources" have small structure (density) and ionization corrections and potentially the most accurate ^{3}He/H abundances. "Complex sources" have density and ionization corrections which *could* be a factor of two or larger. Their abundances could range from the UρS value upwards. Examples of these two classes are shown in Figs. 2 and 3. From the "LTE Uniform" panel of each of these figures we see that sources can be categorized as simple or complex on the basis of whether the observed RRL intensities are well fit by an LTE UρS model. If we restrict ourselves to simple sources we can use UρS abundances and not incur errors more than a few 10's%.

2.3. The H II Region Survey

Since 1995 we have continued our H II region $^3He^+$ survey in Green Bank. The deployment of the new GBT receivers on the 140 ft telescope make it a 2–5 times faster spectrometer. The receivers are also far more stable which leads to greater effective sensitivity per unit time. The survey has been (and will continue to be) directed toward certain specific goals:

Location: (1) The outer galaxy is expected to have less elemental processing and so should have a low ^3He abundance. Yet our earlier results showed that on average the outer galaxy has a lot of ^3He . In the hope of finding nearly primordial abundances we are observing more outer Galaxy sources and already have data for S252, S228, S298, and S212. (2) Our sample needs more sources in the $R_{gal} \sim$ 2–4 kpc region. This zone is critical for pinning down any galactic abundance gradient because the Galactic Center region is known to have odd abundances in many species.

Morphologically Similar Sources: The highest $^3He/H$ in our earlier work was found in "small," bubblelike sources (like W3 and S206) and possibly caused by self-pollution. We have started to test this hypothesis by observing similar sources. Some observations for S76 and S90 have been obtained. Neither show high $^3He/H$.

We can push the precision limit for a few sources: It is important to attain very high precision values for the $^3He/H$ abundance in a few sources. A combination of not so obvious factors can lead to higher precision: (1) a relatively strong $^3He^+$ line; (2) a relatively weak continuum (for less baseline structure); (3) a simple structure; and (4) a well determined distance. **S209**, for example, is a simple outer Galaxy source with low abundance. Early observations were marred by an anomalously large H213ξ line. Further observations and careful re-examination of old data have shown that the anomalous line was due to weak interference during one observing session. Other examples are **S206** and **S311** which appeared to be high abundance ring-like sources. In BBBRW94 we noted an ambiguity in baseline removal. Further observations have resolved the ambiguity and we now find smaller abundances.

Comparison of Related Sources: Physically related sources can be used to assess and constrain source models and address the question of ^3He production or destruction within H II regions. Several H II region complexes have multiple potential targets. Naively we might expect each source within one complex to have initially had the same abundance. We now have observations of G29.9 (W43 complex) shown in Fig. 5, G49.4 (W51), G133.8 (W3), and M16N (M16). The "subregions" typically have much lower density than the central part of the complex, i.e., the continuum peak. They often fall into the "simple"

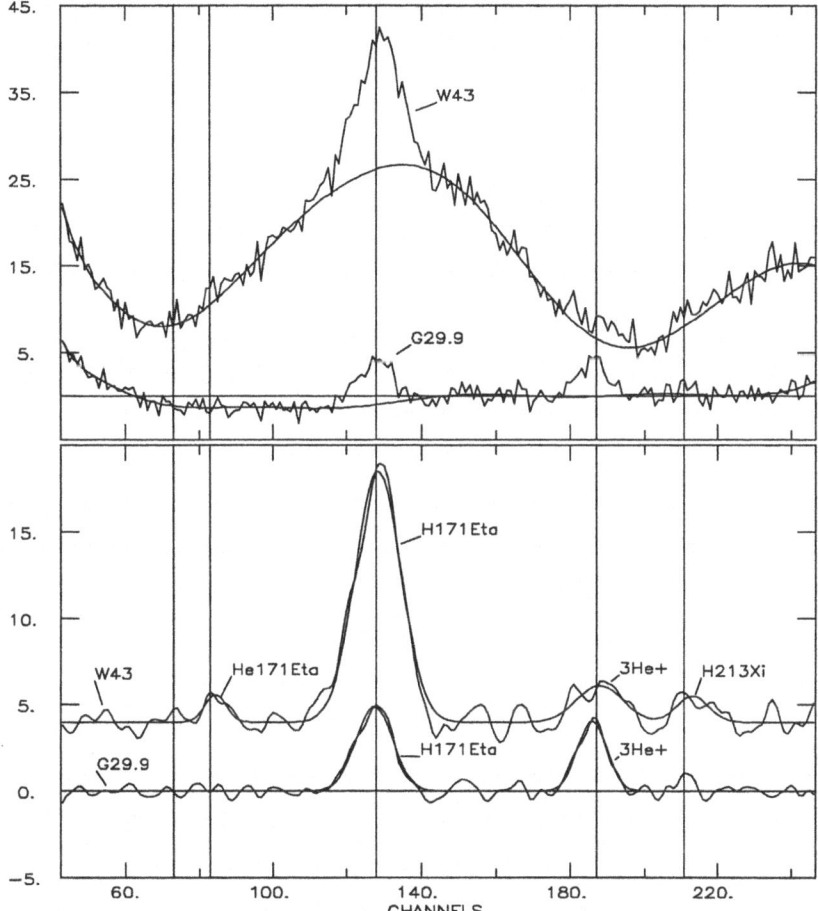

Figure 5. The ^{3}He$^+$ spectrum toward the H II regions W43 and G29.9 after 61.0 hours (30.5 hrs × 2 receivers) and 71 hours of integration respectively. The W43 spectra are on the same intensity scale but offset for clarity. The intensity scale is in mK of antenna temperature. The upper panel shows the spectrum after a linear baseline has been removed together with our adopted polynomial baseline superposed. The baseline structure is due to reflections of source continuum radiation from the telescope feedlegs and other components of the cassegrain system. Because of its higher continuum flux the amplitude of the W43 baseline structure is much larger than that of G29.9. Despite the complexity of the W43 baseline we feel that it can be reliably removed, and our many criteria for establishing baseline reliability are described in detail in BBRW97 §2.4. The lower panel shows the final spectrum with this baseline removed and the data smoothed to a velocity resolution of 8.1 km sec^{-1}. Vertical lines flag the expected positions of the following transitions from left to right: C171η, He171η, H171η, ^{3}He$^+$, and H213ξ. Note the large difference in the H171η/^{3}He$^+$ ratio between W43 and G29.9. This difference arises because W43 has more high density material.

category. To our surprise it has turned out to be much easier to measure the ^{3}He$^+$ line as well. Indeed, it is vastly easier to determine the ^{3}He/H abundance in G29.9 than in W43 itself.

When we establish the robustness of our modeling of "complex" regions, we can test whether the Olive *et al.* (1995) hypothesis that ^{3}He might be depleted in giant H II regions where the ionized gas is significantly polluted by the winds of massive O stars.

It is interesting to contemplate the evolution of our notion of what is an ideal ^{3}He$^+$ target. The only obvious factor is that the gas must be ionized. Given this we started with everybody's favorite H II region, Orion A. It didn't take long to realize the folly* of that idea. We turned to standard catalogs of H II regions, still mostly observing from the "Top 20." Sources like S206 and S209 made it onto our list because there was little else to compete with them in that part of the sky. In the last few years we have discovered that the best targets are obscure H II regions. G29.9 is one such nebula (*cf.*, Fig. 5); S298 is another (*cf.*, Fig. 4). As an H II region, S298 is totally laughable by radio astronomy standards—it has a 6 mK H91α line compared to 4000 mK in Orion A. But the Fig. 4 spectrum shows S298 to be a far more tractable ^{3}He$^+$ target than Orion A. One wonders what the reaction of the telescope TACs would have been if our original proposals had included sources like G29.9 and S298. We now suspect that many good ^{3}He$^+$ targets, including some of the very best, are not included in any catalog of H II regions. Much could be done.

3. Some Preliminary Results

We now have preliminary abundances for a substantial sample of simple H II regions. We are in the process of collecting the necessary continuum data and making the models to get density and ionization structure corrections. For this subsample, we anticipate that the corrections will be small. In Fig. 6 we show the UρS ^{3}He/H abundance as a function of R_{gal}. For the subset of simple H II regions there appears to be no galactic ^{3}He/H abundance gradient.

We also now have enough data to search for a trend in ^{3}He/H as a function of metallicity in much the same spirit as is done for ^{4}He and ^{7}Li. There is no direct measure of metallicity for most of our sample. However, we do have a surrogate. From our RRL and continuum data we can derive the nebular electron temperature, T_e. As has been noted before (e.g., Wink, Wilson and Bieging, 1983; Shaver, 1983) and shown in Fig. 7, T_e varies systematically with R_{gal}. The drop in T_e going toward the center of the Galaxy is generally attributed to more efficient cooling due to higher heavy element abundance. Shaver *et al.* (1983) derived an empirical relationship between T_e and the oxygen abundance [O/H] ($= \log(O/H) - \log(O/H)_{\odot}$). Using this we obtain the relationship shown in Fig. 8 between [O/H] and the ^{3}He/H abundance for the sample of simple sources. The values for the

---

* Orion A is one of the strongest radio continuum sources in the sky. It would have a baseline ripple 6 times larger than that of W43 (shown in the upper panel of Fig. 5). Yet for the same abundance as W43 its ^{3}He$^+$ line would be half the strength. Even worse, its large continuum flux effectively doubles the system temperature, thus quadrupling the required integration time.

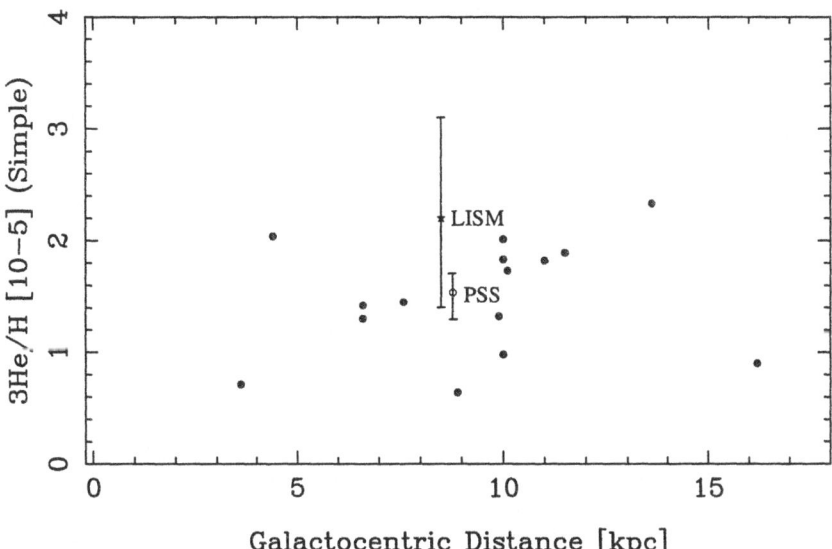

Figure 6. The ^{3}He/H abundance (by number) as a function of R_{gal} for our simple H II region subsample. Based on our past experience and a general assessment of the quality of the data, the statistical error in each abundance is likely to be $\lesssim$ 10% for most sources but ranging up to 20% for a few. The protosolar system value of Geiss (1993) is shown as an open circle labeled "PSS" and the LISM value of Gloeckler and Geiss (1996) is shown as a triangle labeled "LISM."

proto-solar system, LISM, and the PN NGC 3242 are also shown. While the span in metallicity is small, within this range there is no systematic trend. In the spirit of ^{7}Li, we seem to be seeing a " ^{3}He Plateau" for these objects. The fact that the NGC 3242 point lies so far above the plateau is just a reiteration of the "^{3}He problem."

Thus far, our ^{3}He abundance determinations are consistent with the notion that ^{3}He/H abundances are independent of position in the Milky Way and independent of metallicity; they are consistent with that measured in the LISM and in protosolar material. *There is no evidence for substantial ^{3}He enrichment or depletion by any process.* Stars, for reasons not fully understood, neither produce nor destroy ^{3}He. (Note: Most non-standard processes introduced into stellar models, turn stars into non-producers of ^{3}He, not into ^{3}He destroyers.) Our PNe sample must be atypical.

4. Discussion

We have discovered a class of H II regions in which ^{3}He$^+$ is relatively easy to observe and which are "simple" to model. These tend not to be the most well known radio H II regions, and there could be many of them. In this sample there are no trends in ^{3}He/H as a function either of R_{gal} or metallicity. This suggests

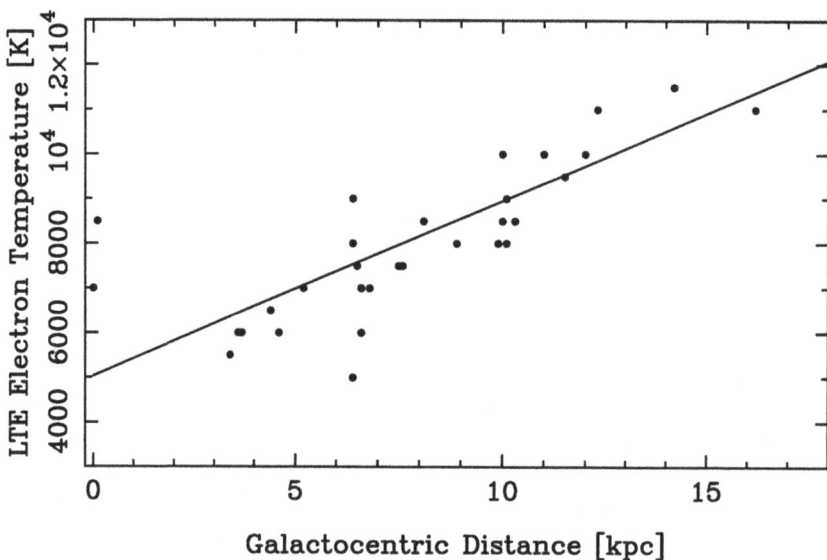

Figure 7. T_e as a function of R_{gal} for our H II region sample. The solid line is a fit to the data (neglecting SgrB2, G1.1, and S298).

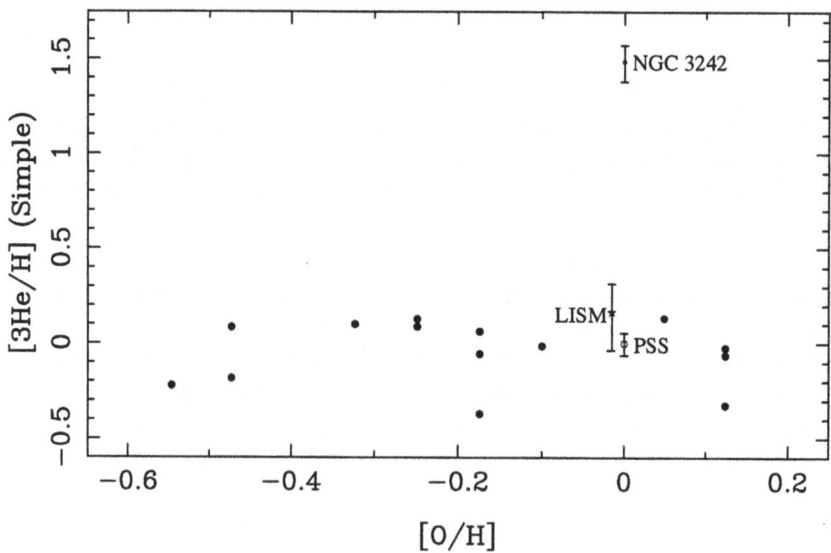

Figure 8. ^{3}He/H abundance for H II regions with small structure (density) and ionization corrections plotted as a function of [O/H] metallicity. The [O/H] abundance is derived from the nebular electron temperature (Shaver *et al.*, 1983). For these sources there appears to be a "^{3}He Plateau."

that neither stellar production or destruction of ^{3}He is very large. Thus, it might be reasonable to interpret the average of our simple source abundances as primordial.

This average is $^3\text{He}/\text{H} = (1.5 \pm 0.5) \times 10^{-5}$. We can only guess at the systematic error. Density and ionization corrections both increase the abundance. For the simple sources we have modeled, the density correction is typically 20%. We have made less progress in modeling the ionization structure. One technique which probably overestimates the effect gives correction factors ranging from 1.05–1.3 for similar sources. Here we crudely recognize the existence of these corrections by doubling the upper error bar giving $^3\text{He}/\text{H} = 1.5^{+1.0}_{-0.5} \times 10^{-5}$. This corresponds to a baryon to photon $\eta = 3.2^{+4.4}_{-1.9} \times 10^{-10}$ based on the BBNS calculations of Kernan, Walker, and Steigman (1994, private communication).

What Can Be Done in the Future? Further ^3He observations and more sophisticated modeling can: (1) confirm stellar production and show to what extent it follows the predictions of stellar nucleosynthesis theory; (2) show local ^3He pollution in a very direct way; (3) lead to improved estimates of systematic error; and (4) determine the global distribution of the ^3He abundance in the Milky Way. The scatter in Figs. 6 and 8 arises from a combination of observational error, the unaccounted for density and ionization corrections, and real abundance differences. As our observations (both in quality and sample size) and modeling improve, we should be able to search for trends in abundance much like others are searching for trends in ^7Li within the Spite plateau.

There are advances in instrumentation which will assist us in our quest: **Green Bank 140 ft Telescope:** We can now make faster, more sensitive observations. As noted above the receiver/spectrometer system is vastly improved over that used for much of our earlier data. E.g., Fig. 5 shows a spectrum of G29.9 obtained over two observing epochs with the new systems. It represents 35.5 hours of telescope time and is comparable to integrations of $\sim$ 165 hours with the old system. **MPIfR 100 m Telescope:** In response to our spectral frequency baseline tests the MPIfR designed, constructed, and recently installed (August 1997) RF spoilers on the feed quadraped of the 100 m telescope. These should reduce non-axial reflections which are particularly problematic. The improvement in baseline structure should enable us to measure ^3He in a larger sample of PNe. **Arecibo Telescope:** The upgraded Arecibo telescope is expected to operate at 3.5 cm. Its high sensitivity and $\sim$ 40 arcsecond expected resolution, which will be a good match to PNe angular sizes, should allow us to expand the PNe ^3He sample. **GBT:** The new off-axis 100 m telescope under construction in Green Bank almost appears to have been designed specifically for this project. Its clear aperture may produce the cleanest, deepest spectral sensitivity in history. We look forward to its completion and hope to get early test time. **VLA:** While the VLA is not well suited to observing ^3He in H II regions, it *is* well suited for sources of small angular size, such as PNe and extragalactic H II regions. Its high-resolution continuum imaging capability is vital for our source modeling.

We can continue to improve the H II region models. As discussed above these require VLA and 100 m continuum data and the high precision recombination line measurements we obtain as a byproduct of the $^3\text{He}^+$ observations. At present only about a quarter of the ^{3}He$^+$ sources have been accurately modeled. We will extend this analysis to the full sample after getting the required VLA and 100 m continuum data. We also plan to refine our models by increasing their physical sophistication.

Indeed there is but one factor which makes us pessimistic—funding. We have no significant funding at this point, and without funding progress will be slow.

Acknowledgements

Having started in 1978, this project is nearing the end of its second decade. The technical support staff and telescope operators at both Green Bank and Effelsberg have contributed greatly to its success. The enthusiasm and support of the international "Light Elements" community has inspired us to continue over these years. Financial support for some of the international travel required for the observations has been supplied by three NATO travel grants. From 1992–1995 the work received support from NSF Grant AST-91-21169.

References

Balser, D. S.: 1994, Ph. D. Dissertation, Boston University.
Balser, D. S., Bania, T. M., Rood, R. T. and Wilson, T. L.: 1995, *ApJS* **100**, 371.
Balser, D. S., Bania, T. M., Brockway, C.J., Rood, R. T. and Wilson, T. L.: 1994, *ApJ* **430**, 667.
Balser, D. S., Bania, T. M., Rood, R. T. and Wilson, T. L.: 1997, *ApJ* **483**, 320.
Balser, D. S., Bania, T. M., Rood, R. T., and Wilson, T. L.: 1998, in preparation.
Bania, T. M., Balser, D. S., Rood, R. T., Wilson, T. L. and Wilson, T. A.: 1997, *ApJ*, in press.
Bania, T. M., Rood, R. T., and Wilson, T. L.: 1987, *ApJ* **323**, 30 [BRW].
Dearborn, D. S. P., Steigman, G., and Tosi, M.: 1996, *ApJ* **465**, 887.
Fields, B. D.: 1996, *ApJ* **456**, 478.
Galli, D., Palla, F., Ferrini, F., and Penco, U.: 1995, *ApJ* **443**, 536.
Galli, D., Stanghellini, L., Tosi, M., Palla, F.: 1997, *ApJ* **477**, 218.
Geiss, J.: 1993, in 'Origin and Evolution of the Elements', eds. N. Prantzos, E. Vangioni-Flam, M. and Cassé (Cambridge University Press, Cambridge), 89.
Gloecker, G. and Geiss, J.: 1996, *Nature* **381**, 210.
Iben, I.: 1967, *ApJ* **147**, 650.
Iben, I. and Truran, J. W.: 1978, *ApJ* **220**, 980.
Olive, K. A., Rood, R. T., Schramm, D. N., Truran, J., and Vangioni-Flam, E.: 1995, *ApJ* **444**, 680.
Prantzos, N.: 1996, *A&A* **310**, 106.
Rood, R. T., Bania, T. M., and Wilson, T. L.: 1984, *ApJ* **280**, 629 [RBW].
Rood, R. T., Bania, T. M., and Wilson, T. L.: 1992, *Nature* **355**, 618.
Rood, R. T., Bania, T. M., Wilson, T. L., and Balser, D. S.: 1995, in 'ESO–EIPC Workshop on the Light Elements', ed. P. Crane (Springer, Heidelberg), 201.
Rood, R. T., Steigman, G., and Tinsley, B. M.: 1976, *ApJ* **207**, L57.
Rood, R. T., Wilson, T. L., and Steigman, G.: 1979, *ApJ* **227**, L97.
Rohlfs, K. and Wilson, T. L.: 1996, 'Tools of Radio Astronomy', 2^{ed}, (Springer, Heidelberg).
Shaver, P. A., McGee, R. X., Newton, L. M., Danks, A. C., and Pottasch, S. R.: 1983, *MNRAS* **204**, 53.
Truran, J. W., and Cameron, A. G. W.: 1971, *ApJS* **14**, 179.
Vangioni-Flam, E. and Cassé, M.: 1995, *ApJ* **441**, 471.
Wink, W. E., Wilson, T. L., and Bieging, J. H.: 1983, *A&A* **127**, 211.

MIXING IN STARS AND THE EVOLUTION OF THE ^{3}HE ABUNDANCE

C. CHARBONNEL

Laboratoire d'Astrophysique de Toulouse, CNRS UMR 5572, 14, av. E.Belin, 31400 Toulouse, France

Space Telescope Science Institute, 3700 San Martin Drive, Baltimore, MD 21218, USA

Abstract. We first recall the observational and theoretical facts that constitute the so-called ^{3}He problem. We then review the chemical anomalies that could be related to the destruction of ^{3}He in red giants stars. We show how a simple consistent mechanism can lead to the destruction of ^{3}He in low mass stars and simultaneously account for the low ^{12}C/^{13}C ratios and low lithium abundances observed in giant stars of different populations. This process should both naturally account for the recent measurements of ^{3}He/H in galactic HII regions and allow for high values of ^{3}He observed in some planetary nebulae. We propose a simple statistical estimation of the fraction of stars that may be affected by this process.

Key words: Nuclear reactions - Nucleosynthesis - Abundances - Stars:Evolution - Interior - Rotation

1. The historical ^{3}He problem

The main features of ^{3}He chemical evolution have long been considered to be simple, dominated by the net production of this light element by low mass stars. In these objects, initial D is processed to ^{3}He during the fully convective pre-main sequence phase. Then, as described by Iben (1967), an ^{3}He peak builds up due to nuclear reactions of the pp-chains on the main sequence, and is engulfed in the envelope of the star during the first dredge-up on the lower red giant branch (RGB). Once in the convective layers of the red giant, ^{3}He is difficult to destroy because of the cool temperature in these regions. So, even if he cautiously noticed that many unknown events could occur to ^{3}He later on, which could affect its abundance, Iben suggested that a large fraction of the ^{3}He in the interstellar medium was perhaps formed in ordinary stars.

It was soon realized that, if this ^{3}He survived the late stages of stellar evolution and was ejected in the ISM, then this source could be sufficient to explain the protosolar abundance. Talbot and Arnett (1973) wrote : "if the low-density big-bang calculations are not relevant, and the initial abundance of D and ^{3}He were zero, then stellar production of ^{3}He is just about the required amount" (... to explain the solar abundance of ^{3}He). Of course, in that case, other difficulties had to be solved, among which the "invention" of a galactic source of D was the most challenging problem.

On the other hand, Reeves *et al.* (1973) showed that primordial nucleosynthesis could be sufficient to account for the protosolar abundance of ^{3}He, provided that enough D is produced. In view of the evolutionary curves of D and ^{3}He obtained with different assumptions for the production/destruction of both elements, Audouze and Tinsley (1974) also had to conclude that ^{3}He production by

Space Science Reviews **84**: 199–206, 1998.

stars was a rather embarrassing process. They suggested then that most of the stellar ^{3}He must be destroyed before the stars lose their envelopes.

Later on, Rood, Steigman and Tinsley (1976) confirmed the large theoretical production of ^{3}He in low mass stars, and raised the question of the actual representativity of the ^{3}He abundance in the solar system compared to the "average" ISM at the time of its formation. They urged for relevant new observations to be done, both in HII regions and in planetary nebulae.

2. The ^{3}He problem in the nineties

Twenty years later, the necessary observations were indeed (at least partly) carried out. We refer to the papers by Geiss, Gloeckler and Rood in this volume, for a complete description of the data, and we simply recall the key numbers :
Pre-solar ^{3}He abundance (Geiss, 1993) :

$$\left(\frac{^3\text{He}}{\text{H}}\right)_\odot = (1.5 \pm 0.3) \times 10^{-5}$$

Abundance of ^{3}He in the local interstellar cloud (Gloeckler and Geiss, 1996):

$$\left(\frac{^3\text{He}}{\text{H}}\right) = 2.2^{+0.7}_{-0.6}(\text{stat}) \pm 0.2(\text{syst}) \times 10^{-5}$$

Abundance of ^{3}He in HII regions (Balser et al., 1994) :

$$\left(\frac{^3\text{He}}{\text{H}}\right) = 1 \text{ to } 5 \times 10^{-5}$$

Abundance of ^{3}He in Planetary Nebulae (Rood et al., 1992; Balser et al., 1997):

$$\left(\frac{^3\text{He}}{\text{H}}\right) \simeq 0.1 - 1.0 \times 10^{-3}$$

The first three sets of observations essentially leave no room for important production ^{3}He in the Galaxy during the last 5 Gyr. On the other hand, the high value of ^{3}He/H reported in a few Planetary Nebulae is however consistent with the idea that some low mass stars do produce this element.

During the last twenty years, computations of standard stellar models with improved microphysics confirmed the production of significant quantities of ^{3}He in low mass stars (Vassiliadis and Wood, 1993; Charbonnel, 1995; Dearborn, Steigman and Tosi, 1996; Weiss, Wagenhuber and Denissenkov, 1996). Meanwhile, independent groups developed different models of chemical evolution. In this volume, Monica Tosi presents the predictions for ^{3}He in a collection of recent chemical evolution models which were computed with various numerical codes, using different approaches and different assumptions on several parameters (see also Tosi, 1996). All these models predict an interstellar medium overenrichment and fail to reproduce both the time evolution of ^{3}He and the current radial distribution of this element. The crucial point is that this common failure is due to the excessively large ^{3}He production in low and intermediate mass stars.

The conventional scenario for chemical evolution of ^{3}He appears thus to be upset, and the actual contribution of low mass stars to the production of this element has to be revised.

3. A nuclear solution?

Galli *et al.* (1994) proposed a very elegant solution to the ^{3}He problem, purely based on nuclear physics. They investigated the effects on the stellar yields of a low-energy resonance in the cross-section of the ^{3}He$+^3$He reaction. The presence of the resonance at energies corresponding to temperatures at which the production of ^{3}He reaches a maximum has important consequences on the internal distribution of the ^{3}He abundance. This leads to a significant reduction of its production in stars of various masses. When using the stellar ^{3}He yields obtained with the models with resonance, the predictions for the Galactic evolution are quite satisfactory when compared to the abundances observed in the solar system and the ISM.

This solution is however in conflict with the measurements of ^{3}He in at least one Planetary Nebulae (Balser *et al.*, 1997). In NGC 3242 indeed, the abundance ^{3}He/H $\simeq 5 - 10 \times 10^{-4}$ is more than one order of magnitude larger than that found in any HII region, the local ISM or the proto-solar system. Such a high abundance, which is in fact in very good agreement with the standard stellar predictions (Galli *et al.*, 1996), rules out the nuclear physics solution.

In order to override the apparent inconsistency, a process has to be found, that destroys ^{3}He in some stars, and preserves it in others.

4. Connection with other chemical anomalies

Rood, Bania and Wilson (1984) suggested that the nonproduction of ^{3}He could be related to other chemical anomalies in red giant stars. During the first dredge-up, the deepening convective envelope of low mass stars reaches only the regions where ^{12}C was processed in favor of ^{13}C and ^{14}N. So basically, the carbon isotopic ratio declines (from 90 to about 20-30), the carbon abundance drops (by about 30%) and nitrogen increases (by about 80%), but oxygen and all other element abundances remain unchanged at the surface of the red giant. Still according to the standard scenario, the surface abundances then stay unaltered as the convective envelope slowly withdraws during the end of the RGB evolution.

However, observational data on the abundance variations of C, N, O, Na, Al in evolved stars reveal a different reality.

- Pop II field and globular cluster giants present $^{12}C/^{13}C$ ratios lower than 10, even down to the near-equilibrium value of 4 in many cases (Sneden *et al.*, 1986; Smith and Suntzeff, 1989; Brown and Wallerstein, 1989; 1992; Bell *et al.*, 1990; Suntzeff and Smith, 1991; Shetrone *et al.*, 1993; Briley *et al.*, 1994; 1997).
- In evolved halo stars, the lithium abundance continues to decrease after the completion of the first dredge-up (Pilachowski *et al.*, 1993).
- A continuous decline in carbon abundance with increasing stellar luminosity along the RGB is observed in globular clusters such as M92 (Carbon *et al.*, 1982; Langer *et al.*, 1986), M3 and M13 (Suntzeff, 1981), M15 (Trefzger *et al.*, 1983), NGC 6397 (Bell *et al.*, 1979; Briley *et al.*, 1990), NGC 6752 and M4 (Suntzeff and Smith, 1991)
- In some globular clusters (M92, Pilachowski, 1988; M15, Sneden *et al.*, 1991; M13, Brown *et al.*, 1991; Kraft *et al.*, 1992; Omega Cen, Paltoglou and Norris, 1989), giants exhibit evidence for O→N processed material.
- In addition to the O versus N anticorrelation, the existence of Na and Al vs N correlations and Na and Al vs O anticorrelations in a large number of globular cluster red giants has been clearly confirmed (Drake *et al.*, 1992; Kraft *et al.*, 1992; 1993; Norris and Da Costa, 1995; Shetrone, 1996).

These observations suggest that, while they evolve on the RGB, low mass stars undergo an extra-mixing in the region situated between the hydrogen burning shell (where the material is processed through the CN-cycle and possibly the ON-cycle) and the deep convective envelope. This extra-mixing adds to the standard first dredge-up to modify the surface abundances. It may well lead to the destruction of ^3He by a large factor in the bulk of the envelope material.

5. When does the extra-mixing occur?

Before speculating on the nature of the extra-mixing which may be responsible for the chemical anomalies described above, one has to determine the evolutionary phase at which this process becomes effective.

Observations of $^{12}C/^{13}C$ ratios in M67 subgiant stars (Gilroy and Brown, 1991) show that the theoretical and the observational first dredge-up are in perfect agreement (Charbonnel 1994). This indicates that the extra-mixing we are looking for is not occuring during the main sequence. Let us insist on this crucial point. Indeed, the idea comes regularly in the literature that mixing in the core of main sequence stars could cut off the production of ^3He, in relation with a possible solution to the solar neutrino problem. In fact, mixing the core of a main sequence star may change both the ^3He and the ^{13}C peak. But with at least two appalling consequences. First, the nice agreement seen for the dredge-up in the less evolved subgiants of M67 would be lost. In addition, as also discussed by Sylvie Vauclair in this volume, mixing in the central core of a solar model is ruled out by helioseismic data (see Richard et al., 1996; Richard and Vauclair, 1997).

In fact, the observations of the $^{12}C/^{13}C$ ratios in M67 strongly suggested that the extra-mixing process is only efficient when the hydrogen burning shell has crossed the discontinuity in molecular weight built by the convective envelope during the first dredge-up (Charbonnel, 1994). This result was recently confirmed in the case of evolved stars with moderate metal deficiencies (Charbonnel, Brown and Wallerstein, 1997). Before this evolutionary point (which corresponds to the so-called "bump" in the RGB luminosity function), the mean molecular weight gradient probably acts as a barrier to the mixing in the radiative zone. Above this point, no gradient of molecular weight exists anymore above the hydrogen burning shell, and extra-mixing is free to act. Importantly, observations of $^{12}C/^{13}C$ in galactic clusters (Gilroy, 1989) clearly indicate that this extra-mixing only occurs in stars with masses lower than about $2M_\odot$. More massive stars ignite He in their central core before reaching the bump.

6. Rotation-induced mixing on the red giant branch

Recently, different groups have simulated extra-mixing between the base of the convective envelope and the hydrogen burning shell in order to reproduce the CNO abundances in RGB stars. Denissenkov and Weiss (1995) modeled this deep mixing by adjusting both the mixing depth and rate in their diffusion procedure. Wasserburg, Boothroyd and Sackman (1995) and Boothroyd and Sackman (1997) used an ad-hoc "conveyor-belt" circulation model, where the depth of the extra-mixed region is related to a parametrized temperature difference up to the bottom of the hydrogen-burning shell.

On the other hand, other authors attempted to relate the extra-mixing with physical processes, among which rotation seems to be the most promising. Sweigart and Mengel (1979) suggested that meridional circulation on the RGB could lead to the low $^{12}C/^{13}C$ observed in field giants. Charbonnel (1995) investigated the influence of such a process on the RGB by taking into account the most recent progress in the description of the transport of chemicals and angular momentum in stellar interiors : Zahn's (1992) consistent theory which describes the interaction between meridional circulation and turbulence induced by rotation. In this framework, the global effect of advection moderated by horizontal turbulence can be treated as a diffusion process. Four important points must be emphasized : 1. The resulting mixing of chemicals in stellar radiative regions is mainly determined by the loss of angular momentum via a stellar wind. 2. Even in the absence of such mass loss, some mixing can take place wherever the rotation profile presents steep vertical gradients. 3. Additional mixing is expected near nuclear burning shells. 4. Due to the stabilizing effect of the composition gradients, the mixing will be efficient on the RGB only when the hydrogen-burning-shell will have crossed the chemical discontinuity created by the convective envelope during the first dredge-up. Since all these conditions are expected to be fulfilled during the non-homologous evolution of low mass stars on the RGB, we suggested that this process could be responsible for the extra-mixing we are looking for. We showed that indeed such a description for the rotation-induced mixing can account for the observed behavior of carbon isotopic ratios and for the Li abundances in Population II low mass giants. Simultaneously, when this extra-mixing begins to act, ^{3}He is rapidly transported down to the regions where it burns by the ^{3}He$(\alpha, \gamma)^7$Be reaction. This leads to a decrease of the surface value of ^{3}He/H, confirming the predictions by Hogan (1995).

The description for extra-mixing described above provides an attractive solution of several discrepancies between the predictions of "classical" stellar models and the observed abundances, not only on the red giant branch, but for stars of various spectral types in different evolutionary states. For example, it was used to sucessfully reproduce the slight over (under) abundances of C (N) observed in B-type stars (Talon *et al.*, 1997) and the Li dip in galactic cluster F-stars (Talon and Charbonnel, 1997). For what concerns the red giant stars, a self-consistent analysis of the effect of the deep mixing on the energy production in the hydrogen shell and on the horizontal branch morphology (see Sweigart, 1997) is still lacking. Work is in progress in this direction (Charbonnel and Talon, 1997).

In any case, all these "non-standard" models, which consider the effect of an extra-mixing process (related or not to a physical process) on the red giant branch, lead to the same conclusion : The mechanism which is responsible for chemical anomalies on the RGB must also affect the ^{3}He abundance. The important question now is : What fraction of stars are affected by this mechanism?

7. Statistical significance of the Planetary Nebulae sample

As already discussed, a solution of the inconsistency between the galactic constraints and the high ^{3}He abundance in Planetary Nebulae may be related to a process that preserves ^{3}He in some stars, while destructing it in others. In the case of the rotation-induced mixing described previously, stars with different rotation and mass loss histories should actually suffer different mixing efficiency and thus display different chemical anomalies.

We suggest that a crude estimation of the percentage of stars which destroy their ^{3}He can be done thanks to the carbon isotopic ratio data. The observations in stars of different populations show that 80 to 90% of the giants for which data exist show low ^{12}C/^{13}C ratios. One can say that the same percentage of red giants destroy efficiently the ^{3}He they produced while on the main sequence (see Dias and Charbonnel, 1997 for a complete statistical study). This value is in perfect agreement with the one needed by Galli *et al.* (1996) to fit the galactic constraints.

In this framework, one has to raise the question of the statistical significance of Balser's *et al.* sample. As discussed by the authors, this source sample is indeed highly biased, due to selection criteria (see also Robert Rood's paper in this volume). These Planetary Nebulae should then belong to the 10 to 20% of stars which do not suffer from extra-mixing on the red giant branch. They should also show "normal" carbon isotopic ratios. This crucial test has already been verified for one Planetary Nebulae of Balser's *et al.* sample : NGC 6720 shows a ^{12}C/^{13}C ratio of 23 (Bachiller *et al.*, 1997), in perfect agreement with the "standard" predictions. Further observations are on their way.

References

Audouze, J., Tinsley, B.M.: 1974, *ApJ* **192**, 487.
Bachiller R., Forveille, T., Huggins P.J., Cox, P.: 1997, *A&A*, preprint.
Balser, D.S., Bania, T.M., Brockway, C.J., Rood, R.T., Wilson T.L.: 1994, *ApJ* **430**, 667.
Balser, D.S., Bania, T.M., Rood, R.T., Wilson T.L.: 1997, *ApJ*, in press.
Bell, R.A., Briley, M.M., Smith, G.H.: 1990, *AJ* **100**, 187.
Bell, R.A., Dickens, R.J., Gustafsson B.: 1979, *ApJ* **229**, 604.
Boothroyd, A., Sackman, I.: 1997, preprint.
Briley, M.M., Bell, R.A., Hoban, S., Dickens, R.J.: 1990, *ApJ* **359**, 307.
Briley, M.M., Smith, V.V., King, J., Lambert, D.L.: 1997, *AJ* **113**, 1.
Briley, M.M., Smith, V.V., Lambert, D.L.: 1994: *ApJLetters* **429**, 119.
Brown, J.A., Wallerstein, G.: 1989, *AJ* **98**, 1643.
Brown, J.A., Wallerstein, G.: 1992, *AJ* **104**, 1818.
Brown, J.A., Wallerstein, G., Oke J.B.: 1991, *AJ* **101**, 1693.
Carbon, D.F., Langer, G.E., Butler, D., Kraft, R.P., Trefzger, C.F., Suntzeff, N.B., Kemper, E., Romanishin, W.: 1982, *ApJS* **49**, 207.
Charbonnel, C.: 1994, *A&A* **282**, 811.
Charbonnel, C.: 1995, *ApJLetters* **453**, 41–44.
Charbonnel, C., Brown, J.A., Wallerstein, G.: 1997, *A&A*, submitted.
Charbonnel, C., Talon, S.: 1997, *A&A*, in preparation.
Drake, J.J., Smith, V.V., Suntzeff, N.B.: 1992, *ApJLetters* **395**, 95.
Dearborn, D.S., Steigman, G., Tosi, M.: 1996, *ApJ* **465**, 887–897.
Denissenkov, P. A., Weiss, A.: 1995, *ApJ* **308**, 773.
Dias, J., Charbonnel, C.: 1997, in preparation.
Galli, D., Palla, F., Straniero O., Ferrini F.: 1994, *ApJLetters* **432**, 101–104.
Galli, D., Stanghellini, L., Tosi, M., Palla, F.: 1997, *ApJ* **477**, 218.
Geiss, J.: 1993, in *Origin and Evolution of the Elements*, ed. Prantzos, Vangioni-Flam and Cassé (Cambridge Univ. Press), 89.
Gilroy, K. K.: 1989, *ApJ* **347**, 835.
Gilroy, K. K., Brown, J.A.: 1991, *ApJ* **371**, 578.
Gloeckler, G., Geiss, J.: 1996, *Nature* **381**, 210–212.
Hogan, C.J.: 1995, *ApJLetters* **441**, 17–20.
Iben, I.: 1967, *ApJ* **143**, 642.
Kraft, R.P., Sneden, C., Langer, G.E., Prosser, C.F.: 1992, *AJ* **104**, 645.
Kraft, R.P., Sneden, C., Langer, G.E., Shetrone, M.D.: 1993, *AJ* **106**, 1490.
Langer, G.E., Kraft, R.P., Friel, E., Oke, J.B.: 1986, *PASP* **97**, 373.
Norris, J.E., Da Costa, G.S.: 1995, *ApJLetters* **441**, 81.
Pilachowski, C.A.: 1988, *ApJLetters* **326**, 57.
Pilachowski, C.A., Sneden, C., Booth, J.: 1993, *ApJ* **407**, 713.
Paltoglou, G., Norris, J.: 1989, *ApJ* **336**, 185.

Reeves, H., Audouze, J., Fowler, W.A., Schramm, D.N.: 1973, *ApJ* **179**, 909–930.
Richard, O., Vauclair, S.: 1997, *A&A*
Richard, O., Vauclair, S., Charbonnel, C., Dziembowski W.A.: 1996, *A&A* **312**, 1000–1011.
Rood, R.T., Bania, T.M., Wilson, T.L., 1984, *ApJ* **280**, 629–647.
Rood, R.T., Bania, T.M., Wilson, T.L., 1992, *Nature* **355**, 618.
Rood, R.T., Steigman, G., Tinsley, B.M.: 1976, *ApJLetters* **207**, 57–60.
Shetrone, M.D.:1996, *AJ*
Shetrone, M.D., Sneden, C., Pilachowski, C.A., 1993, *PASP* **195**, 337.
Smith, V.V., Suntzeff, N.B.: 1989, *AJ* **97**, 1699.
Sneden, C., Kraft, R.P., Prosser, C.F., Langer, G.E.: 1991, *AJ* **102**, 2001.
Sneden, C., Pilachowski, C.A., VandenBerg, D.A.: 1986, *ApJ* **311**, 826.
Suntzeff, N.B., Smith, V.V.: 1991 *ApJ* **381**, 160.
Suntzeff, N.B.: 1981, *ApJS* **47**, 1.
Sweigart, A. V.: 1997, *ApJ* **474**, L23.
Sweigart, A. V., Mengel, J. G.: 1979, *ApJ* **229**, 624.
Talbot, R.J., Arnett W.D.:1973, *ApJ* **186**, 51–67.
Talon, S., Charbonnel, C.: 1997, *A&A*, submitted.
Talon, S., Zahn, J.P., Maeder, A., Meynet, G.:1997, *A&A* **322**, 209.
Trefzger, C.F., Carbon, D., Langer, G.E., Suntzeff, N.B., Kraft, R.P.: 1983, *ApJ* **266**, 144.
Tosi, M.: 1996, in *From stars to galaxies : The impact of stellar physics on galaxy evolution*, eds.
    Leitherer, von Alvensleben, Huchra, 299–310.
Vassiliadis, E., Wood, P.R.: 1993, *ApJ* **413**, 641–657.
Wasserburg, G. J., Boothroyd, A. I., Sackman, I.-J.: 1995, *ApJLetters* **447**, 37.
Weiss, A., Wagenhuber, J., Denissenkov, P.A.: 1996, *A&A* **313**, 581.
Zahn, J.P.: 1992, *A&A* **265**, 115.

Address for correspondence: Corinne.Charbonnel@obs-mip.fr

GALACTIC EVOLUTION OF D AND ^{3}HE

MONICA TOSI

Osservatorio Astronomico di Bologna, Via Zamboni 33, I-40126 Bologna, Italy

Abstract. The most recent chemical evolution models for D and ^{3}He are reviewed and their results compared with the available data.

Models in agreement with the major galactic observational constraints predict deuterium depletion from the Big Bang to the present epoch smaller than a factor of 3 and therefore do not allow for D/H primordial abundances larger than $\sim 5 \times 10^{-5}$. Models predicting higher D consumption do not seem to be able to reproduce other observed features of our galaxy (e.g. SFR, abundances, abundance ratios and/or gradients of heavier elements, metallicity distribution of G-dwarfs).

Observational and theoretical ^{3}He abundances can be reconciled with each other if the majority of low mass stars experience in the red giant phase a deep mixing allowing the consumption of most of the ^{3}He produced during core-hydrogen burning.

Key words: Galaxy: evolution, nucleosynthesis, abundances

1. Introduction

It has been amply described in this meeting that deuterium and ^{3}He are measurable only in a few types of objects (see e.g. the contributions by Geiss and Gloeckler, 1998, Linsky, 1998, Rood *et al.*, 1998, Tytler, 1998, and Vidal-Madjar *et al.*, 1998, in this volume). D abundances are derived from observations of the solar system, of the local interstellar medium (ISM) and of a few high-redshift absorbers on the line of sight of distant quasars. ^{3}He is measured in the solar system, in the ISM (not only local but at all galactocentric distances) and in a few Planetary Nebulae (PNe). All these objects, with the only exception of high-redshift clouds, are relatively young (the oldest being the sun with an age of 4.5 Gyr) and have therefore formed from an ISM whose chemical composition had inevitably been modified by the previous stellar generations. To infer the primordial abundances of the light elements from those derived from these measurements, it is thus necessary to take into account the effects of the galactic chemical evolution. This is done by means of theoretical chemical evolution models which consider the various cycles of gas astration and gas return, and the variations of the chemical composition in the ISM due to stellar nucleosynthesis and gas accretion taking place up to the time when the observed objects have formed.

Several parameters are involved in the computation of a chemical evolution model, the most important ones being the star formation rate (SFR), the initial mass function (IMF), the gas flows in and/or out of the galactic region being studied. To avoid misleading conclusions, it is therefore of primary importance to always compare the model predictions with all the available observational constraints. This has not yet led to a unique solution to the galactic evolution problem, but has at least reduced the range of possible values for the adjustable parameters. In the

Space Science Reviews **84**: 207–218, 1998.

following, we will examine the predictions for D and ^{3}He of chemical evolution models taking into account all the major constraints.

2. D Evolution

The galactic evolution of D is relatively straightforward. Since any D which enters a star is immediately burnt into ^{3}He already during the pre-main sequence phase (Reeves *et al.*, 1973), the gas returned to the ISM by the stars is always D-free. For this reason, the only fraction of ISM gas still containing some D, at any epoch and galactic location, is the fraction which has not been through stars yet. This virgin fraction clearly contains the primordial D abundance. This implies that the D survival factor $X_2(t)/X_{2,p}$ (where $X_{2,p}$ and $X_2(t)$ represent the primordial deuterium abundance and the value at time t, respectively) is equal to the fraction of gas which has never been through stars.* As a consequence:

a) $X_2(t)/X_{2,p}$ is independent of $X_{2,p}$, and

b) $X_2(t)/X_{2,p}$ depends mostly on the SFR and the gas flows which have affected the examined region up to time t.

After the pioneering study by Audouze and Tinsley (1974), many models have been computed to predict the galactic evolution of deuterium. In the following I will separate these models into two groups:

i) those that were originally computed with the aim of reproducing the major observed features of our Galaxy to better understand its evolution, with no special assumptions for the evolution of the light elements; and

ii) those that were computed to study in detail the light element evolution, with no special attention to the whole set of other galactic constraints.

I will refer to group i) as the *standard* models and to group ii) as *non-standard* models.

2.1. STANDARD MODELS

Figure 1 shows the evolution of deuterium predicted by those standard models which are currently in better agreement with the major galactic constraints. These models have been computed by different authors (Tosi, 1988; Ferrini *et al.*, 1994; Chiappini *et al.*, 1997; Prantzos, 1996), with different assumptions on the SFR, IMF and gas infall, and with different numerical codes. Despite their differences (see Tosi, 1996, for a comparison of their assumptions and results) their predictions are consistent with the observed current value and radial distribution of the SFR, of the gas density, of the gas/total mass ratio, and of the element abundances. They reproduce the empirical age-metallicity relation in the solar neighbourhood (and also at different galactocentric distances, although such data are poor and sparse), and the metallicity distribution derived from observations of local G-dwarfs.

---

* Notice that X_2 stands for the D mass fraction and D/H for its number ratio, the two quantities being related by $X_2 = 2 \times X \times D/H$, where X is the hydrogen mass fraction.

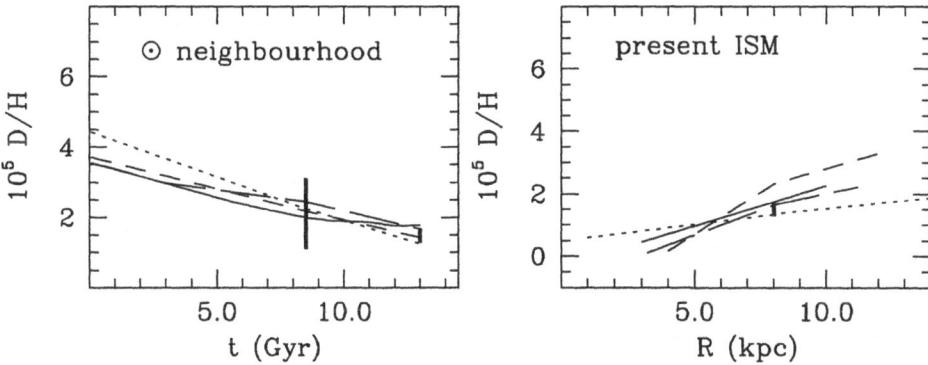

Figure 1. Left hand panel: Time behaviour of deuterium as predicted in the solar neighbourhood by the standard models mentioned in section 2.1 when applied to the light elements (short-dashed line: Galli *et al.*, 1995, and Ferrini and Mollà, 1996; solid line: Dearborn *et al.*, 1996 (hereinafter DST96); dotted line: Prantzos, 1996; long-dashed line: Matteucci and Chiappini, 1996). The vertical bars show the 2-σ ranges of values derived from local ISM and solar system observations (Linsky, 1998; Geiss and Gloeckler, 1998). Right hand panel: Corresponding radial distribution at the present epoch; the vertical bar at 8 kpc represents the local ISM value as in the left hand panel.

When applied to deuterium, these standard models predict time behaviors in the solar neighbourhood (left hand panel in Figure 1) in striking agreement with each other and with the D abundances observationally attributed to the local region at the present epoch and 4.5 Gyr ago. The two vertical bars represent the 2-σ ranges of values derived respectively from Linsky (1998; see also Linsky *et al.*, 1993) and Geiss and Gloeckler (1998; see also Geiss, 1993), assuming that the local ISM and the pre-solar nebula are representative of the average conditions in the solar neighbourhood at these two epochs. In fitting these data all the standard models examined (and see also Copi *et al.*, 1995; Pilyugin and Edmunds, 1996) predict that a fraction between 1/3 and 2/3 of the primordial D is still present in the local ISM. Notice that even if one assumes as representative of the local ISM the extreme case with maximum $D/H = 4 \times 10^{-5}$ derived by Vidal-Madjar *et al.* (1998) in the gas toward one of their target stars, the primordial D/H resulting from these models cannot exceed 12×10^{-5}. We recall, however, that standard chemical evolution models predict the average conditions of the studied regions at any epoch; to predict any fluctuation from the mean trend one should consider much more complicated models including stellar dynamics and gas hydrodynamics, which are still beyond the capabilities of standard computers.

The right hand panel of Fig. 1 shows the D distribution as a function of galactocentric distance predicted by the same models at the present epoch. The vertical bar at 8 kpc corresponds again to the 2-σ range of values from Linsky's data in the local ISM. There are no reliable data available for the D abundances at different galactic radii, but such data should hopefully arrive from the FUSE satellite in a couple of years. All the models predict similar radial gradients at the present epoch, with positive slope as obvious consequence of the fact that the inner galactic regions have and/or have had more SFR (and therefore higher D consumption) than the outer ones.

From the predictions of those standard models in better agreement with the largest set of observational constraints one can derive an upper limit of 3 and a lower limit of 1.3 to the D destruction factor during the Galaxy lifetime. These limits imply a primordial deuterium in the range $2 \lesssim 10^5 (D/H)_p \lesssim 5$ which, according to the standard theory for Big Bang Nucleosynthesis, corresponds to a baryon/photon ratio $\eta \simeq (4-7) \times 10^{-10}$.

2.2. NON-STANDARD MODELS

The results described above refer to models computed with the goal of satisfying as far as possible all the observed galactic features. Alternative models have however been computed with the explicit purpose of testing the possibility of higher D destruction (e.g. Vangioni-Flam et al., 1994; Scully and Olive 1995; Scully et al., 1997). As first pointed out by Vangioni-Flam and Audouze (1988), the major drawback of these non standard models is their tendency to overestimate the metallicity, due to the fact that the higher SFR (often coupled with IMFs skewed in favour of massive stars) required to destroy D also implies a higher production of heavy elements.

To verify whether or not a self-consistent scenario can be found for galactic chemical evolution with high D destruction, C.Chiappini, F.Matteucci, G.Steigman and myself have computed a large number of models with varying assumptions on the parameters that most directly affect the D evolution, namely SFR, IMF, gas accretion and gas loss (Tosi et al., 1997). The predictions of these non-standard models have then been compared with the corresponding observational constraints to check on their consistency.

Let's first examine the effect on D evolution of the assumption for its abundance in the infalling gas. The disk of the Galaxy has accreted gas from outside since the earliest epochs, the two most probable origins for such gas being the halo collapse and the intergalactic medium (e.g. the Magellanic Stream made of gas stripped by the Galaxy from the two Magellanic Clouds). The direct evidence for the current accretion comes mainly from observations of Very High Velocity Clouds, which show almost exclusively negative velocities (i.e. motions toward the disk). These clouds, when seen on the line of sight of bright background objects (e.g. DeBoer and Savage, 1983; West et al., 1985) reveal an overall metallicity $Z \lesssim 0.2 Z_\odot$.

The best standard model by DST96 assumes in fact an infall metallicity $Z_{inf} = 0.2 Z_\odot$ and a D mass fraction in the infalling gas $X_{2,inf} = 0.8 X_{2,p}$, which is the D value resulting at that Z from reasonable models. The D evolution in the solar neighbourhood for this model is shown by the solid line in the left hand panel of Fig. 2.

The predictions of the same model for the present oxygen abundances at different galactocentric distances are shown (solid curve) in the right hand panel of Fig. 2.

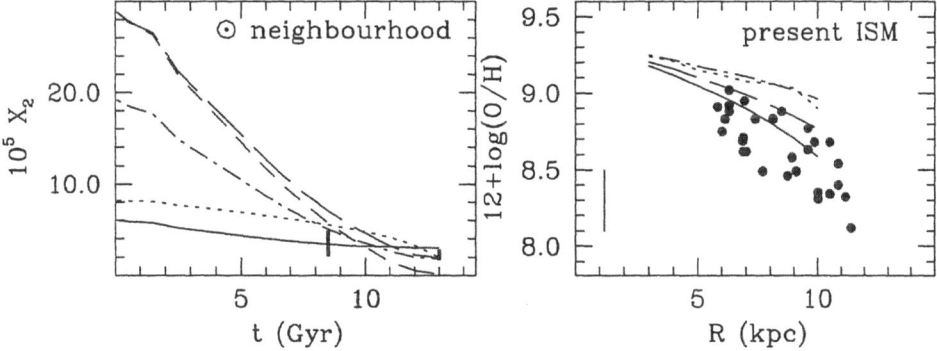

Figure 2. Left hand panel: Time behaviour of deuterium as predicted in the solar neighbourhood by DST96's best model with varying assumptions on the D abundance in the infalling gas (dotted line: no infall after disk formation; solid line: $X_{2,inf} = 0.8X_{2,p}$ as in DST96; dash-dotted line: $X_{2,inf} = 0.2X_{2,p}$; long-dashed line: $X_{2,inf} = 0.1X_{2,p}$; short-dashed line: $X_{2,inf} = 0$). The vertical bars are as in Fig. 1 Right hand panel: Oxygen radial distribution predicted at the present epoch by DST96's model, with varying assumptions on the infall metallicity (dotted line: no infall after disk formation; solid line: $Z_{inf} = 0.2Z_\odot$ as in DST96; long-dashed line: $Z_{inf} = 0.5Z_\odot$, dash-dotted line: $Z_{inf} = Z_\odot$). The dots represent the values derived by Shaver *et al.* (1983) from HII region observations, whose average uncertainty is shown by the vertical line.

In this diagram the dots represent the values derived by Shaver *et al.* (1983) from observations of HII regions. The solid line fits very well the observed distribution.

The other curves in the left hand panel of Fig. 2 show the effect on the same model of adopting a different D abundance in the infalling gas. If the accreted gas is D-free (short-dashed line), the D destruction is so strong that starting from a high primordial D, the model ends up at the present epoch with almost no D left in the local ISM. However, if the infall gas contains as little D as $0.1X_{2,p}$ (long-dashed line) the current abundance is reproduced, but the solar abundance is overestimated. The maximum primordial abundance allowed to fit both the D data is $X_{2,p} \leq 20 \times 10^{-5}$ and is achieved if $X_{2,inf} = 0.2X_{2,p}$ (dash-dotted line). Such choice however implies that the infall metallicity should be at least solar, since the stellar processes needed to destroy D to 1/5 of its primordial value, in the absence of galactic winds, inevitably produce a large amount of heavy elements. Unfortunately, a solar infall metallicity leads at the present epoch to a predicted oxygen abundance (dash-dotted line in the right hand panel) higher and with a flatter gradient than observed in HII regions. Besides, it is hard to imagine a galactic or extragalactic location where such high metallicity could have been produced, since both the halo and the closest galaxies don't reach solar metallicities.

Also assuming that the disk has been completely formed 13 Gyr ago and has not accreted any gas ever since, one can fit the D data with slightly higher deuterium destruction than in the standard models. In this case (dotted line in Fig. 2) the maximum primordial value compatible with the observed abundances in the pre-solar nebula and in the local ISM is $X_{2,p} = 8 \times 10^{-5}$. Models with no infall are

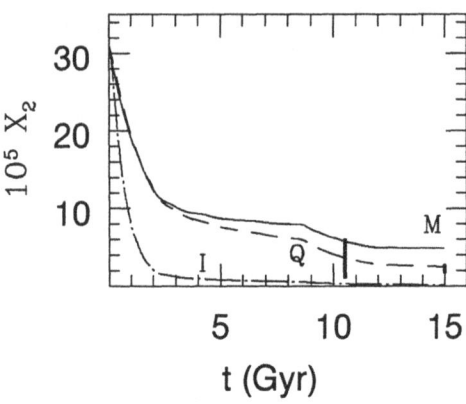

Figure 3. Evolution in the solar ring of the deuterium abundance as predicted by Tosi *et al.* (1997) halo-disk models with galactic winds. The vertical bars are as in Fig. 1. The abscissa allows for 2 Gyr of halo evolution alone plus 13 Gyr for disk evolution.

however known to be inconsistent with many of the galactic constraints (first of all the metallicity distribution of G-dwarfs in the solar neighbourhood). Also the resulting oxygen abundances do not fit the HII region data, as shown in the right hand panel of Fig. 2.

It is thus evident that one cannot reconcile high D destruction with the other galactic features by simply changing its abundance in the accreted gas.

Alternatively, we have considered higher SFRs. It has been shown (e.g. Twarog, 1980; Scalo, 1986; Prantzos, 1998) that the past SFR in the solar neighbourhood cannot have been more than a few times its current value, and this does not allow for high D destruction, as demonstrated by the standard models discussed in section 2.1. Hence, the much higher SFR required to deplete D significantly must have occurred in the halo. We (Tosi *et al.*, 1997) have therefore computed models for both the disk and the halo, of the kinds described by Matteucci and François (1989) and by Chiappini *et al.* (1997).

Since high SFR implies high metal production, but halo and old disk stars don't show high metallicities, a mechanism must be invoked to remove the excess of metals. This mechanism is presumably related to the possible galactic winds triggered by the many simultaneous explosions of supernovae occurring a few Myr after the peak of star formation activity (e.g. Scully *et al.*, 1997; Fields *et al.*, 1998). Since the wind may affect differently the ISM enrichment of elements produced by stars in different mass ranges, it is of crucial importance to test the predictions of such models with the observed element ratios.

The dashed curve in Fig. 3 shows the best solution for D evolution with high depletion of our many cases. It refers to a model *à la* Matteucci and François, with the halo phase occurring in the first 2 Gyr and the disk phase in the remaining 13 Gyr; it assumes an early SFR (in the halo) 20 times higher than at the present epoch (in the disk) and galactic winds expelling preferentially the elements synthetized

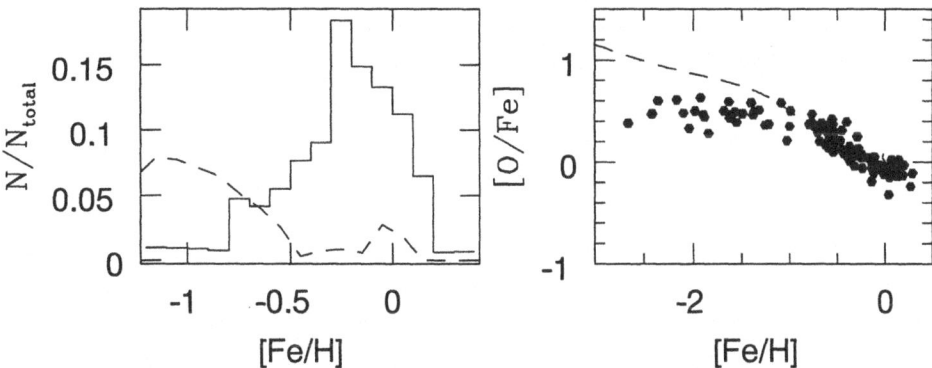

Figure 4. Left hand panel: G-dwarf metallicity distribution in the solar vicinity predicted by Tosi *et al.* (1997) model Q. The histogram refers to the data collected by Rocha-Pinto and Maciel. Right hand panel: oxygen abundance ratio as a function of the iron abundance predicted by model Q. The data are taken from the compilation by Gratton *et al.* (1997).

by SNeII. Fig. 4 shows the predictions of the same model (dashed curve) for the metallicity distribution of G-dwarfs in the solar neighbourhood (left hand panel) and for the oxygen/iron ratio as a function of the iron abundance (right hand panel). In both cases, the model fails to reproduce the observed distributions, because, on the one hand, it predicts too many metal poor G-dwarfs and too few metal rich ones and, on the other hand, it overpredicts oxygen with respect to iron in halo stars. Both inconsistencies are the consequence of the high initial SFR. Adopting an early IMF without low mass stars, one could obviously solve the G-dwarf inconsistency by simply not allowing them to form when the metallicity is low. This however inevitably implies either to assume a SFR much lower than that derived from observations, or to severely overproduce the oxygen abundance of old stars.

Despite the large number of models we tested (with and without winds, with and without varying IMF, see Tosi *et al.*, 1997, for details), in no other case have we been able to find predictions consistent with the data for models with high D destruction. Either they are inconsistent with the abundance data, or with the observational ranges allowed for the SFR or the gas or the total mass of the Galaxy. The only models in agreement with the observational constraints have been those similar to the standard ones, with lower SFR and therefore D depletion factors smaller than 3.

3. ^{3}He Evolution

The stellar nucleosynthesis of ^{3}He was first described 30 years ago (Iben, 1967), finding that this element is produced by all stars during core hydrogen burning; it is subsequently burnt into heavier elements in massive stars but preserved and then ejected into the ISM by low mass stars. This *standard* view for the amount

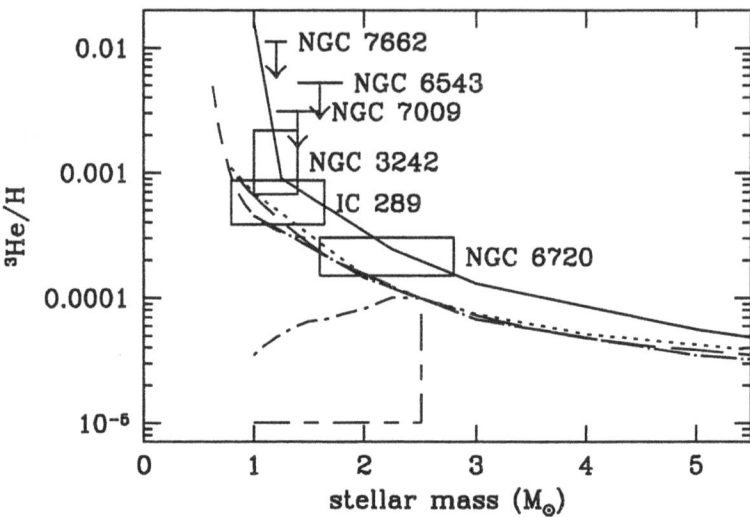

Figure 5. ³He ejected by stars into the ISM as a function of the stellar initial mass. The nucleosynthetic predictions are from: Iben (1967, solid line), Rood *et al.* (1976, short-dashed line), DST96 (dotted line), Boothroyd (1996, long-dashed line for standard prescriptions, dash-dotted line for deep-mixing prescriptions), Hogan (1995, short-dash-long-dashed line). The three boxes and the three arrows show the position derived by GSTP97 for the PNe observed by Rood *et al.* (1995).

of ³He ejected by stars of different initial mass is shown in Fig. 5, where the solid curve represents Iben's predictions and the other curves, roughly parallel to it, other *standard* predictions, all consistent with the idea that low mass stars are net producers and high mass stars are net destroyers of this element. In the last few years, however, several authors (see Charbonnel, 1995; 1998) have independently suggested that a mixing deeper than previously thought may occur during the red giant phase of low mass stars and bring the ³He previously synthetized into regions where it is significantly destroyed. In these conditions, the amount of ³He that low mass stars can eject in the ISM is significantly lower, for instance as represented in Fig. 5 by the dash-dotted curve (Boothroyd, 1996) and the short-dash-long-dashed curve (Hogan, 1995).

The position in Fig. 5 of the six PNe observed by Rood *et al.* (1995) (Galli *et al.*, 1997, hereinafter GSTP97) definitely supports the ³He standard predictions. However, their abundances are almost two orders of magnitude higher than those measured in the ISM (Rood *et al.*, 1995; Gloeckler and Geiss, 1996) and this makes it quite difficult to reconcile all the data with each other (see also Rood *et al.*, 1998). In fact, none of the chemical evolution models computed so far with standard nucleosynthesis prescriptions is able to reproduce the observed ISM ³He abundances. The left hand panel of Fig. 6 shows the time behaviour in the solar neighbourhood predicted by the best standard models described in Section 2.1 (same symbols as in Fig. 1. The lower solid line represents the only model with non-standard nucleosynthetic prescriptions). All the models with standard

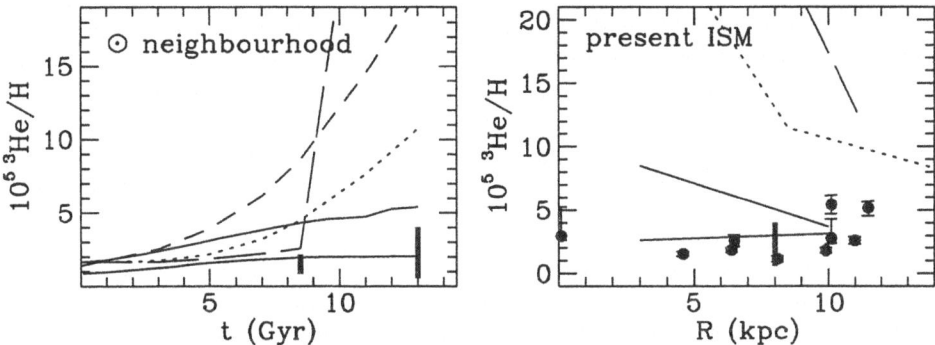

Figure 6. Left hand panel: Time behaviour of ^{3}He as predicted in the solar neighbourhood by the best standard models mentioned in section 2.1. Symbols are as in Fig. 1. All the models assume the standard nucleosynthesis of ^{3}He, except the model represented by the lower solid line (see text and GSTP97 for details). The vertical bars show the 2-σ ranges of values derived from local ISM and solar system observations (Gloeckler and Geiss, 1996, and Geiss, 1993, respectively). Right hand panel: Corresponding radial distribution at the present epoch. The data points corresponds to the HII region values derived by Rood *et al.* (1995) and the vertical bar at 8 kpc to the local ISM value, as in the left hand panel.

nucleosynthesis overpredict the ^{3}He abundance already by the time the sun formed, and by even more at the present epoch, thus being completely inconsistent with the data. The same results are obtained, independently of the galactic model, by any other author adopting standard stellar models (see e.g. Vangioni-Flam *et al.*, 1994; Galli *et al.*, 1995; Palla *et al.*, 1995; Scully *et al.*, 1996).

The right hand panel of Fig. 6 shows the predictions of the same models for the ^{3}He radial distribution at the present epoch. All the models with standard nucleosynthesis predict too much ^{3}He at all galactocentric distances and a steep negative gradient. A negative gradient is typical of all the heavy elements and is the natural consequence of the fact that the inner galactic regions have experienced more star formation activity (and hence more chemical enrichment) than the outer ones. Contrary to what happens for heavy elements, though, the negative gradient of ^{3}He is at odds with the observed distribution in HII regions, which appears fairly flat.

Several hypotheses have been suggested to eliminate the model predicted over-production of ^{3}He (e.g. Galli *et al.*, 1994), but none has so far provided satisfactory results, except the deep mixing mentioned above and described by Charbonnel (1998). The advantage of this mechanism is that it depends on the ambient con-ditions and doesn't therefore apply necessarily to all the low mass stars, thus allowing some stars to still have high ^{3}He yield in agreement with the PNe data (e.g. Lattanzio and Boothroyd, 1997)

To check quantitatively what percentage of low mass stars must experience the deep mixing and have a moderate ^{3}He yield to reproduce the abundances observed in the ISM and in the solar system, GSTP97 have recomputed the best models by

DST96, assuming a varying fraction of stars with and without deep mixing. The lower solid curve in the two panels of Fig. 6 show the model results when 20% of low mass stars follow the standard nucleosynthetic prescriptions (DST96) and 80% follow the deep mixing results by Boothroyd (1996). The strong reduction in the ^{3}He yield resulting from this choice brings the upper solid lines in the two panels of Fig. 6 down to the lower solid line, which fits perfectly all the available observational data.

If Hogan's (1995) more drastic reduction of the ^{3}He yield is adopted, GSTP97 find that a fraction of 70% of low mass stars experiencing the deep mixing is sufficient to solve the ^{3}He problem. Very similar results are obtained by Olive et al. (1997) with chemical evolution models for the solar neighbourhood.

In this context, the PNe with high ^{3}He content should originate from the 20–30% of the whole low mass population which have standard stellar evolution, a fraction apparently compatible with the selection criteria applied by Rood et al. in choosing their targets. To verify this consistency, a good method should be (see Palla et al., 1998) to measure the ^{12}C/^{13}C isotopic ratio both in the six PNe by Rood et al. (1995) with high ^{3}He and in a statistically significant sample of other planetaries. In fact, it has been shown (e.g. Charbonnel, 1995; 1998) that the low mass stars experiencing extra mixing are predicted not only to have lower ^{3}He but also lower ^{12}C/^{13}C than standard stars (namely a carbon isotopic ratio around 5 rather than 30), a circumstance which should be easily verifiable with millimetric spectral observations. If the suggestion by GSTP97 based on chemical evolution arguments is correct, about 80% of the planetaries should show low carbon isotopic ratios and only about 20% should show ratios of the order of 30.

4. Conclusions

It has been shown that for a safe use of chemical evolution models it is of primary importance to compare their predictions with all the available observational constraints. The results of the predicted galactic evolution of D and ^{3}He can be summarized as follows:

- Chemical evolution models in agreement with the major observed features of our Galaxy predict a depletion factor smaller than 3 from the primordial to the present D abundance.
- We haven't been able to find models with higher D destruction which are able to satisfy all the other galactic constraints.
- The ^{3}He abundances derived from observations of PNe, HII regions and local ISM can be reconciled with each other and with the model predictions if the vast majority ($\sim$ 80%) of low mass stars experience deep mixing in red giant phase and burn most of the ^{3}He previously produced.

Acknowledgements

The results described here originate from the contributions of several people: I wish to thank Cristina Chiappini, Daniele Galli, Francesca Matteucci, Francesco Palla, Letizia Stanghellini and Gary Steigman for the nice collaboration. I also thank Nikos Prantzos for interesting conversations. I'm grateful to Johannes Geiss, Vittorio Manno and Ruedi von Steiger for the warm hospitality and the efficient organization of this workshop.

References

Audouze, J. and Tinsley, B.M.: 1974, Astrophysical Journal**192**, 487.
Boothroyd, A.I.: 1996, (private communication).
Charbonnel, C.: 1995, Astrophysical Journal**453**, L41.
Charbonnel, C.: 1998, Space Science Reviews, this volume.
Chiappini, C., Matteucci, F., and Gratton, R.: 1997, Astrophysical Journal**477**, 765.
Copi, C.J., Schramm, D.N. and Turner, M.S.: 1995, *Science* **267**, 192.
Dearborn, D.S.P., Steigman, G. and Tosi, M.: 1996, Astrophysical Journal**465**, 887. (DST96)
DeBoer, K.S. and Savage, B.D.: 1983, Astrophysical Journal**265**, 210.
Ferrini, F., Mollà, M., Pardi, C., and Díaz, A.I.: 1994, Astrophysical Journal**427**, 745.
Ferrini, F. and Mollà, M.: 1996, private communication.
Fields, B., Mathews, G.J., and Schramm, D.N.: 1998, Space Science Reviews, this volume.
Galli, D., Palla, F., Ferrini, F., and Penco, U., 1995 Astrophysical Journal**443**, 536.
Galli, D., Palla, F., Straniero, O., and Ferrini, F.: 1994 Astrophysical Journal**432**, L1.
Galli, D., Stanghellini, L., Tosi, M., and Palla F.: 1997 Astrophysical Journal**477**, 218. (GSTP97)
Geiss, J.: 1993, in *Origin and evolution of the elements*, eds N.Prantzos, E.Vangioni-Flam, M.Cassé (Cambridge University Press, UK), p. 89.
Geiss, J. and Gloeckler, G.: 1998, Space Science Reviews, this volume.
Gloeckler, G. and Geiss, J.: 1996, *Nature* **381**, 210.
Gratton, R., Carretta, E., Matteucci, F., and Sneden, C.: 1997, *Astron.Astrophys.* , submitted.
Hogan, C.J.: 1995, Astrophysical Journal**441**, L17.
Iben, I.: 1967, Astrophysical Journal**147**, 650.
Lattanzio, J.C. and Boothroyd, A.I.: 1997, preprint (astro-ph/9705186).
Linsky, J.L.: 1998, Space Science Reviews, this volume.
Linsky, J.L., Brown, A., Gayley, K., Diplas, A., Savage, B.D., Ayres, T.R., Landsman, W., Shore, S.W., and Heap, S.: 1993 Astrophysical Journal**402**, 694.
Matteucci, F. and Chiappini, C.: 1996, private communication.
Matteucci, F. and François, P.: 1989, Monthly Notices of the RAS **239**, 885.
Olive, K.A., Schramm, D.N., Scully, S.T., and Truran, J.W.: 1997, Astrophysical Journal**479**, 752.
Palla, F., Galli, D., and Silk, J.: 1995, Astrophysical Journal**451**, 44.
Palla, F., Galli, D., Bachiller, R., and Pérez Gutiérrez, M.: 1998, Space Science Reviews, this volume.
Pilyugin, L.S. and Edmunds, M.G.: 1996, *Astron.Astrophys.* 313, 792.
Prantzos, N.: 1996, *Astron.Astrophys.* **310**, 106.
Prantzos, N.: 1998, Space Science Reviews, this volume.
Reeves, H., Audouze, J., Fowler, W.A., and Schramm, D.N.: 1973, Astrophysical Journal**179**, 979.
Rocha-Pinto, H.J. and Maciel, W.J.: 1996, Monthly Notices of the RAS **279**, 447.
Rood, R.T., Steigman, G., and Tinsley, B.M.: 1976, Astrophysical Journal**207**, L57.
Rood, R.T., Bania, T.M., Wilson, T.L., and Balser, D.S.: 1995, in *The light elements abundances*, ed. P.Crane (Springer, De), p. 201.
Rood, R.T., Bania, T.M., Balser, D.S., and Wilson, T.L.: 1998, Space Science Reviews, this volume.
Scalo, J.M.: 1986, *Fund.Cosmic Phys.* **11**, 1.
Scully, S. and Olive, K.A.: 1995, Astrophysical Journal**446**, 272.

Scully, S., Cassé, M., Olive, K.A., Schramm, D.N., Truran, J., and Vangioni-Flam, E.: 1996, Astrophysical Journal**462**, 960.

Scully, S., Cassé, M., Olive, K.A., and Vangioni-Flam, E.: 1997, Astrophysical Journal**476**, 521.

Shaver, P.A. *et al.*, 1983, Monthly Notices of the RAS **204**, 53.

Tosi, M.: 1988, *Astron.Astrophys.* **197**, 33.

Tosi, M.: 1996, in *From stars to galaxies*, eds C.Leitherer, U.Fritze-von Alvensleben, J.Huchra, ASP Conf.Ser. **98**, 299.

Tosi, M., Steigman, G., Matteucci, F., and Chiappini, C.: 1997 Astrophysical Journal, submitted (astro-ph/9706114).

Twarog, B.A.: 1980, Astrophysical Journal**242**, 242.

Tytler, D.: 1998, Space Science Reviews, this volume.

Vangioni-Flam, E. and Audouze, J.: 1988, *Astron.Astrophys.* **193**, 81.

Vangioni-Flam, E., Olive, K.A. and Prantzos, N.: 1994, Astrophysical Journal**148**, 3.

Vidal-Madjar, A., Ferlet, R., and Lemoine, M.: 1998, Space Science Reviews, this volume.

West, K.A., Pettini, M., Penston, M.V., Blades, J.C., and Morton, D.C.: 1985, Monthly Notices of the RAS **215**. 481.

HALO WHITE DWARFS AND BARYONIC DARK MATTER

B.D. FIELDS* and G.J. MATHEWS
Department of Physics, University of Notre Dame, Notre Dame, IN 46556, USA.

D.N. SCHRAMM
University of Chicago, Chicago, IL 60637, USA.

Abstract. We describe the formation of hot intergalactic gas along with baryonic remnants in galaxy halos. In this scenario, the mass and metallicity of the hot intracluster and intragroup gas relates directly to the production of baryonic remnants during the collapse of galactic halos. We construct a schematic but self consistent model in which early bursts of star formation lead to a large remnant population in the halo, and to the outflow of stellar ejecta into the halo and ultimately the Local Group. We consider local as well as high redshift constraints on this scenario. This study suggests that the microlensing objects in the Galactic halo may predominantly be $\sim 0.5 M_\odot$ white dwarfs, assuming that the initial mass function for early star formation favored the formation of intermediate mass stars with $m \gtrsim 1 M_\odot$. However, the bulk of the baryonic dark matter in this scenario is associated with the ejecta of the white dwarf progenitors, and resides in the hot intergalactic medium.

Key words: galaxies:evolution — galaxies:halos — intergalactic medium — dark matter

1. Introduction

Big bang nucleosynthesis provides one of the best available determinations of the cosmic baryon density, and requires that the universe contain baryonic dark matter (Schramm, 1997). We summarize here a model (Fields *et al.*, 1997; hereafter FMS97) which relates recent observations of (presumably baryonic) dark matter in the Galactic halo, and hot intergalactic gas in groups and clusters of galaxies.

On the one hand, dark halos have long been inferred to exist in other galaxies, and almost certainly our own as well. Recently, the presence of dark matter in our Galactic halo has been directly confirmed by microlensing observations towards the LMC (Alcock *et al.*, 1996; Aubourg *et al.*, 1993). These remarkable experiments have detected massive compact halo objects (MACHOs), and thus are the most direct detection to date of dark matter in the universe. Current estimates of the lensing object's mass, $m = 0.5^{+0.3}_{-0.2} M_\odot$ (Alcock *et al.*, 1996) are suggestive of white dwarfs.

On the other hand, hot intergalactic gas is found to be ubiquitous in rich clusters, and has recently been observed in galaxy groups (Mulchaey *et al.*, 1996; Davis *et al.*, 1996). This X-ray gas is metal enriched: $Fe/Fe_\odot \sim 0.3$ in clusters, and $Fe/Fe_\odot \sim 0.1 - 0.2$ in groups. The presence of these heavy elements implies that some of the gas has undergone significant processing through massive stars, and subsequently was ejected.

---

* Current address: School of Physics and Astronomy, U. Minnesota, Minneapolis, MN 55655, USA.

Our model is an attempt to connect these observations; our account here is a condensed account of the detailed work of FMS97. To have a dark halo composed of a significant population of white dwarfs demands that extensive stellar processing occurred in the past. White dwarf production is accompanied by the ejection of planetary nebulae whose masses are significantly larger than the compact remnants themselves. Thus, if white dwarfs are a significant component of the halo mass, then the gas released by the progenitor stars (1) contains a large fraction of the initial Galactic baryonic mass, and (2) must be removed from the Galaxy, simply to maintain a reasonable total Galactic mass budget. These implications are encouraging, since hot gas is indeed observed to be a significant component of galaxy aggregates. Furthermore, the presence of metals in the hot gas underscores the need for stellar processing, as well as a mechanism for the ejecta to escape into the intergalactic medium (IGM). Therefore, we posit a scenario in which there were strong bursts of star formation in the early Galaxy. Most of the stars are now dead: the remnants are MACHOs; the ejecta was lost in galactic wind, and becomes a part of the IGM. In FMS97, important local constraints on this scenario (e.g. Ryu *et al.* 1990) are discussed. Here, we note as well (§5) some high-redshift constraints, which could test and/or point out necessary refinements to our schematic picture.

2. Observations and Constraints

Our model should apply generally to groups and clusters, but here we focus on the Local Group. This is dominated by two spiral galaxies, each having both luminous and dark components. The visible portion of our Galaxy consists in a disk of mass $\sim 6 \times 10^{10} M_\odot$, and a central bulge with about $10^{10} M_\odot$. The halo mass is uncertain because it is not clear how far the halo extends. Nevertheless, the halo seems to contain at least $5 \times 10^{11} M_\odot$. The Local Group itself is estimated (Fich and Tremaine, 1991) to have a total mass of $3 - 5 \times 10^{12} M_\odot$.

We will assume that the Local Group's X-ray properties are similar to those of similar groups. In groups which clearly evince X-ray gas (Mulchaey *et al.*, 1996), the temperatures are typically very close to $T \sim 1$ keV, with metallicity that is poorly determined (Davis *et al.*, 1996) but of order 10–20% solar. The gas masses inferred for these groups vary more than the temperature does, and directly correlates with the content of early-type galaxies (Mulchaey *et al.*, 1996). Recent observations (Ponman *et al.*, 1996) have complicated this issue, suggesting that some spiral-dominated groups might still have a large amount of gas that is relatively cool ($T \sim 0.3$ keV, at the limit of the ROSAT sensitivity). We will, therefore, regard the X-ray data as allowing the presence of relatively cool X-ray gas.

3. The Model

We model (FMS97) the Local Group evolution schematically, in the spirit of a hierarchical clustering picture. Namely, we establish a hierarchy of three mass scales: (1) protogalactic clouds, (2) galactic halos, and (3) the group itself. Each mass scale corresponds to a primordial density fluctuation which must overcome the cosmological expansion. Thus, the dynamics of each component is a gravitational collapse, specifically that of a spherical overdensity (Mathews and Schramm, 1993).

The different mass scales are self-similar. At the smallest scale, the clouds contain stars, as well as both hot and cold gas. The halos include the clouds as sub-components, as well as their own component of hot gas and stars ejected from clouds. The group includes the two galaxies as a sub-component, as well as hot gas ejected from the halos. While the clouds contain only baryons, we introduce a component of non-baryonic dark matter at the halo and group levels.

Star formation is a key ingredient of the model; stars provide significant heating, gas recycling, as well as nucleosynthesis products. Star formation occurs only in the cold gas components of the clouds, and has both a quiescent and merger-induced term, with the former dominant (Mathews and Schramm, 1993). We include the production of metals and helium using the yields of Maeder (1992), and we follow the luminosity evolution of the stars in the halo.

The model components interact via several mechanisms. One of these is merging, which reduces the number of clouds while increasing their average mass. Initially, there are 10^6 clouds, each with mass $10^6 M_\odot$. These ultimately coalesce to become one massive object, the progenitor of the disk and bulge. We do not distinguish these components nor follow the disk evolution, as we are only interested in the halo. We also allow for stellar enrichment of the gas (predominantly via supernovae), which contributes heat and metals. The gas may also be heated or cooled via pdV work, evaporation and radiative cooling. At each mass scale, the heating leads to evaporation of the hottest gas particles with thermal velocities above the local escape velocity. For the hot gas, evaporation is very efficient: there is a significant wind at all levels, with mass loss even from the group itself (this leads to the presence of hot intergroup gas as a significant component of the dark baryons). The winds also function to remove material after a single stellar processing and so prevents recycling which would otherwise lead to the overproduction of metals and helium.

The evolution depends sensitively on the choice of the initial mass function (IMF) for the halo. A remnant-rich halo scenario such as ours requires that the halo IMF differed significantly from that of the disk. Whereas in the disk, stars with $m < 1 M_\odot$ are by far the most numerous, they must be rare in the halo. Otherwise, these very long-lived stars would still be shining and lead to an unacceptably bright halo. Thus we must choose an IMF biased away from low mass ($\lesssim 1 M_\odot$) stars; arguments for such an IMF in the early Galaxy are presented by Silk (1993). We parameterize the IMF as a log-normal form (Adams and Laughlin, 1996), characterized by a centroid m_c and and dimensionless width σ. In fact, we are able to obtain acceptable results for a whole spectrum of IMF parameters; here we present results with $m_c = 2.3 M_\odot$ (the value suggested by Adams and Laughlin 1996) and $\sigma = 1.6$ (the present-day value; Miller and Scalo 1979).

4. Results

Our results, and their dependence on the model parameters, are summarized in detail in FMS97. The model inputs are a set of parameters, describing: the IMF, star formation rate, initial masses, wind strength, and gas mixing. The outputs are the time evolution of the gas mass, temperature, and composition at each level in the hierarchy. We also compute total stellar and remnant masses, and the X-ray luminosity of the hot gas. In the space allotted here, we will sketch the basic results for our "best-fit" model, which maximizes the white dwarf component of the halo while satisfying the constraints we have adopted.

For our best-fit model, the mass and metal budgets for the various hierarchy scales are as follows. (1) Initially, there are 10^6 protogalactic clouds, each with a mass of $10^6 M_\odot$. These merge to form a single object, the proto-disk and bulge, with a final mass of $8.1 \times 10^{10} M_\odot$. (2) The halos begin with a mass $1.35 \times 10^{12} M_\odot$, of which 81% is baryonic (mostly in the form of the clouds). The final mass is $5.0 \times 10^{11} M_\odot$, of which 50% is baryonic. Thus there is a net loss of $1.7 \times 10^{12} M_\odot$ of gas from the galaxies into the IGM. (3) The group itself begins with a mass of $5.6 \times 10^{12} M_\odot$, and ends with a mass $4.3 \times 10^{12} M_\odot$, 17% of which is baryonic.

Of the baryonic group mass, each galaxy contains $1.7 \times 10^{11} M_\odot$ in the form of remnants, which we predict are the MACHOs; these are mostly white dwarfs, with 20% neutron stars. An additional $2.9 \times 10^{11} M_\odot$ of dark baryons reside as $T = 0.25$ keV intragroup gas, an amount below but close to the ROSAT limits for diffuse emission. The group as a whole loses $1.4 \times 10^{12} M_\odot$ of gas to the intergalactic medium; thus about 64% of the initial baryonic mass in the group is ejected later into intergalactic space. This amount of hot (ionized) material is consistent with Gunn-Peterson limits on the intergalactic medium.

Several key constraints on halo white dwarfs are considered in detail in FMS97. Aside from nucleosynthesis issues, there are various luminosities to consider, arising from: long-lived stars in halo (Ryu *et al.*, 1990); the now-cooled halo white dwarfs; and the diffuse radiation from the (presumably universal) occurrence of the starbursts themselves (Charlot and Silk, 1995). We find all of these luminosities to be acceptably low, but the white dwarf luminosity, and perhaps the diffuse background are near current limits and thus are potentially detectable.

5. High Redshift Constraints

Previous discussions of halo white dwarf constraints has focussed on local observables (Ryu *et al.*, 1990), or on integrated diffuse backgrounds (Charlot and Silk, 1995); these constraints are discussed in detail in FMS97. In addition to these, very recent observations of high redshift systems could provide additional and possibly decisive constraints on the early epochs we model. One should, however, bear in mind that high redshift observations include objects of all mass scales and not just poor groups such as the Local Group. That is, even if our Local Group model is typical of evolution for objects of this mass scale, other scales which evolve differently will contribute to the cosmic average properties at a given epoch. Nevertheless, properly interpreted high–z data can provide crucial information about halo white dwarf scenarios.

One potential constraint comes from the cosmic star formation history. As determined by ground-based observations and by the HST (see, e.g., Madau 1997), cosmic star formation increases with redshift up to $z \sim 1 - 2$, when the mean comoving star formation density was about an order of magnitude higher than the present. The inferred star formation behavior at still higher redshift is less clear, and depends strongly on assumptions one makes about absorption corrections. If one can sort out issues such as this, one can begin to put strong constraints on all chemical evolution scenarios, including ours.

Another key high-redshift constraint comes from quasar absorption line systems. Observations of these systems provide metal abundances as a function of redshift. These abundances show a possible floor at a level of about 10^{-2} of solar. This could be signature of a first burst of star formation in our scenario. Related to the quasar absorbers is the Gunn-Peterson determination of an ionized intergalactic medium. That this is a significant component of baryons is encouraging for our scenario, which requires that most baryons are in the form of hot intergalactic gas.

6. Discussion and Conclusions

We have summarized work (FMS97) which schematically illustrates how a plausible model of galaxy evolution can relate the hot gas seen in galaxy aggregates to a dark halo population of remnants. Since we link these phenomena, confirmation of one in the Local Groups would imply, in our model, the existence of the other. Furthermore, signatures of the model are within reach. For example, X-ray gas may be observable in the Local Group or other groups (and perhaps has been; Ponman *et al.* 1996). Also promising are the signatures of white dwarfs. Aside from the microlensing experiments, these objects should be directly detectable with pencil beam and wide angle observations. Also, it is intriguing that several edge-on galaxies have an observed IR halo, as one would expect from cooled white dwarfs (e.g., Lehnert and Heckman 1996, and references therein). Indeed, even if white dwarf scenarios such as this one can be ruled out, we would be led to demand that the Galaxy be made of something stranger still.

Acknowledgements

It is a pleasure to acknowledge useful discussions with K. Freese, D. Graff, D. Bennett, J. Truran, and M. Turner.

References

Adams, F.C., and Laughlin, G.: 1996, *ApJ* **468**, 586–597.
Alcock, C., *et al.* (MACHO Collaboration): 1996, *ApJ* **461**, 84–103.
Aubourg, E., *et al.*: 1993, *Nature* **365**, 623.
Charlot, S., and Silk, J.: 1995, *ApJ* **445**, 124–132.
Davis, D.S., Mulchaey, J.S., Mushotzky, R.F., and Burstein, D.: 1996, *ApJ* **460**, 601–611.
Fich, M., and Tremaine, S.: 1991, *ARAA* **29**, 409–445.
Fields, B.D., Mathews, G.J., and Schramm, D.N.: 1997, *ApJ* **483**, 625–637 (FMS97).
Lehnert, M.D., and Heckman, T.M.: 1996, *ApJ* **462**, 651–671.
Madau, P., *et al.*: 1997, *MNRAS* **283**, 1388–1404.
Maeder, A.: 1992, *A&A* **264**, 105–120.
Mathews, G.J., and Schramm, D.N.: 1993, *ApJ* **404**, 468–475.
Miller, G.E., and Scalo, J.M.: 1979, *ApJS* **41**, 513.
Mulchaey, J.S., Davis, D.S., Mushotzky, R.F., and Burstein, D.: 1996, *ApJ* **456**, 80.
Ponman, T.J., Bourner, P.D.J., Ebeling, H., and Böhringer, H.: 1996, *MNRAS* **283**, 690.
Ryu, D., Olive, K.A., and Silk, J.: 1990, *ApJ* **353**, 81.
Schramm, D.N.: this volume.
Silk, J.: 1993, *Phys. Reports* **227**, 143–148.

Address for correspondence: School of Physics and Astronomy, 116 Church Street SE, University of Minnesota, Minneapolis, MN 55455, USA.

THE CHEMICAL EVOLUTION OF THE MILKY WAY DISK

Some certainties, several open questions and
implications for "cosmic chemical evolution"

N. PRANTZOS

Institut d'Astrophysique de Paris

Abstract. A brief review is presented of our current understanding of the evolution of the Milky Way disk and of its relevance to "cosmic chemical evolution" studies. The implications of this understanding for the evolution of deuterium are emphasized.

Key words: Galaxy: evolution, nucleosynthesis, abundances

1. Introduction

Despite more than 30 years of intense theoretical and observational studies, galactic chemical evolution is not yet a mature astrophysical discipline (compared e.g. to stellar evolution). One can identify two main reasons for that situation:

• lack of a galactic equivalent of the HR diagram, that would show unambiguously the evolutionary status of galaxies;

• lack of an understanding of the main motor engine of galactic evolution, namely the creation of stars out of galactic gas (compared to our fairly good understanding of the motor of stellar evolution, namely nuclear reactions).

Progress in the theory of galactic chemical evolution has been very slow (almost 20 years later, Tinsley's (1980) review continues to be probably the best on the subject). Since the number of model parameters is, in general, larger than the number of observables, one may sometimes feel that she/he is only constrained by her/his own imagination. This may be the case for most extragalactic systems, but it certainly does not apply to the case of our Galaxy: the wealth of available data, especially in the solar neighborhood, constrain seriously the parameters of simple models of chemical evolution and point to a rather well defined history for the local disk (LD).

In the following, we present a brief review of the observational data for the LD and the hints they reveal as to the past history of that region. As shown in Tosi (1998) that history allows only for a small depletion of deuterium (D), less than a factor of 3 from its pregalactic value. The observational data for the rest of the Milky Way disk are much less constraining for the models. They suggest, however, that a much larger astration (and, hence, D depletion) has taken place in the inner Galaxy; the resulting D gradient, measurable by the future *FUSE-LYMAN* mission (see Vidal-Madjar *et al.*, 1998) should provide invaluable information as to the past history of the disk. Finally, assuming that our Galaxy is a typical spiral,

Space Science Reviews **84**: 225–236, 1998.
© 1998 *Kluwer Academic Publishers*

one can calculate the properties of disk galaxies as a function of redshift (in the framework of a given cosmological model) and compare to the observed properties of the extragalactic universe: global star formation rate, gas content and metal abundances in gas clouds. Preliminary conclusions of such a comparison appear in Sect. 4.

2. The past history of the solar neighborhood

In the solar neighborhood (defined as a cylinder of ~ 1 kpc radius at a distance $R_\odot = 8.5$ kpc from the galactic center), the main observables relevant to chemical evolution are (Prantzos and Aubert, 1995; Pagel, 1997):

• the current surface densities of gas, stars and total amount of matter (~ 10, 40, and 55 $M_\odot$ pc^{-2}, respectively), leading to a current gas fraction of $\sigma_G \sim 20\%$; also, the current star formation rate (SFR), of $1 - 3$ $M_\odot$ pc^{-2} Gyr^{-1} (e.g. Rana 1991 and references therein).

• the metallicity at solar birth ($Z_\odot$) and today (Z_0). Notice that measurements of abundances in the local interstellar medium (ISM) and in young stars in nearby Orion show $Z_0 < Z_\odot$ (at least for CNO elements), suggesting that $Z_\odot$ is, perhaps, not representative of the local ISM 4.5 Gyr ago. If this turns out to be true, it would have major implications for chemical evolution studies; indeed, all models at present are required to reproduce solar elemental and isotopic abundances at the time of solar system formation.

• the oxygen vs. Fe (O-Fe) relationship, showing a steady decline from O/Fe $\sim$ const. ~ 3 times solar at [Fe/H] < -1 (i.e. during the halo phase), to O/Fe $\sim$ solar at [Fe/H] ~ 0; this is usually attributed to the delayed (~ 1 Gyr) appearance of SNIa, producing 2/3 of the galactic Fe, while 1/3 of Fe and the totality of O are produced by short lived massive stars exploding as SNII. Notice that the delayed onset of SNIa activity is dictated by our understanding of the halo formation timescale, not by any understanding of the SNIa occurrence (despite the use of some "sophisticated" formulae, the SNIa rate is essentially a free parameter of the models, unlike the SNII rate). In particular, it is not at all clear why the end of the halo phase coincides with the onset of the decline in the O/Fe ratio.

• the age-metallicity (Z-t) relationship, traced by the Fe abundance of long-lived, F-type stars. The data (Fig. 1b) show a monotonous increase in the mean metallicity and a considerable dispersion at any age; the latter, if real, may result from imperfect mixing of the ISM, orbit diffusion of stars coming from galactic regions with different metallicity, or both. Simple, one-zone, chemical evolution models always yield a unique age-metallicity relationship and cannot reproduce that scatter without further assumptions and free parameters.

• the metallicity distribution of long-lived G-type stars (Fig. 1a), showing that very few of them were formed at [Fe/H] < -0.7 (1/5 solar). This is the most important constraint, never satisfied in models forming rapidly the local disk (like

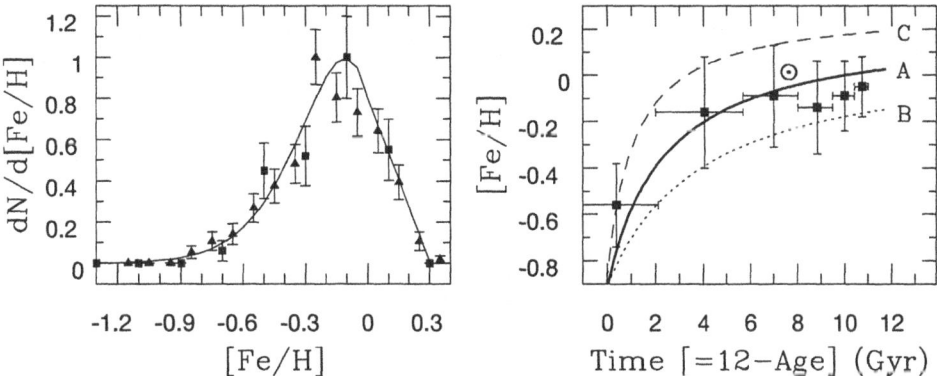

Figure 1. Observables in the solar neighborhood.
Left: Metallicity distribution of G-type stars, from 2 recent surveys (Rocha-Pinto and Maciel, 1996; Wyse and Gilmore, 1995). The solid line is a fit to the data.
Right: Age-metallicity relationship of F-stars (Edvardsson *et al.*, 1993). The data (182 stars) are binned in groups of 0.2 in log(Age/Gyr); instead of Age, Time = 12 − Age is plotted here. Metallicities are for stars with galactocentric distances evaluated to 8 − 9 kpc (i.e. appropriate to local disk) and volume corrected, where the vertical error bars represent 1σ dispersion in metallicity for each age group (Col. 6 of Tables 14 and 15, respectively, in Edvardsson *et al.*, 1993). The 3 cases A,B and C trace schematically the mean metallicity and the observed dispersion.

closed-box or assuming infall on time-scales much shorter than $\sim$ 5 Gyr). The alternative solution, namely that the disk started with an initial metallicity of [Fe/H] $= -0.7$ (corresponding to [O/H] ~ -0.4, or $\sim 1/3$ solar) is rather implausible, since neither the galactic halo nor the bulge are massive enough to have polluted the disk to such an extent.

The last two observables allow to recover the past SFR history of the local disk (Rana, 1991; Prantzos and Silk, 1997), as:

$$\text{SFR} \propto \frac{dN_G}{dt} = \frac{dN_G}{d[\text{Fe/H}]} \frac{d[\text{Fe/H}]}{dt} \qquad (1)$$

where the assumption of a constant IMF is necessary for the proportionality between the SFR and the creation rate of G-stars to hold. $dN/d[\text{Fe/H}]$ is the observed differential metallicity distribution and $d[\text{Fe/H}]/dt$ is the derivative of the observed Z-t relationship w.r.t. time. The scatter in the latter relationship does not allow to recover unambiguously the SFR. For illustration purposes, we adopt the 3 curves of Fig. 1b, which have considerably different early and late slopes and cover most of the observed dispersion. When the resulting SFR histories are normalised to the current local star surface density $\Sigma_* \sim 42\ M_\odot\ \text{pc}^{-2}$ over the age $T = 12$ Gyr of the galactic disk ($\int_0^T \text{SFR}(t)dt = \Sigma_*$), only cases A and B lead to an acceptable current SFR ($\sim 1 - 3\ M_\odot\ \text{pc}^{-2}\ \text{Gyr}^{-1}$); case C, as well as all other Z-t relationships with large early slopes, lead to a large early star formation and to too low current SFR (Fig. 2a).

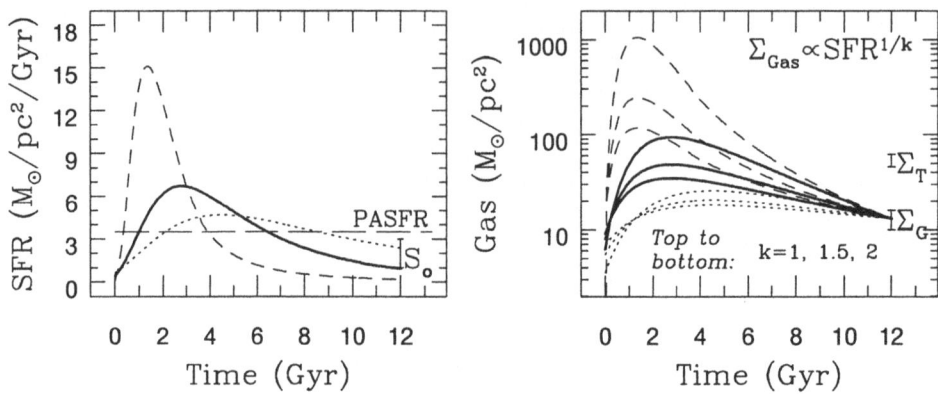

Figure 2. Observationally derived Star Formation Rate in the solar neighborhood.
Left: Curves correspond to cases A, B and C of Fig. 1b ; they are all normalised to $\int_0^T SFR(t)dt = 42M_\odot$ pc^{-2}, with $T = 12$ Gyr. Case C leads to unacceptably low current SFR $\sim 0.1M_\odot$ pc^{-2} Gyr^{-1}, much lower than the observed $S_0 \sim 1 - 3M_\odot$ pc^{-2} Gyr^{-1}. The current SFR is comparable to the Past Average SFR (PASFR).
Right: Past history of the gas surface density, assuming SFR $\propto \Sigma_G^k$. The 3 types of curves (*solid, dotted, dashed*) correspond again to those of Fig. 1b and 2a. Case C violates the constraint $\Sigma_G < \Sigma_T$ for all plausible values of k.

There is another indication against C-type cases. Observations suggest that in disk galaxies SFR $\propto \Sigma_G^k$, with $k = 1 - 2$ (Kennicutt, 1989). Knowing the local SFR history from eq. (1), one can then recover the gas history as $\Sigma_G \propto SFR^{1/k}$, normalising to the current $\Sigma_G \sim 10$ $M_\odot$ pc^{-2}. It can be seen from Fig. 2b that C-type cases should be excluded, since they lead to past Σ_G larger than the current total surface density Σ_T (only in the case of an important, but quite implausible, outflow from the disk would this conclusion be invalidated).

The above results clearly indicate that the local disk was formed on timescales of many Gyr. Models with slow infall of primordial composition reproduce reasonably well the above constraints (in particular, the last one in the list above). Such models (Prantzos, 1996; Tosi *et al.*, 1997) deplete D only by factors $\sim 2 - 3$ (Fig. 3) at least for reasonable stellar Initial Mass Functions (IMF), like the one of (Kroupa *et al.*, 1993). Notice, however, that the main reason for the small D depletion is the currently large local gas fraction ($\sim 20\%$), showing that a rather large fraction of the ISM has not been astrated yet. These results point to low values for pregalactic D/H$\sim 3 - 4 \times 10^{-5}$, i.e. compatible with those advocated by Tytler (1998). Those values also alleviate (but do not solve) the problem of ^{3}He overproduction, which is definitively a stellar nucleosynthesis, not a galactic evolution, problem (see Charbonnel, 1998; Galli *et al.*, 1997, for possible solutions).

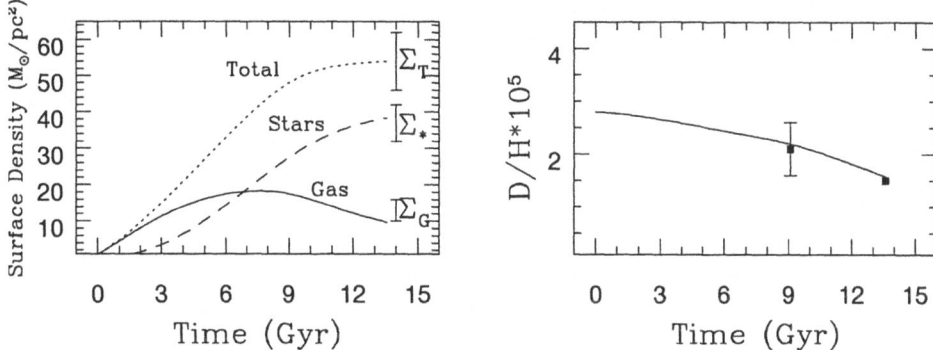

Figure 3. Results of a model for the chemical evolution of the local disk.
Left: Evolution of the gaseous, stellar and total mass and comparison of the final results to observations of current local Σ_G, Σ_* and Σ_T, respectively. The adopted SFR is proportional to $\Sigma_G^{1.5}$, while the infall rate is a broad gaussian with $\Delta\tau = 6$ Gyr, i.e. the local disk is formed very slowly. Such models reproduce reasonably well the observational constraints of Sec. 2 and Figs. 1 and 2a.
Right: Corresponding evolution of the D abundance, compared to observations at solar system formation (Geiss and Gloeckler, 1998) and in the local ISM (Linsky, 1998). Only small factors of D depletion $(2 - 3)$ are obtained in this kind of models, with slow infall of primordial composition.

3. The evolution of the Milky Way disk

Contrary to the case of the solar neighborhood, the available observations for the Milky Way disk offer information mainly about its current status, not its past history. The main relevant observables are (Prantzos and Aubert, 1995):

i) The total mass of gas and stars in the disk ($\sim 8 \times 10^9 \, M_\odot$ and $\sim 5 \times 10^{10} \, M_\odot$, respectively); the total current SFR ($\sim 3 - 5 \, M_\odot \, yr^{-1}$) and the current supernova rates (~ 2 SNII/century and ~ 0.4 SNIa/century, respectively, as suggested by observations of external spirals).

ii) The current gas profile, dominated by the molecular ring at galactocentric distance $R \sim 4 - 5$ kpc and by HI of roughly constant surface density at distances $6 - 14$ kpc.

iii) The stellar profile, exponentially decreasing outwards with a scale-length of ~ 3 kpc. The combination of observables (ii) and (iii) leads to a gas fraction profile steeply decreasing in the inner disk, suggesting that the star formation efficiency has been larger in those regions than in the outer disk.

iv) The current SFR profile (traced by the surface density of pulsars and super-nova remnants or the H_α emissivity profile), strongly decreasing outwards. Notice that the SFR profile does not follow the molecular or the total (molecular+atomic) one, i.e. the SFR is not simply proportional to some power of the gaseous profile.

v) The current metallicity profile, usually traced by oxygen observed in HII regions and planetary nebulae and showing a gradient of $d[O/H] \sim -0.07$ dex/kpc (Shaver *et al.*, 1983). Observations of O abundances in young, B-type stars in the early 90ies suggested a surprising flat oxygen abundance profile; the controversy seems to be settled by the recent non-LTE treatment of those objects (Smartt and Rolleston, 1997), confirming the gradient of HII regions.

vi) Current abundance profiles of other elements (e.g. N, S, etc.) and isotopic ratios (like $^{12}C/^{13}C$, $^{16}O/^{17}O$ etc.). However, the constraints imposed by those observables are either similar to those imposed already by the oxygen profile or much more sensitive to the adopted nucleosynthesis prescriptions (concerning e.g. the primary and/or secondary origin of N, ^{13}C; the stellar production of ^3He; etc) than to the chemical evolution model itself.

Since there are essentially no constraints on the past history of the Milky Way disk (i.e. no age-metallicity relations or metallicity distributions are available for other regions) there is much more freedom in its modelisation than in the case of the solar neighborhood. Still, it is meaningful to construct models, insofar as the number of parameters used is considerably smaller than the constraints (i-v) above.

These constraints make the Milky Way disk a "testbed" for theories of star formation in disk galaxies. Indeed, several ideas have been put forward over the years for a radial dependence of the star formation efficiency, most of them on the basis of various instability criteria for gaseous disks (Talbot and Arnett, 1975; Wyse and Silk, 1989). It should be noted that star formation theories exist and may be tested mainly for disk galaxies, not for e.g. ellipticals or irregulars.

Up to now, these theoretical prescriptions have been used or tested only in the framework of extremely schematic models, considering the disk as a system of independent (one-zone) rings. This (over)simplification ignores the possibility of radial inflows in gaseous disks, resulting e.g. by viscosity (inducing friction between adjacent gas layers) or by infalling gas with specific angular momentum different from the one of the underlying disk; in both cases, the resulting redistribution of angular momentum leads to radial mass flows. The magnitude of the effect is difficult to evaluate, because of our poor understanding of viscosity and our ignorance of the kinematics of the infalling gas. At the present stage of our knowledge, introduction of radial inflows in the models (Lacey and Fall, 1985) would imply even more free parameters and make impossible the study of a radial variation in the efficiency of the SFR.

Notice that a radial variation of the infall timescale (in the framework of the simple, independent ring, model) may also play a similar role. To give an example, a disk galaxy may be formed inside-out in either of two ways: a) with a constant efficiency of the SFR and an infall much more rapid in the inner zones than in the outer ones; b) with the same infall timescale all over the disk and a radially decreasing outwards SFR efficiency. In other terms, the possibility of a radially varying infall timescale makes it difficult to test unambiguously theories of radial variation of the SFR efficiency (and vice versa).

Despite the degeneracy of the problem, simple toy-models can often give useful insights on the above questions. Also, once a "successful" (albeit non-unique) model is obtained, it can be used to test some nucleosynthesis prescriptions (resulting

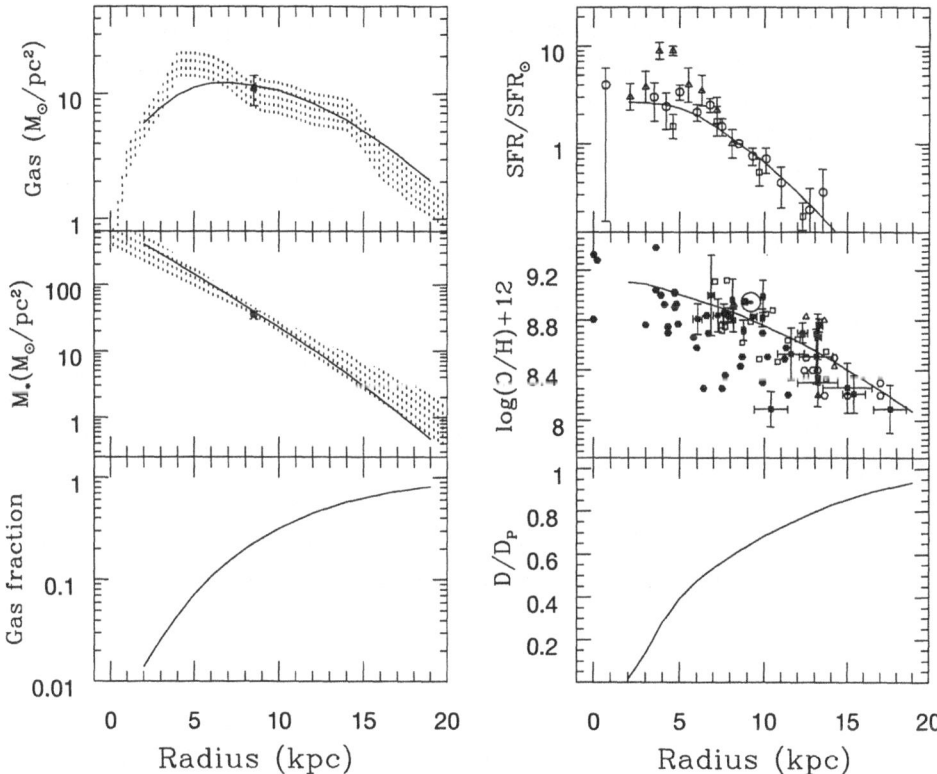

Figure 4. Results of a simple (independent ring) model for the chemical evolution of the Milky Way disk at a galactic age $T = 13$ Gyr, and comparison to observations. The adopted SFR is SFR $\propto \Sigma_G^{1.5}/R$ and the adopted infall rate is gaussian in time with $\Delta\tau = 6$ Gyr in all the zones.
Left, from top to bottom: final profiles of the surface density of gas, stars and of the gas fraction, respectively. *Solid lines*: model results; *shaded regions* correspond to observations for the disk, and data points at $R = 8.5$ kpc to solar system values.
Right, from top to bottom: current SFR, O and D profiles, respectively. SFR is normalised to its local value and D to its primordial one D_P. Data for O are from HII regions [*open symbols*, from (Shaver *et al.*, 1983; Fich and Silkey, 1991)] and B-stars [*filled symbols*, from (Smartt and Rolleston, 1997)].

in radial variations of the primary/secondary elemental or isotopic ratios) or even to make further predictions..

The results of such a model appear on Fig. 4. The adopted SFR $\propto \Sigma_G^{1.5}/R$ is based on the idea that stars are formed in spiral galaxies when the interstellar medium with angular frequency $\Omega(R)$ is periodically compressed by the passage of the spiral pattern, having a frequency $\Omega_P = $ const. $\ll \Omega(R)$. This leads to SFR $\propto \Omega(R) - \Omega_P$ and, for disks with flat rotation curves, to SFR $\propto R^{-1}$ (Wyse and Silk, 1989). On the other hand, a gaussian infall with a half-width $\Delta\tau = 6$ Gyr is adopted all over the disk (such as to account for the G-dwarf problem in the solar neighborhood). The radial variation in the SFR efficiency is the only new parameter introduced in the model, all the others been already fixed by the modelisation of the solar neighborhood (Sect. 2). It turns out that with this simple parametrisation the model reproduces reasonably well the constraints (i-v); it can be used then with some confidence for making further predictions.

The most important result of this type of models (in the context of this Conference) concerns the evolution of deuterium. It is found that its final abundance profile shows an important gradient, much more important than (and anticorrelated to) the one of oxygen. The reason is that all stars deplete deuterium (especially the numerous low-mass ones) but only massive stars produce oxygen. In regions with large early SFR, low mass stars become the major actors in the late stages of chemical evolution, ejecting D free and O free material; as a result, the abundance of D continues decreasing, while the one of O saturates (and may even decrease, if not compensated by sufficient production from the relatively few massive stars that are born lately). Thus, the D abundance profile, not subject to saturation effects, is the most sensitive tracer of the past SFR history of the disk (Prantzos, 1996), especially in the inner regions. Measurements of the D profile with the ISO satellite and the forthcoming FUSE-LYMAN mission (Vidal-Madjar et al., 1998) should give invaluable information about the past star formation activity in the disk.

4. Implications for "cosmic chemical evolution"

In recent years, observations of high redshift systems started revealing some views of the early phases of chemical evolution in the Universe. These observations concern the "global" SFR and HI content of the Universe (i.e. both averaged over sufficiently large volumes), as well as the abundances of various elements in gaseous systems, in the line of sight of quasars. To these one should add, in principle, observations of D abundances (Tytler, 1998), although the current status of those observations does not allow to establish a trend as a function of redshift. More specifically, the available observations concern:

(a) The comoving *luminosity density of the Universe*, from the present epoch (redshift $z = 0$) back to $z \sim 4$, in several broad pass-bands (Madau, 1997). The data, obtained from various deep spectroscopic and photometric surveys with ground based telescopes and the HST, show a rather steep rise in the luminosity density from $z = 0$ to $z \sim 1.5$ (by a factor of ~ 10), then a slow decline; in fact, the data points at high redshift should be considered only as tentative, since possible obscuration by dust could "hide" even larger amounts of UV emitting stars. In principle, the UV luminosity traces the underlying massive star population and (assuming a universal stellar IMF) the corresponding SFR. In practice, the method has several limitations (Madau, 1997), but it can still give a first idea of the evolution of the SFR on a cosmic scale, at least at late times.

(b) The comoving *density of neutral gas* in Lyα absorbers, taking into account the observed column density distribution of those systems as a function of redshift (Storrie-Lombardi et al., 1996). The data show an increase in the gas density back to $z \sim 2$, then a flattening at a value roughly comparable to the one of the stellar density in the local Universe. This could indicate that the bulk of star formation took place after $z \sim 2$, a conclusion corroborated by observable (a). Notice, however, the absence of information on the amounts of molecular and ionised gas, which may modify that conclusion.

(c) The *gas phase abundances of several elements* in a few dozens of damped Lyα absorbers at various redshifts (Pettini *et al.*, 1997; Lu *et al.*, 1996). There is an overall trend of decreasing metallicity with increasing redshift, but also a large dispersion at all redshifts. Values range between 1/10 and 1/300 solar, while an average metallicity of [Fe/H] $\sim$ 1/30 solar is derived in the redshift range $2 < z < 3$. However, the possibility of dust depletion does not allow to use most of those metals as reliable tracers of the metal content of the gas, with the possible exception of Zn (less refractory than most of the other observed elements).

Observables (a-c) have prompted several studies of "cosmic chemical evolution", e.g. (Pei and Fall, 1995). In most cases, a simple model of chemical evolution is adopted (i.e. one zone model assuming an IMF and a SFR, as well as the possibility of inflows and outflows), which is assumed representative of a sufficiently large volume of the Universe; obviously, the total amount of gas in the box (processed in the system *and* participating in inflows/outflows) corresponds to the total baryon content of that representative volume. The predictions of the model concerning the gas and metal content as a function of time are then translated into functions of the redshift (in the framework of a given cosmological model) and compared to observables (b) and (c).

Notice however that, although observables (a) and (b) can be considered as truly "global" (if the possibility of, up to now undetected, amounts of molecular and ionised gas is neglected), observable (c) is rather "local": indeed, it traces the metallicity of only those systems that have still retained some gas at a given redshift. The metals produced e.g. early on in ellipticals and already incorporated in their stars or expelled as ionised gas in the intergalactic medium of galaxy clusters, do not appear in the "census" of (c). The predictions of the "global" model for the evolution of the metallicity cannot then be directly compared to the observations in (c).

Since the only system with data sufficient to constrain its history is the Milky Way, it may be interesting to see how its evolution, modelled in Sect. 3 and translated in a cosmological framework, compares to the observables (a-c) of "cosmic evolution". A favourable comparison would imply that the Milky Way is a really "average" galaxy, but this would be a rather improbable result. The aim of that comparison is rather to see to what extent the Milky Way evolution differs from the "average" one (Prantzos and Silk, 1997).

The results are shown in Fig. 5. It can be seen that the Milky Way SFR does not increase between $z = 0$ and $z \sim 1$ as steeply as indicated by observations, although it peaks at about the right redshift. Milky Way type spirals can then account only partially for the observed evolution of the cosmic SFR; this is a

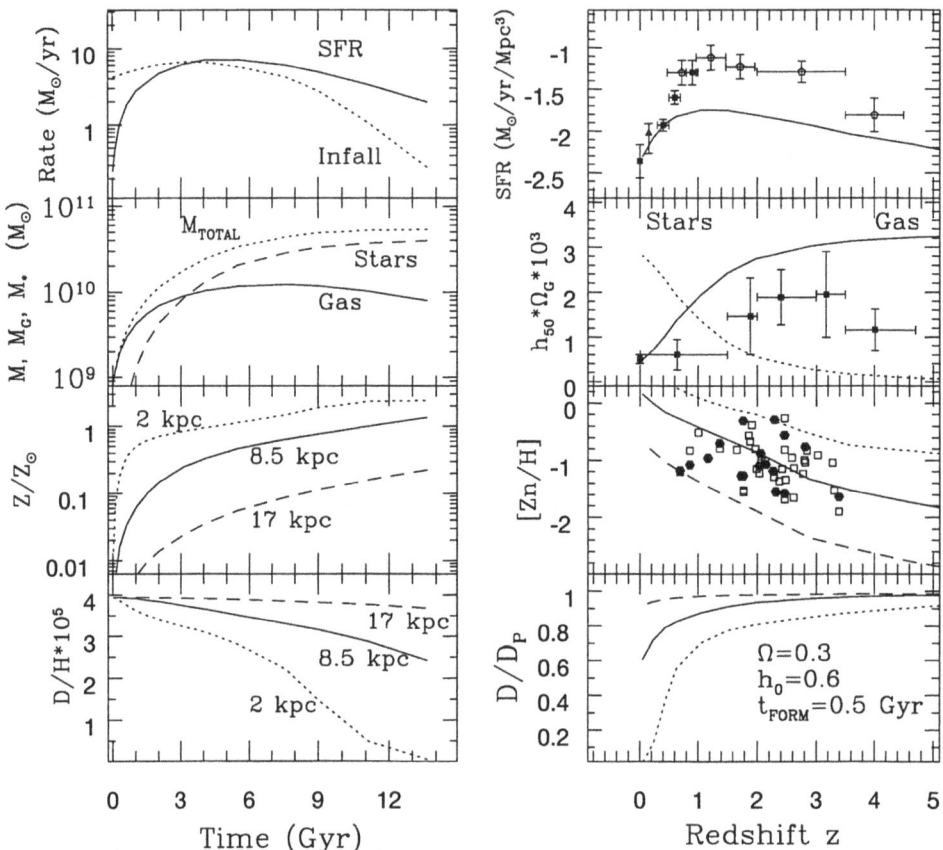

Figure 5. History of the Milky Way disk as a function of time (*left*) and of redshift (*right*), assuming that our galaxy is a typical spiral; a cosmological model with $\Omega = 0.3$, $h_0 = 0.6$ and galaxy formation starting 0.5 Gyr after the Big Bang is adopted.

Left, from top to bottom: History of a) SFR and infall rate; b) gaseous, stellar and total mass; c) overall metallicity in three different zones, at distances of 2, 8.5 and 17 kpc from the galactic center; d) D abundance in the same regions.

Right, from top to bottom: a) the "cosmic" SFR of disk galaxies,when normalised to the current local value ($z = 0$), does not show the steep observed increase back to $z \sim 1$ (although it peaks at $z \sim 1$, as observational data do); other galaxy types (ellipticals ?) should account for the discrepancy between theory (*solid line*) and observations; data from Madau (1997). b) Evolution of gas and stars; data for HI gas density from Storrie-Lombardi *et al.* (1996). c) The evolution of Zn/H in various regions of spiral disks (only three are shown here, corresponding to those on the left) brackets well the observed abundances of Zn/H in Lyα absorbers; data from Pettini *et al.* (1997) and Lu *et al.* (1996); those systems may well be (proto)galactic disks. d) The corresponding evolution of D shows that considerable depletion may take place in the inner disk regions, but only at low redshifts; abundances measured at high redshifts should be close to the primordial value.

reasonable conclusion, since other galaxy types (ellipticals? mergers? earlier type spirals? dwarfs?) are also expected to have a large contribution to the cosmic SFR at $z > 1$.

The interesting result of that exercise (i.e. placing the Milky Way in a cosmological setting) is that the evolution of Zn in the various disk zones brackets well the observations as a function of the redshift. It may well be indeed, that most of the lines of sight to quasars intercept various parts of (proto)galactic disks. This conclusion is opposite to the one reached in Lu *et al.* (1996), which was based on a comparison of the data to the age-metallicity relationship of the LD only. Notice that for geometrical reasons, the probability of detecting outer disk regions is higher, favouring systematically lower abundances than those spanned during the LD history.

Finally, in the framework of the same model, it turns out that large D depletion may indeed take place in inner galactic disks, but only at low redshifts. D abundances observed at high redshifts in disk galaxies should always be close to the primordial value. If the observed values differ indeed by the large factors reported in this meeting (Tytler, 1998; Vidal-Madjar *et al.*, 1998), then at least one of the corresponding systems has evolved in a radically different way from a typical spiral.

5. Conclusions

The basic points of this short overview may be summarised as follows:

i) Observational data in the solar neighborhood allow in principle to derive the past SFR history of that region. In practice (because of large dispersion in the age-metallicity relationship) they allow only to exclude high early SFRs, favouring a very slow building of the local disk. If this is achieved through infall of primordial composition (or of material slightly astrated, i.e. with metallicity $< 0.2\ Z_\odot$) , only small depletion factors of D ($\sim 2 - 3$) are obtained, favouring low primordial values of D.

ii) Observations of the Milky Way disk suggest a high SFR efficiency in its inner regions. Simple toy-models that reproduce reasonably well those observations, predict an important D gradient in the disk; when measured by forthcoming experiments, the D profile will provide invaluable information on the past history of the disk.

iii) The SFR evolution of the Milky Way is presumably smoother than the observed "cosmic" one. However, the range of metal abundances observed in Lyα absorbers corresponds well to the range spanned by the disk of our Galaxy during its evolution; those systems may well be (proto) galactic disks. Finally, according to these simple models, D can be considerably depleted in inner disk zones, but only at low redshifts.

Acknowledgements

I am grateful to our hosts of ISSI in Bern, J. Geiss and R. von Steiger, for their hospitality, and to B. Pagel and M. Tosi for their comments on this work.

References

Charbonnel, C.: 1998, *Space Sci. Rev.*, this volume.
Edvardsson, B., Andersen, J., Gustafsson, B., Lambert, D., Nissen, P., and Tomkin, J.: 1993, *A&A* **275**, 101.
Fich, M., and Silkey, M.: 1991, *ApJ* **366**, 107.
Galli, D., Stanghellini, L., Tosi, M., and Palla, F.: 1997, *ApJ* **477**, 218.
Geiss, J., and Gloeckler, G.: 1998, *Space Sci. Rev.*, this volume.
Kennicutt, R.: 1989, *ApJ* **344**, 685.
Kroupa, P., Tout, C., and Gilmore, G.: 1993, *MNRAS* **262**, 545.
Linsky, J.: 1998, *Space Sci. Rev.*, this volume.
Madau, P.: 1997, in *7th International Origins Conference*, in press.
Lacey, C. G., and Fall, M.: 1985, *ApJ* **290**, 154.
Lu, L., Sargent, W. and Barlow, T.: 1996, *ApJ Suppl.* **107**, 475.
Pagel, B.: 1997 *Nucleosynthesis and Chemical Evolution of Galaxies*, CUP.
Pei, Y., and Fall, M.: 1995, *ApJ* **454**, 69.
Pettini, M., Smith, L., King, D., and Husntead, R.: 1997, astro-ph/9704102.
Prantzos, N., and Aubert, O.: 1995, *A&A* **302**, 69.
Prantzos. N.: 1996, *A&A* **310**, 106.
Prantzos, N., and Silk, J.: 1997, in preparation.
Rana, N.: 1991, *ARAA* **29**, 129,
Rocha-Pinto, H., and Maciel, W.: 1996, *MNRAS* **279**, 447.
Shaver, P., McGee, R., Newton, L., Danks, A. and Pottasch, S.: 1983, *MNRAS* **204**, 53.
Smartt, S., and Rolleston, W.: 1997, *ApJ* **481**, L47.
Storri-Lombardi, L., McMahon, R., and Irwin, M.: 1996, *MNRAS* **283**, L79.
Talbot, R., and Arnett, D.: 1975, *ApJ* **197**, 551.
Tinsley, B.: 1980, *Fund. Cosm. Phys.* **5**, 287.
Tosi, M.: 1998, *Space Sci. Rev.*, this volume.
Tosi, M., Steigman, G., Matteucci, F. and Chiapinni, C.: 1997, *ApJ*, submitted.
Tytler, D.: 1998, *Space Sci. Rev.*, this volume.
Vidal-Madjar, A., Ferlet, R., and Lemoine, M.: 1998, *Space Sci. Rev.*, this volume.
Wyse, R., and Silk, J.: 1989, *ApJ* **379**, 700.
Wyse, R., and Gilmore, G.: 1995, *AJ* **110**, 2771.

V: SOLAR NEBULA

ABUNDANCES OF DEUTERIUM AND HELIUM-3 IN THE PROTOSOLAR CLOUD

J. GEISS

International Space Science Institute, Hallerstrasse 6, 3012 Bern, Switzerland

G. GLOECKLER

Department of Physics, University of Maryland, College Park MD 20742, USA, and Department of Atmospheric, Oceanic and Space Sciences, University of Michigan, Ann Arbor, MI 48109, USA

Abstract. The mass spectrometric determinations of the isotopic composition of helium in the solar wind obtained from (1) the Apollo Solar Wind Composition (SWC) experiment, (2) the Ion Composition Instrument (ICI) on the International Sun Earth Explorer 3 (ISEE-3), and (3) the Solar Wind Composition Spectrometer (SWICS) on Ulysses are reviewed and discussed, including new data given by Gloeckler and Geiss (1998). Averages of the ^{3}He/^{4}He ratio in the slow wind and in fast streams are given. Taking account of separation and fractionation processes in the corona and chromosphere, ^{3}He/^{4}He $= (3.8 \pm 0.5) \times 10^{-4}$ is derived as the best estimate for the present-day Outer Convective Zone (OCZ) of the sun. After corrections of this ratio for secular changes caused by diffusion, mixing and ^{3}He production by incomplete H-burning (Vauclair, 1998), we obtain (D $+$ ^{3}He)/H $= (3.6 \pm 0.5) \times 10^{-5}$ for the Protosolar Cloud (PSC). Adopting ^{3}He/H $= (1.5 \pm 0.2) \times 10^{-5}$ for the PSC, as is indicated from the ^{3}He/^{4}He ratio in the 'planetary gas component' of meteorites and in Jupiter (Mahaffy *et al.*, 1998), we obtain (D/H)$_{\text{protosolar}} = (2.1 \pm 0.5) \times 10^{-5}$. Galactic evolution studies (Tosi, 1998) show that the measured D and ^{3}He abundances in the Protosolar Cloud and the Local Interstellar Cloud (Linsky, 1998; Gloeckler and Geiss, 1998), lead to (D/H)$_{\text{primordial}} = (2-5) \times 10^{-5}$. This range corresponds to a universal baryon/photon ratio of $(6.0 \pm 0.8) \times 10^{-10}$, and to $\Omega_b = 0.075 \pm 0.015$.

1. Introduction

So far, primordial abundances of deuterium and helium-3 are deduced from observations and measurements in three samples of cosmic material. (1) The present-day galaxy, (2) the Protosolar Cloud (PSC), and (3) clouds with very low contamination from stellar nucleosynthesis, using metallicity as an indicator. In this paper, we present estimates for the abundances of D and ^{3}He in the Protosolar Cloud that represents a sample of galactic material with a nucleosynthetic age of approximately 4.6 Gy. Since D was converted into ^{3}He in the early sun, the abundance of ^{3}He in the Outer Convective Zone (OCZ) of the sun corresponds approximately to the abundance of D $+$ ^{3}He in the PSC. We summarise and review the ^{3}He/^{4}He abundance measurements in the solar wind, from which we then obtain an estimate of the ^{3}He/^{4}He ratio in the present-day OCZ. After a discussion of the processes that have changed the He/H and ^{3}He/^{4}He ratios in the OCZ over the lifetime of the sun, we give our best estimate for (D $+$ ^{3}He)/H in the PSC. The protosolar D/H ratio is then obtained by combining the solar wind ^{3}He/^{4}He data with the ^{3}He/^{4}He ratio in the PSC that is estimated from the values in the 'planetary gas component' of meteorites and in Jupiter. Finally, by comparing the D and ^{3}He abundances in

Space Science Reviews **84**: 239–250, 1998.

the PSC and in the Local Interstellar Cloud (LIC) we discuss constraints on the evolution of these two nuclei in the galaxy and on their primordial abundances.

2. The ^{3}He/^{4}He Ratio in the Solar Wind

The ^{3}He/^{4}He ratio in the solar wind has been measured by several spaceborne instruments. Since the results clearly show that this ratio is not constant, a comprehensive database and at least some understanding of the causes for the changes in ^{3}He/^{4}He are needed for obtaining the best estimate for the ^{3}He/^{4}He ratio in the present-day OCZ. Comprehensive results have been obtained from three investigations: (1) the Apollo Solar Wind Composition (SWC) experiments, using solar wind collection in foils with subsequent analysis by laboratory mass spectrometry, (2) the Ion Composition Instrument (ICI) on the International Sun Earth Explorer 3 (ISEE-3) using an electromagnetic mass spectrometer allowing unambiguous measurement of the mass/charge ratio of the ions, and (3) the Solar Wind Ion Composition Spectrometer (SWICS) on Ulysses, a time-of-flight system giving the mass/charge ratio as well as the mass of the ions. The results obtained with these three techniques are given in Table I. True variations in the ^{3}He/^{4}He ratio were found by all three experiments. However, for comparable solar wind flows (cf. the averages in the slow wind) the agreement between the results obtained over 25 years by these three completely different techniques is very remarkable. Data not explicitly given in Table I include those obtained by electrostatic analysers (Bame et al., 1968; Grünwaldt, 1976) and a result from a collection experiment flown with a Luna mission (Boltenkov et al., 1972). Some of these data represent special events, emphasizing the variability in the ^{3}He/^{4}He ratio, but they do not indicate a disagreement with the more general results given in Table I.

Prior to the Ulysses mission, all ^{3}He/^{4}He data were taken in the ecliptic plane, where the low speed solar wind dominates, although some data were obtained in fast streams and during CME events. The SWC-Apollo foils collected solar wind particles mainly during slow wind conditions. Since the technique employed by this experiment gave very small systematic errors, we assign a high weight to the SWC-Apollo results in calculating the average ^{3}He/^{4}He ratio in the slow wind from the Apollo, ISEE-3 and Ulysses data. Thanks to the polar orbit of the spacecraft and the large energy range of the instrument, SWICS-Ulysses has allowed the first systematic investigation of the helium isotopes in the high speed streams coming out of large coronal holes. The result is that ^{3}He/^{4}He is lower in the fast streams than it is in the average slow solar wind. The uncertainties in the absolute ^{3}He/^{4}He ratios obtained in the fast streams and in the slow wind are larger than the uncertainties in the difference. We have taken this into account in our estimate for the average of ^{3}He/^{4}He in fast streams. As Table I shows, Bodmer et al. (1995) reported a smaller difference than Gloeckler and Geiss (1998). In part, this is due to a different choice of detector combinations in the SWICS instrument.

Table I

Results of mass spectrometric measurements of ^{3}He/^{4}He in the solar wind.

	^{3}He/^{4}He [10^{-4}]	
SWC Apollo 11-16 (1969–1972)		
Weighted average of 5 missions	4.26 ± 0.21	a
Average over flux and time	4.36 ± 0.25	a
ICI ISEE-3 (1978–1982)		
Long-time average	4.88 ± 0.48	b,c,d
Average of ratios	4.37 ± 0.5	d
Slow Wind ($V < 500$ km/s)	4.8 ± 0.48	c,d
SWICS Ulysses (1991–1996)		
Slow Wind ($V < 500$ km/s)	4.6 ± 0.6	e
Fast Stream ($V > 700$ km/s)	4.4 ± 0.6	e
Slow Wind ($V < 500$ km/s)	4.08 ± 0.25	f
Fast Stream ($V > 700$ km/s)	3.3 ± 0.3	f
Slow Wind, Adopted Average	4.3 ± 0.4	
Fast Stream, Adopted Average	3.6 ± 0.5	

(a) Geiss *et al.*, 1970a; Geiss *et al.*, 1972;

(b) Ogilvie *et al.*, 1980;

(c) Coplan *et al.*, 1984;

(d) Bochsler, 1984; the error for the average of ratios was adopted from the error of the long-time average;

(e) Bodmer *et al.*, 1995; the given errors include an estimate of the systematic error;

(f) Gloeckler and Geiss, 1998.

Bodmer (1996) has since compared three detector combinations and found ratios between ^{3}He/^{4}He in the slow wind and in fast streams ranging from 1.07 to 1.25. In determining the ^{3}He/^{4}He ratio representative for fast streams, we have given more weight to the new data (Gloeckler and Geiss, 1998), because they represent a broader range of solar latitudes than the earlier result of Bodmer *et al.* (1995).

3. ^{3}He/^{4}He in the Outer Convective Zone of the Sun

Elemental and isotopic abundances in the solar wind are not exactly the same as they are in its source reservoir, the Outer Convective Zone (OCZ) of the sun. Separation processes in the chromosphere and in the corona produce a general variability in solar wind ion abundances as well as systematic differences between solar wind and OCZ composition.

In this section, we present two independent derivations of the isotopic composition of helium in the OCZ, the one based on the ^{3}He/^{4}He ratio in the fast streams, the other on the ^{3}He/^{4}He ratio in the slow wind.

Table II

Derivation of the ^{3}He/^{4}He abundance ratio in the Outer Convective Zone of the sun. Corrections to the observed solar wind ^{3}He/^{4}He ratios.

	^{3}He/^{4}He Correction
Fast Streams, ^{3}He/^{4}He $= (3.6 \pm 0.5) \times 10^{-4}$	
Chromospheric Separation	$(0 \pm 4)\%$
Corona / Solar Wind Acceleration	
Slow Wind, ^{3}He/^{4}He $= (4.3 \pm 0.4) \times 10^{-4}$	
Chromospheric Separation	$(0 \pm 4)\%$
Corona / Solar Wind Acceleration	$-(8 \pm 5)\%$
Outer Convective Zone	
from Fast Stream Data	^{3}He/^{4}He $= (3.6 \pm 0.5) \times 10^{-4}$
from Slow Wind Data	^{3}He/^{4}He $= (4.0 \pm 0.5) \times 10^{-4}$

An ion-atom separation process operating in or near the chromosphere causes an overabundance of elements with low first ionisation potential (the "FIP-Effect") and an underabundance of helium in the plasma that is fed into the corona. Specifics concerning the dynamics and geometry of the separation mechanism remain uncertain (cf. Hénoux and Somov, 1992; von Steiger et al., 1997). However, solar flare particle and solar wind data (cf. Garrard and Stone, 1994; von Steiger et al., 1997), covering more than a dozen species with very different atomic properties, indicate that the FIP effect results from a competition between ionisation time in the outer chromosphere and a characteristic ion-atom separation time. This observation allows the fractionations of the ^{3}He/^{4}He and He/H ratios to be linked.

In fast streams, the He/H ratio is remarkably constant at 0.05 (Bame et al., 1977). With He/H = 0.084 in the OCZ (Pérez Hernández and Christensen-Dalsgaard, 1994), this gives a depletion of ^{4}He by 40%. If atom-ion separation in the chromosphere is controlled by diffusion, the separation lengths would scale with $m^{1/4}$. Since the ionisation times for ^{3}He and ^{4}He are the same (cf. von Steiger and Geiss, 1989), a depletion of ^{3}He by 42% is calculated, i.e. the chromospheric process would cause a decrease in the ^{3}He/^{4}He ratio by $\sim 3.5\%$. This estimate agrees very well with the decrease obtained by the numerical models of von Steiger and Geiss (1989). On the other hand, if the helium depletion in the gas fed into the corona should be caused by some sort of gravitational settling (cf. Barraclough et al., 1996), an increase in ^{3}He/^{4}He of a similar magnitude could result. We adopt $0 \pm 4\%$ (cf. Table II) for the ^{3}He/^{4}He separation in the chromosphere or below.

As compared to the slow wind, the fast streams are more steady and homogenous, variations in elemental abundances and charge states are smaller, and the expansion geometry is fairly well known. Therefore, the fast streams are much better represented by existing steady-state models than the slow wind. Theories and numerical calculations show that much of the kinetic energy in the flow is supplied by wave heating and momentum transfer from waves (Munro and Jackson,

1977; Bürgi and Geiss, 1986; McKenzie *et al.*, 1997; Banaszkiewicz *et al.*, 1998) which, if non-resonant, gives the same thermal velocity or acceleration to all ion species (Hollweg, 1974; Bürgi and Geiss, 1986). Ion heating by waves is directly supported by the SOHO-UVCS results which show that the ion temperatures in coronal holes are much higher than the electron temperatures, and they increase with increasing mass (Kohl *et al.*, 1997). Iron has the highest mass and mass/charge ratio among the elements monitored in the solar wind, but we have as yet not seen a depletion of the Fe/Mg ratio in the fast streams. Since there is no evidence for ion fractionation in the process of solar wind acceleration out of coronal holes, we assume that the only correction to be applied to the ^{3}He/^{4}He ratio in fast streams is the chromospheric correction (Table II).

The He/H ratio in the slow wind is quite variable, daily averages vary typically between 0.035 and 0.05. Observation and theory (cf. Bürgi and Geiss, 1986) indicate that these variations are due to helium settling in the corona and losses in the solar wind acceleration process (see below). The typical upper value of 0.05 for the daily averages of He/H indicates that the chromospheric effect on the He/H ratio is similar for the fast streams and the slow wind. Therefore, we adopt the $0 \pm 4\%$ correction also for the slow wind (Table II).

If we have He/H = 0.05 in the plasma coming up from the chromosphere, the depletion below 0.05 must occur in the corona. There, helium is fully ionised, and thus the dynamical behaviour of ^{3}He^{++} lies somewhere between H$^+$ and ^{4}He^{++}. The slow wind represents a mixture of flow types, and since expansion geometries and wave fields are not well known, we do not have an adequate theoretical description. Temporary excursions of the ^{3}He/^{4}He ratio to high values in the slow wind were observed by several instruments (Bame *et al.*, 1968; Geiss *et al.*, 1970; Grünwaldt, 1976; Coplan *et al.*, 1984; Gloeckler *et al.*, 1997; Gloeckler and Geiss, 1998). Certainly, these high ratios show that, on average, ^{3}He/^{4}He in the slow wind is enhanced above the OCZ abundance ratio, although there is as yet no adequate theoretical explanation for these excursions. However, even under relatively quiet slow wind conditions some degree of helium isotope fractionation must be expected. Steady-state models give some information on the relative behaviour of ion species, and they show that ^{4}He^{++} can be much more depleted than ^{3}He^{++} (Geiss *et al.*, 1970; Bürgi and Geiss, 1986). In the absence of significant momentum transfer from waves, the relative dynamical behaviour of heavy ions in a proton-electron gas is characterised by a drag-factor which in its most simple form is given by $\Gamma = Z^2/(2A - Z - 1)$, where Z is the charge state, and A the mass number of the ion species (Geiss *et al.*, 1970). Thus, for ^{3}He^{++}, ^{4}He^{++}, ^{16}O^{6+} and ^{56}Fe^{11+} the Γ-factors are 1.33, 0.80, 1.44 and 1.21, respectively. These figures show that ^{4}He^{++} is particularly disadvantaged when escape from the solar gravitational field depends strongly on the momentum transfer from the proton-electron gas.

The average ^{4}He/H ratio in the slow wind is 0.038, corresponding to a reduction of $\sim 25\%$ caused by coronal separation processes. Considering the large difference in the drag factors for the two helium isotopes, the depletion of ^{3}He/H in the slow wind could be considerably less. We take this and an occasional, inadvertant admixture of plasma with anomalously high ^{3}He abundance (Gloeckler *et al.*, 1997) into account by assigning a large error to the increase of ^{3}He/^{4}He in the slow wind, which we adopt to be $(8 \pm 5)\%$ (Table II).

The corrections here lead to the two values for the ^{3}He/^{4}He ratio in the OCZ given at the bottom of Table II. Giving the two determinations equal weight, we obtain for the Outer Convective Zone

$$(^3\text{He}/^4\text{He})_{\text{OCZ}} = (3.8 \pm 0.5) \times 10^{-4} \tag{1}$$

We have discussed here in some detail the corrections needed to derive ^{3}He/^{4}He in the OCZ from the solar wind data. The aim was to show that these corrections can be estimated by straightforward physical arguments, using the observed solar wind He/H abundance pattern, independent of very particular assumptions which go into model calculations. We emphasize that the net correction is relatively minor, the average of the ^{3}He/^{4}He ratios in the fast streams and the slow wind being only 4% higher than the $(^3\text{He}/^4\text{He})_{\text{OCZ}}$ value given above (cf. Tables I and II).

4. (D+^{3}He)/H, D/H and ^{3}He/H in the Protosolar Cloud

When 4.6 Gy ago a fragment of an interstellar cloud collapsed to form the solar system, the sun was largely formed by direct infall (Tscharnuter, 1987) implying that the material going into the sun was representative of the Protosolar Cloud. In the early sun, D was converted into ^{3}He which has not been further processed in the material of the Outer Convective Zone (OCZ), as can be surmised from the existence there of beryllium (Geiss and Reeves, 1972). ^{9}Be is indeed much faster destroyed than ^{3}He at all temperatures (Figure 1). Thus, the ^{3}He/^{4}He ratio in the OCZ basically represents the protosolar (D + ^{3}He)/H ratio. There are, however, two processes that could have changed the ^{3}He/^{4}He ratio in the Outer Convective Zone during the lifetime of the sun.

Solar seismic data and solar models show that He/H in the OCZ is 16% lower than it was in the PSC (Pérez Hernández and Christensen-Dalsgaard, 1994; Bahcall and Pinsonneault, 1995). The difference is interpreted as being due to settling of helium out of the OCZ into deeper layers of the sun. ^{3}He settles more slowly than ^{4}He, and model calculations (Gautier and Morel, 1997) show an increase of a few percent of the ^{3}He/^{4}He ratio in the present-day OCZ.

The second possible change of $(^3\text{He}/^4\text{He})_{\text{OCZ}}$ over solar history is due to solar mixing. Well below the OCZ, at intermediate levels in the sun, the p-p reaction slowly produces ^{3}He, which is not converted into ^{4}He, because a much higher temperature would be required (cf. Figure 1). Some of the ^{3}He produced by this incomplete H-burning could have been convected upwards during solar history, thus increasing ^{3}He/^{4}He in the OCZ (cf. Schatzman and Maeder, 1981; Bochsler *et al.*, 1990). Using mixing models to various solar depths, Vauclair (1998, 1998a)

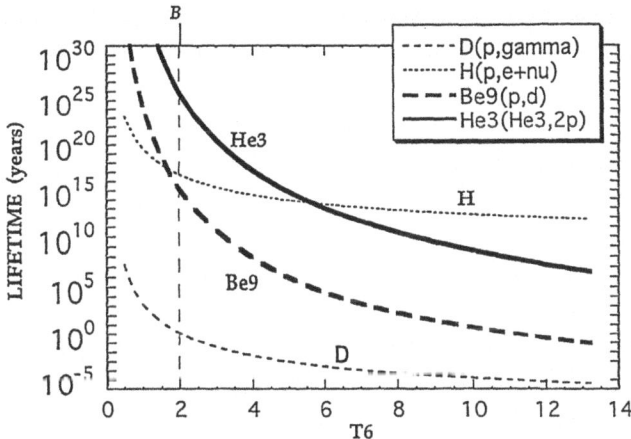

Figure 1. Lifetimes of light nuclei as a function of temperature. Given is the lifetime due to the fastest destructive reaction of the species (cf. insert). Density and composition were normalised to the present conditions at the bottom of the Outer Convective Zone (OCZ). The dashed line (B) is the present temperature at the bottom of the OCZ.

has shown that in order to deplete lithium by two orders of magnitude, as is required, some p-p produced ^{3}He might be added to the OCZ, possibly increasing ^{3}He/^{4}He by several percent.

On the lunar surface, we have a record of solar wind irradiation that goes back to about 4 Gy. In most materials, the ^{3}He/^{4}He ratio of the old solar wind samples has been affected by strong diffusive losses of helium. The loss is least severe in ilmenite, and from the results of an on-line etching technique applied to this mineral, Wieler *et al.* (1992) have deduced that the change in the solar wind ^{3}He/^{4}He ratio over the last 3 Gy was less than 10 percent. Since the peak of ^{3}He increases over solar history and moves slowly outward, we expect a large fraction of the contamination of the OCZ with ^{3}He from incomplete H-burning to have occurred during the last 3 billion years of solar history. Thus settling of helium out of the OCZ and solar mixing could not have increased the ^{3}He/^{4}He ratio in the OCZ by much more than 10%. We thus adopt a correction of $(5 \pm 3)\%$ and, using He/H = 0.10 for the PSC (Bahcall and Pinsonneault, 1995), we obtain

$$[(D + {}^3\mathrm{He})/H]_{\mathrm{Protosolar}} = (3.6 \pm 0.5) \times 10^{-5} \qquad (2)$$

from the ^{3}He/^{4}He ratio in the present-day OCZ.

The solar wind data give only the protosolar abundance of the sum of D and ^{3}He. For distinguishing between the two species, an estimate for the ^{3}He/^{4}He ratio in the PSC is needed. In the past, ^{3}He/^{4}He = 1.5×10^{-4}, as measured in the 'planetary gas component' of meteorites (Black, 1972; Eberhardt, 1974) was widely used as the best approximation to the protosolar value. In 1996, a Jovian ^{3}He/^{4}He value of $(1.1 \pm 0.2) \times 10^{-4}$ was obtained by the Galileo Probe Mass Spectrometer (Niemann *et al.*, 1996). This result was somewhat surprising, because one would expect the 'planetary component' to be depleted, rather then augmented in ^{3}He, because most trapping and/or loss mechanisms would tend to decrease the ^{3}He/^{4}He ratio in the helium that is retained in the solid phase. Since the new Jupiter value (Mahaffy *et al.*, 1998; Niemann *et al.*, 1998) is higher than the meteoritic value, this puzzle is now resolved, and we adopt here $(1.5 \pm 0.2) \times 10^{-4}$ for the protosolar ^{3}He/^{4}He ratio, which is compatible with the presently available Jovian and meteoritic data. Thus we obtain from (2)

$$(D/H)_{\text{protosolar}} = (2.1 \pm 0.5) \times 10^{-5} \qquad (3)$$

The error limits in (2) and (3) are 1σ-uncertainties. They include experimental errors, uncertainties in the correction for chromospheric and coronal effects on the solar wind composition, and the uncertainty resulting from helium settling and admixture of p-p produced ^{3}He to the OCZ.

Concerning the best estimate for $(D/H)_{\text{protosolar}}$, we give in this paper priority to the value derived from the solar ^{3}He/^{4}He abundance ratio. Determinations of Jovian D/H by different methods have given values which disagree by a factor of two and more (cf. Niemann *et al.*, 1996). We note, however, that the D/H ratio obtained from ISO infrared spectra of molecular hydrogen (2.2×10^{-5}; Encrenaz *et al.*, 1996) and the new ratio from the Galileo Probe Mass Spectrometer (Mahaffy *et al.*, 1998; Niemann *et al.*, 1998) are close to the $(D/H)_{\text{protosolar}}$ value derived here, cf.(3). Furthermore, gas losses from the protoplanetary disc or from the Jovian sub-nebula could in two ways have increased the D/H ratio in the planet: the escape of gas under the influence of gravity would (1) favour H_2 over DH, and (2) enhance the grain component which is probably enriched in D, leading to an increased D/H ratio in the material going into Jupiter.

The method of deriving the protosolar deuterium abundance from the ^{3}He/^{4}He ratio in the solar wind has been used for twenty-five years (Geiss and Reeves, 1972: $(D/H)_{\text{protosolar}} = (2.5 \pm 1.0) \times 10^{-5}$). As additional solar wind data (cf. Table II), new estimates of He/H in the OCZ and PSC, other data for the protosolar ^{3}He/^{4}He ratio, and revised assumptions or model results on solar mixing and fractionation processes in the solar atmosphere became available, several authors have used the same method for deriving $(D/H)_{\text{protosolar}}$ (Reeves *et al.*, 1973: $(2.6 \pm 1.0) \times 10^{-5}$; Anders and Grevesse, 1989: $(3.4 \pm 0.1) \times 10^{-5}$; Geiss, 1993: $(2.6 \pm 1.0) \times 10^{-5}$; Gautier and Morel, 1997: $(3.01 \pm 0.17) \times 10^{-5}$). All these results, as well as our result of $(2.1 \pm 0.5) \times 10^{-5}$ (cf. 3) lie within the error bars given in 1972, indicating the robustness of the method.

5. Implications for Galactic Evolution and Cosmology

With the Protosolar Cloud (PSC) and the Local Interstellar Cloud (LIC) we have two galactic samples, differing in nucleosynthetic age by 4.6 Gy, for which we have reliable data on the isotopic abundances of both hydrogen and helium. D/H in the

LIC (Linsky, 1998) is lower than it was in the PSC. On the other hand, ^{3}He/^{4}He in the LIC (Gloeckler and Geiss, 1998) is higher than it was in the PSC. The direction of these changes is as expected. Because D is destroyed but not produced by stars, the D/H ratio ought to decrease monotonously with time. ^{3}He, on the other hand, is both destroyed and produced by stars. The observed increase in ^{3}He from the PSC to the LIC value is mainly due to p-p production in small stars (cf. Tosi, 1998), a stellar population that began to leave the main sequence and to lose material only relatively late in galactic history.

The LIC is of course not a direct descendent of the PSC. However, since (a) we have good D *and* ^{3}He abundance data for both, (b) they evolved at roughly the same distance from the galactic centre and (c) their age difference is not small, but corresponds to > 30% of the age of the universe, a comparison of the two clouds provides us with unique information on galactic evolution. The PSC is the sample with the best defined nucleosythetic status. Our knowledge on elemental and isotopic abundances in the LIC is still scarce, but will be growing, thanks to refined spectroscopic methods, and to direct measurements on interstellar grains and gas components that pass through the heliosphere. At the present time, the data do not allow a determination of the difference in the He/H ratio or in metallicity between LIC and PSC.

Thus, in Figure 2 we plot the abundance ratios as a function of nucleosynthetic age, with the age of the universe taken as $\sim$ 14 Gy (cf. Tammann, 1998). Since interstellar deuterium is destroyed and not produced by stars and since production of D by cosmic rays is minor, the curve labelled "Minimum D/H" gives the lower limit of primordial D/H ($\sim 2 \times 10^{-5}$). The curve labelled "Steep Decrease D/H" connects the D/H ratios in the LIC and the PSC with the primordial deuterium abundance of D/H = 5×10^{-5} given by Tosi (1998) as the upper bound derived by galactic evolution models. We note that the low D/H values obtained in high-z clouds given by Burles and Tytler (1998) are consistent with this range of primordial D/H ratios. Higher values are difficult to reconcile with the D/H and ^{3}He/^{4}He abundance ratios measured in the PSC and LIC.

From the primordial deuterium abundance given here, $(D/H)_{\text{primordial}} = (2 - 5) \times 10^{-5}$ (cf. Figure 2), the theory of Standard Big Bang Nucleosynthesis (Walker *et al.*, 1991) allows the determination of ranges for other cosmologically relevant parameters: $(^3He/D)_{\text{primordial}} = 0.29 - 0.54$; the baryon/photon ratio in units of 10^{-10}, $\xi_{10} = 6.0 \pm 0.8$; the present baryonic density $\rho_{b,0} = (4.7 - 3.6) \times 10^{-31}$ g/cm^3; and (with $H_0 = 55 \pm 7$ km/s/Mpc, cf. Tammann, 1998) the present relative baryonic density $\Omega_b = 0.075 \pm 0.015$. This range of Ω_b values derived from galactic data is compatible with the lower limit of $\Omega_b > 0.06$ obtained from Lyman absorption lines (Rauch *et al.*, 1997; Tammann, 1998) and confirms that the baryonic matter density is insufficient to close the universe as has been known for more than two decades (cf. Geiss and Reeves, 1972; Reeves *et al.*, 1973). There is mounting evidence (Tammann, 1998) that the observed total density Ω_{tot}

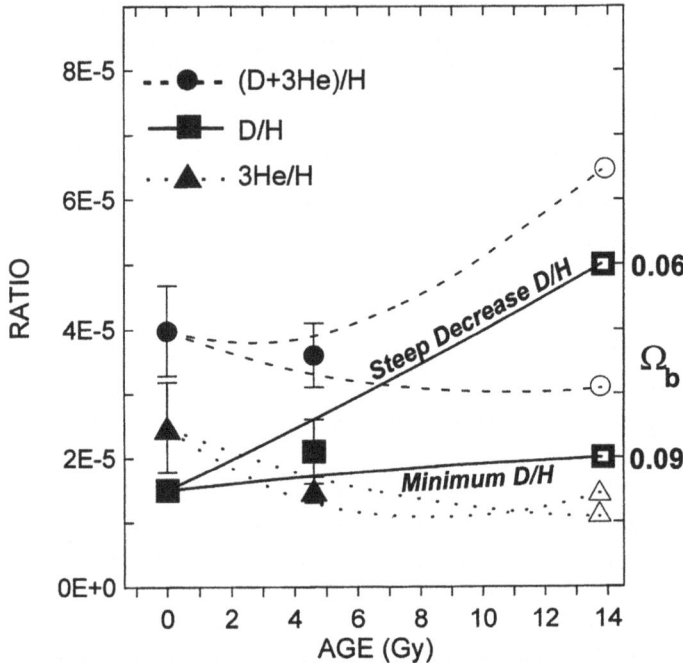

Figure 2. Illustration of the evolution of D and ³He in the "solar ring" of the galaxy. The measured abundance data for the Local Interstellar Cloud (LIC) at age 0 are from Linsky (1998) and Gloeckler and Geiss (1998), those for the Protosolar Cloud (PSC) at age 4.6 Gy are from this work and from Mahaffy *et al.* (1998). The open symbols at ∼ 14 Gy represent primordial abundances obtained from extrapolations of these galactic data. The curve labelled "Minimum D/H" leads to the minimum primordial abundance ratio inferred from the PSC data. The curve labelled "Steep Decrease D/H" leads to $(D/H)_{primordial} = 5 \times 10^{-5}$, the upper value obtained from galactic evolution studies by Tosi (1998). The dotted lines connect the measured ³He/H ratios with the theoretical primordial value, the dashes lines connect the measured (D+³He)/H ratios with the theoretical primordial (D+³He)/H ratio.

is higher by a factor of 2 to 10 than the baryonic density of $\Omega_b = 0.075 \pm 0.015$, the difference probably being made up of weakly interacting matter. A welcome corollary of this finding is that a dominant component of weakly interacting matter could better than baryonic matter explain the observed inhomogeneities of matter and radiation in the universe (cf. Bennett *et al.*, 1997).

Acknowledgements

We have very much benefitted from discussions with Thomas Donahue, Hubert Reeves, Gustav Andreas Tammann, Monica Tosi, Sylvie Vauclair, Heinz Völk and Rudolf von Steiger and we thank Ursula Pfander for her help with the manuscript. This work was supported by the ISSI-Foundation, the European Space Agency and the National Aeronautics and Space Administration (NASA/JPL contract 955460).

References

Anders, E. and Grevesse, N.: 1989, *Geochim. and Cosmochim. Acta* **53**, 197–214.

Bahcall, J.N. and Pinsonneault, M.H.: 1995, *Rev. Modern Physics* **67**, 781.

Bame, S.J., Hundhausen, A.J., Asbridge, J.R. and Strong, I.B.: 1968, *Phys. Rev. Lett.* **20**, 393.

Bame, S.J., Asbridge, J.R., Feldman, W.C. and Gosling, J.T.: 1977, *J. Geophys. Res.* **82**, 1487–1492.

Banaszkiewicz, M., Czechowski, Axford, W.I., McKenzie, J.F., and Sukhoruva, G.V.: 1998, 'Proc. 31st ESLAB Symposium', in press.

Barraclough, B.L., Feldman, W.C., Gosling, J.T., McComas, D.J., Phillips, J.L. and Goldstein, B.E.: 1996, in 'Solar Wind Eight' *AIP Conf. Proc.* **382**, (eds. Winterhalter, D., Gosling, J.T., Habbal, S.R., Kurth, W.S., and Neugebauer, M.), AIP Press, Woodbury, N.Y., 277–280.

Bennett, L., Turner, M.S., White, M.: 1997, *Physics Today* **November 1997**, 32–38.

Black, D.C.: 1972, *Geochim. Cosmochim. Acta* **36**, 347.

Bochsler, P.: 1984, 'Helium and Oxygen in the Solar Wind', Habilitation Thesis, University of Bern.

Bochsler, P., Geiss, J. and Maeder, A.:1990, *Solar Phys.* **128**, 203.

Bodmer, R., Bochsler, P., Geiss, J., von Steiger, R. and Gloeckler, G.: 1995, *Space Sci. Rev.* **72**, 61.

Bodmer, R.: 1996, 'The Helium Isotopic Ratio as a Test for Minor Ion Fractionation in the Solar Wind Acceleration Process: SWICS/ULYSSES Data Compared with Results from a Multifluid Model', Ph.D. Thesis, University of Bern.

Boltenkov, B.S., Gartmanov, V.N., Kocharov, G.E., Naidenov, V.O. and Starbunov, Ju.N.: 1972, *Isvestia Akad. Nauk USSR, Ser. Phys.* **34**, 2319.

Bürgi, A. and Geiss, J.: 1986, *Solar Phys.* **103**, 347.

Burles, S. and Tytler, D.: 1998, *Space Sci. Rev.*, this volume.

Coplan, M.A., Ogilvie, K.W., Bochsler, P. and Geiss, J.: 1984, *Solar Phys.* **93**, 415.

Eberhardt, P.: 1974, *Earth Planet. Sci. Lett.* **24**, 182.

Encrenaz, Th. *et al.*: 1996, *Astron. Astrophys.* **315**, L397–402.

Garrard, T.L. and Stone, E.C.: 1994, *Adv. Space Res.* **14**, 589–598.

Gautier, D. and Morel, P.: 1997, *Astron. Astrophys.* **323**, L9–L12.

Geiss, J., Hirt, P. and Leutwyler, H.: 1970, *Solar Physics* **12**, 458.

Geiss, J., Eberhardt, P., Bühler, F., Meister, J. and Signer, P.: 1970a, *J. Geophys. Res.* **75**, 5972.

Geiss, J. and Reeves, H.: 1972, *Astron. Astrophys.* **18**, 126–132.

Geiss, J., Bühler, Cerutti, H., Eberhardt, P. and Filleux, Ch.: 1972 *'Apollo 16 Preliminary Science Report', NASA SP-315* **Section 14**.

Geiss, J.: 1993, *'Origin and Evolution of the Elements'* (eds. Prantzos, N., Vangioni-Flam, E., and Cassé, M.), Cambridge University Press, 89–106.

Gloeckler, G., *et al.*: 1997, *EOS Trans. Am. Geophys. Union* **78**, 438.

Gloeckler, G. and Geiss, J.: 1998, *Space Sci. Rev.*, this volume.

Grünwaldt, H.: 1976, *Space Research* **XVI**, 681–684.

Hénoux, J.-C. and Somov, B.V.: 1992, *ESA SP-348*, pp. 325–330.

Hollweg, J.V.: 1974, *J. Geophys. Res.* **79**, 1539.

Kohl, J. L., *et al.*: 1997, 'UVCS/SOHO Empirical Determinations of anisotropic velocity distribution in the solar corona' *Harvard College Observatory* **Preprint Series No. 4630**.

Linsky, J.L.: 1998, *Space Sci. Rev.*, this volume.

Mahaffy, P. R., *et al.*: 1998, *Space Sci. Rev.*, this volume.

McKenzie, J.F., Sukhorukova, G.V. and Axford, W.I.: 1997, *Astron. Astrophys.*, in press.

Munro, R.H. and Jackson, B.V.: 1977, *Astrophys. J.* **213**, 874.

Niemann H.B. *et al.*: 1996, *Science* **272**, 846–849.

Niemann, H.B. *et al.*: 1998, *J. Geophys. Res.*, in press.

Ogilvie, K.W., Coplan, M.A., Bochsler, P. and Geiss, J.: 1980, *J. Geophys. Res.* **85**, 6021.

Pérez Hernández, F. and Christensen-Dalsgaard, J.: 1994, *MNRAS* **269**, 475.

Rauch, M. *et al.*: 1997, *Astrophys. J.* **489**, 7.

Reeves, H. Audouze, J., Fowler, W.A. and Schramm, D.N.: 1973, *Astrophys. J.* **179**, 909.

Schatzman, E. and Maeder, A.: 1981, *Astron. Astrophys.* **96**, 1.

von Steiger, R. and Geiss, J.: 1989, *Astron. Astrophys.* **225**, 222–238.

von Steiger, R., Geiss, J. and Gloeckler, G.: 1997, in 'Cosmic Winds and the Heliosphere' (eds. Jokipii, J.R., Sonett, C.P. and Giampapa, M.S.), Tucson (Conference 1973), University of Arizona Press, pp. 581–616.
Tammann, G.A.: 1998, *Space Sci. Rev.*, this volume.
Tosi, M.: 1998, *Space Sci. Rev.*, this volume.
Tscharnuter, W.M.: 1987, *Astron. Astrophys.* **188**, 55–73.
Vauclair, S.: 1998, *Space Sci. Rev.*, this volume.
Vauclair, S.: 1998a, *Space Sci. Rev.*, in press.
Walker, T.P., Steigman, G., Schramm, D.N., Olive, K.A. and Kang, H.S.: 1991, *Astrophys. J.* **376**, 51.
Wieler, R., Baur, H. and Signer, P.: 1992, *Lunar. Planet. Sci.* **XXIII**, 1525–1526.

GALILEO PROBE MEASUREMENTS OF D/H AND ^{3}HE/^{4}HE IN JUPITER'S ATMOSPHERE

P.R. MAHAFFY

Goddard Space Flight Center, Greenbelt, MD 20771, USA

T.M. DONAHUE and S.K. ATREYA

Department of Atmospheric, Oceanic and Space Sciences, University of Michigan, 2455 Hayward Street, Ann Arbor, MI 48109, USA

T.C. OWEN

Institute for Astronomy, University of Hawaii, 2680 Woodlawn Drive, Honolulu, HI 96822, USA

H.B. NIEMANN

Goddard Space Flight Center, Greenbelt, MD 20771, USA

Abstract. The Galileo Probe Mass Spectrometer measurements in the atmosphere of Jupiter give

$$D/H = (2.6 \pm 0.7) \times 10^{-5}$$

$$^3He/\,^4He = (1.66 \pm 0.05) \times 10^{-4}$$

These ratios supercede earlier results by Niemann *et al.* (1996) and are based on a reevaluation of the instrument response at high count rates and a more detailed study of the contributions of different species to the mass peak at 3 amu. The D/H ratio is consistent with Voyager and ground based data and recent spectroscopic and solar wind (SW) values obtained from the Infrared Spectroscopic Observatory (lSO) and Ulysses. The $^3He/^4He$ ratio is higher than that found in meteoritic gases $(1.5 \pm 0.3) \times 10^{-4}$. The Galileo result for D/H when compared with that for hydrogen in the local interstellar medium $(1.6 \pm 0.12) \times 10^{-5}$ implies a small decrease in D/H in this part of the universe during the past 4.55 billion years. Thus, it tends to support small values of primordial D/H – in the range of several times 10^{-5} rather than several times 10^{-4}. These results are also quite consistent with no change in $(D+^3He)/H$ during the past 4.55 billion years in this part of our galaxy.

1. Introduction

Measurements of the relative abundances of primordial light nuclei can provide fundamental insights into conditions prevailing in the early universe. They have implications in particular for the baryon-to-photon density ratio at that time and thus for the fraction of dark matter that can be ordinary matter. Following the Big Bang, the ratio of deuterium to hydrogen (D/H) in the universe should have decreased monotonically, since destruction of D in stars far outstrips any known processes that create it. The situation with ^{3}He is less simple, since it can be destroyed as well as created in stars. To constrain models of galactic evolution of deuterium and ^{3}He it is necessary to know the primordial ratio of D to H and how it and the ^{3}He to H ratio have evolved. Unfortunately, measurements of D/H in extragalactic, very highly red-shifted hydrogen clouds that would give us the primordial ratio have ranged over a full order of magnitude (Songaila, *et al.*, 1997; Tytler *et al.*, 1996; Webb *et al.*, 1997). Consequently we are far from realizing this objective. On the

Space Science Reviews **84**: 251–263, 1998.
© 1998 *Kluwer Academic Publishers*

other hand rather precise measurements of ^{3}He/H and D/H in the local interstellar medium are now available.

The values of these ratios 4.55 billion years ago in the region of the interstellar medium where the solar system was formed can be determined if they can be measured at suitably chosen places in solar system objects today. One such measurement involves determining the ratio of ^{3}He to ^{4}He in the solar wind and subtracting the contribution of protosolar ^{3}He to obtain the protosolar D to He ratio (Geiss and Gloeckler, 1998). All of the sun's deuterium was converted to ^{3}He during its pre-main sequence phase. To derive the D/H ratio in the primitive solar nebula itself (the protosolar value) from these measurements it is necessary to correct the solar wind value. These corrections account for changes in isotopic composition that occur in the chromosphere and corona and then for changes of ^{3}He/H and ^{4}He/H in the outer convective zone that have occurred because of transport processes and incomplete H burning. Until the measurements reported here were made, meteorites have been the sole sources of helium that might yield the protosolar nebular ^{3}He/^{4}He ratio. They can do so only if the unknown process by which helium was incorporated in the bodies from which the meteorites formed did not fractionate the isotopes.

Another method of determining the values of D/H and ^{3}He/^{4}He in the primitive solar nebula is to measure these quantities in the atmosphere of Jupiter. This procedure is valid if all but a negligible portion of the isotopes of the light gases found in the Jovian atmosphere today came directly from the gaseous nebula with no important contributions from sources of fractionated hydrogen or helium, such as icy planetesimals. This assumption is supported by the fact that the D/H in comet Halley (Balsiger et al., 1995; Eberhardt et al., 1995), Hyakutake (Bocklee-Morvan et al., 1998) and Hale-Bopp (Meier et al., 1998) is at least an order of magnitude greater than the Jovian value derived in this paper. It also supposes that the process that depletes He in the atmosphere of Jupiter is not mass dependent; that no significant fractionation of hydrogen isotopes had occurred in the solar nebula by the time and in the place where Jupiter formed; and that the appropriate He to H ratio is known.

It is the purpose of this paper to report measurements of D/H and ^{3}He/^{4}He in the Jovian atmosphere by the Galileo Probe Mass Spectrometer (GPMS) and to discuss their implications. The new analysis presented here is based on extensive laboratory calibration and replaces our earlier result (Niemann et al., 1996). These results are compared with remote sensing observations of Jupiter, which give a D/H measurement, and meteoritic measurements of ^{3}He/^{4}He, which give an estimate of the protosolar value of this ratio.

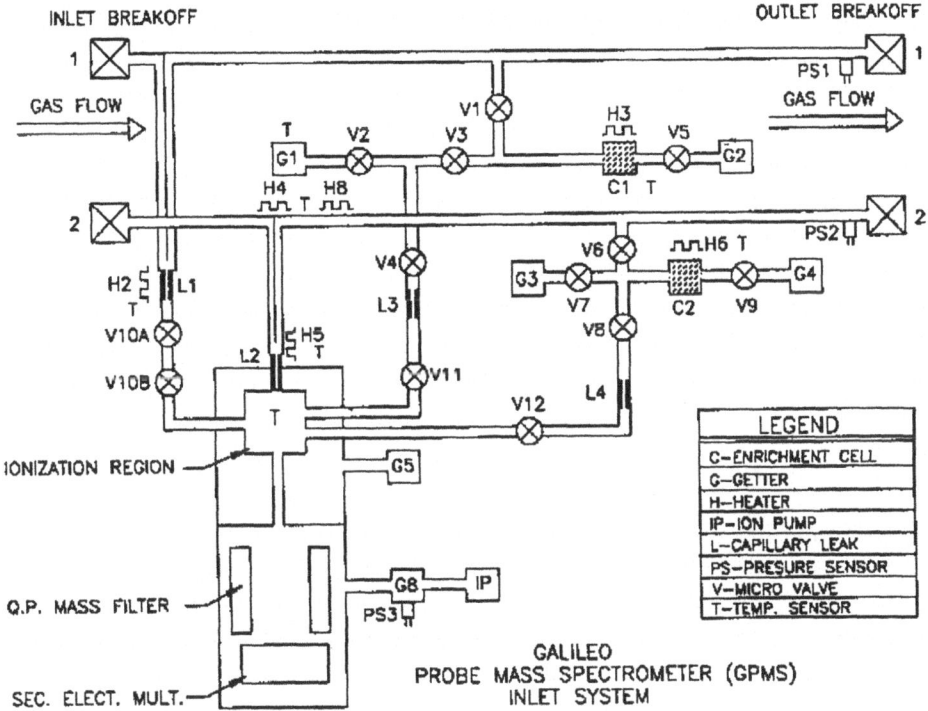

Figure 1. Schematic diagram of the Galileo Probe Neutral Mass Spectrometer

2. Galileo Probe Mass Spectrometer Measurements

2.1. THE INSTRUMENT AND ITS OPERATION

The GPMS has been described in detail by Niemann *et al.* (1992), and a report shortly after the Probe encounter of results obtained during its descent through the Jovian atmosphere has been published (Niemann *et al.*, 1996). A schematic diagram is shown in Figure 1. The mass analyzer was a quadrupole mass filter with a secondary electron multiplier detector. The instrument scanned in integral steps from 2 to 150 amu, devoting 0.5 sec to integrate the detector response at each step. The dynamic range was 10^8. Chemical getter pumps pumped the ionization and detector chambers of the GPMS, and a sputter ion pump backed the chamber getter pump. Atmospheric gases were admitted to the ionization chamber through a system of capillary leaks. On their high-pressure sides these leaks were exposed to gas flowing through tubing open to the atmosphere or to enrichment chambers in which these gases had been processed. At pressures between 0.5 and 4 bar a relatively high conductance leak (L1), sampled the atmosphere. During a portion of this time, gas was also circulated through an enrichment cell (EC1) to collect a portion of the heavier species. After L1 had been sealed, the contents of a volume next to the enrichment cell from which hydrogen and other chemically active species had been eliminated by getter pumps was admitted to the ionization chamber through leak L3. Methane and the noble gases were sampled in this experiment designated the noble gas (NG) sequence. Gas released from EC1 by heating this cell was then introduced through L3 to the ionization chamber. After an instrument background measurement, which extended to an atmospheric pressure of 8.6 bar, the direct atmospheric sampling resumed with a lower conductance leak (L2). A second enrichment cell sequence (EC2) took place between 12.1 and 15.6 bar and the cell contents introduced to the source through leak L4. Direct atmospheric sampling then continued until the end of the mission at approximately 22 bar. The portions of the L2 measurement sequence before and after the EC2 sequence are designated L2a and L2b respectively.

2.2. D/H AND ^{3}HE/ ^{4}HE FROM GPMS

The D/H ratio has to be derived from measurements of the ratio of HD to H_2, which is twice the D/H ratio. This requires separating the contributions of other ions appearing at 3 amu (^{3}He and H_3^+) from HD. The H_3^+ is manufactured in the ion source in two ways, one by ion-molecule reactions involving H_2 and the other by dissociative ionization of methane:

$$e + CH_4 \rightarrow CH + H_3^+ + e \tag{1}$$

Thus, if [3], [2], and [16] denote the number of counts per step at 3, 4 and 16 amu respectively,

$$[3] = a(HD/H_2)[2] + b(^3He/^4He)[4] + k[16] + c(H_3^+) \tag{2}$$

where HD/H2 is the relative abundance of HD and H2 in the atmosphere, ^{3}He/^{4}He the relative abundance of ^{3}He and ^{4}He, and a and b are factors that relate relative counting rates to relative atmospheric densities The ion source density depends on the flow characteristics through the capillary inlet into the source and the pumping speeds with which species are removed from the ionization chamber. The factor k is the rate constant for creation of H_3^+ from CH_4 by dissociative ionization multiplied by an instrument efficiency factor, and c (H_3^+) is the contribution of the other sources of H_3^+ to [3]. The values of these factors in the pressure regions of interest are obtained from the descent data, from calibration data obtained from the GPMS Flight Unit (FU) in 1985 prior to launch, and from dedicated experiments carried out after the Galileo Probe encounter using a refurbished Engineering Unit (EU) that duplicates the performance of the FU.

The sequence of measurements at 3 amu and at 2 and 4 amu before these species saturate, is illustrated in Figure 2 where the 3 amu counts per integration period are plotted from step 92 at 0.523 bar to step 6538, 54 minutes later at the 20.8 bar level. Between steps 1810 and 2160 and between steps 3100 and 3485 the mass spectrometer was isolated and background count rates were measured. It should be

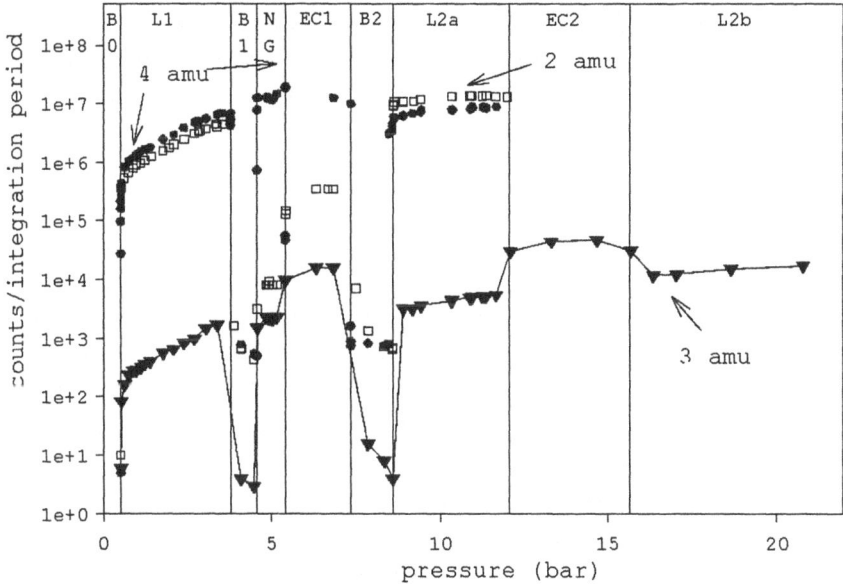

Figure 2. Log of counts per step at 3 amu versus pressure.

noted that, since the gas emerging from capillary leak L2 beams directly into the ionization region, the relative responses of L1 and L2 to different species is not the same. This is evident from the difference in the [4]/[2] ratios in the two leaks.

The relative contributions of CH_4 and ^3He to [3] can be determined from the measurement sequence involving EC1. In this sequence the chemical getters have largely removed the hydrogen and the ratio of CH_4 to He changes in different parts of this sequence as the enrichment cell is heated. The contribution of CH_4 to [3] has also been independently verified by recently obtained EU data. After the contributions of CH_4 and He to [3] have been subtracted from the signal in the L1 and L2 regions, the remaining 3 amu counts contain contributions from H_3^+ and from HD. The H_3^+ contribution can be quantified using 1985 FU calibration experiments where a descent sequence was carried out over the full range of pressures that were encountered during the Probe experiment. Following subtraction of the H_3^+ signal the residual [3]/[2] ratio together with instrument efficiency for detection of the HD and H_2 gave the HD/H_2 ratio from which the Jovian D/H ratio was derived.

2.2.1. *Contribution of CH_4 to [3]*
The change in the mixing ratio of helium and methane in EC1 on heating this cell can be used to determine the relative contributions of methane and helium to the 3 amu counts. The GPMS mass spectrum derived from measurements during the noble gas experiment is shown in Figure 3. The getter pump has removed much of the hydrogen and other active gases leaving methane and the noble gases as the primary constituents of the volume sampled by the mass spectrometer. The

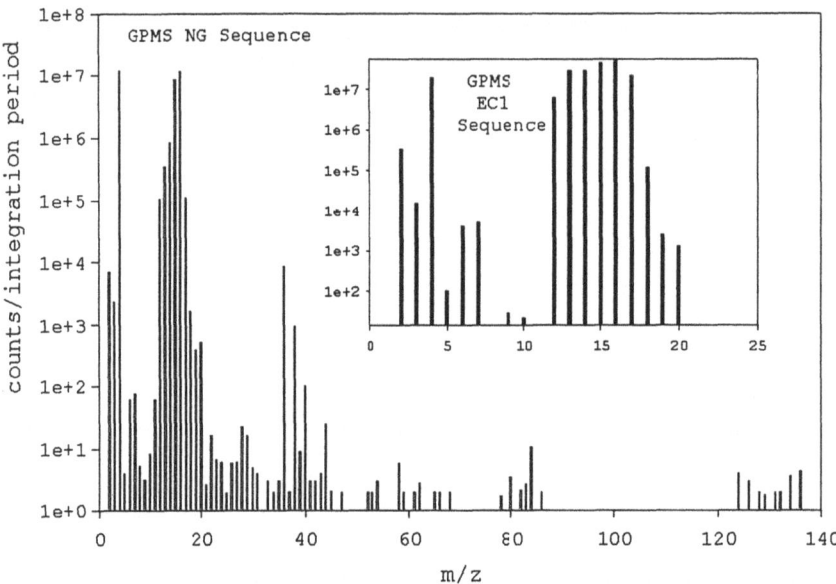

Figure 3. GPMS spectra in the NG portion of the descent sequence show counts per step *vs.* the m/z value. The full spectrum is derived from the signal expected near the center of this sequence based on a polynomial fit extrapolation of the signal at each m/z value to step 2400 in the NG sequence. The inset shows a portion of a similar spectrum extracted from the EC1 sequence. During this period, EC1 is heated to release gas trapped on the enrichment cell. At the mass values of interest for this study, there is a negligible cross talk between adjacent mass channels.

H_2-derived H_3^+ will not contribute substantially in this case to [3] and the terms associated with factors a and c in equation (2) are zero. The inset in this plot shows a portion of the EC1 sequence spectrum resulting from heating the enrichment cell to release its trapped gas. The signal at 16 amu and several methane fragment peaks saturate the detector sufficiently in the EC1 spectrum that it is difficult to derive accurate corrected count rates at these masses. The 12 amu signal, however, is not saturated and can be used to retrieve the relative increase in the methane mixing ratio from the noble gas sequence. The counts at 12 amu come almost entirely from methane in this spectrum since the relative abundances of other carbon containing species are very low.

Polynomial fits to [3], [4], and [16] in the NG region give values of 2.266×10^3, 1.239×10^7, and 1.027×10^7 respectively for the counts per period at these masses at step 2400. A similar exercise in EC1 gives $[3] = 1.583 \times 10^4$, $[4] = 1.856 \times 10^7$, and $[16] = 6.877 \times 10^8$ where the latter value is derived from the ratio to [12]. Solution of the simultaneous equations for the contributions to [3] give the $^3He/^4He$ ratio and the contribution of CH_4 to [3] of $[3]/[16] = 1.85 \times 10^{-5}$. A recent exercise on the EU with pure CH_4 agrees quite closely with the latter result giving directly $[3]/[16] = 2.2 \times 10^{-5}$. The EU exercise furthermore demonstrated that there are no large variations in this ratio over the pressure range of interest for the FU NG and EC1 sequences.

2.2.2. The ^{3}He/^{4}He Ratio

The [3]/[4] ratio derived from the FU NG and EC1 [3], [4], and [16] data as described above is 1.67×10^{-4}. Additional EU experiments show that the instrument response to the two helium isotopes introduced to the ion source in the molecular flow regime is flat for ^{3}He and ^{4}He so this count ratio represents a measure of the Jovian ^{3}He/^{4}He ratio. Using the EU derived [3]/[16] ratio gives a Jovian ^{3}He/^{4}He ratio of 1.65×10^{-4}. The adopted ^{3}He/^{4}He is an average of these two giving

$$^3\text{He}/^4\text{He} = (1.66 \pm 0.04) \times 10^{-4}.$$

2.2.3. The D/H Ratio

The EC1 data allow the contributions of helium and methane to be subtracted in the direct leaks using ratios to [4] and [16]. Figure 4 shows the result of this subtraction together with the subsequent subtraction of other contributions to [3] on the [3]/[2] ratio in the L1 and L2a portions of the descent sequence. The counts from ^{3}He in the L1 sequence region are designated $[3]_{\text{He}}$ and have a value of $1.66 \times 10^{-4}[4]$. The residual 3 amu counts due to H_3^+ and HD are labeled "([3] $-$ $[3]_{\text{He}})/[4]$" in Figure 4. In both this and the EC1 sequence, the transport both into and out of the ionization region is in the molecular flow regime and no pumping speed corrections are required. In the higher pressures of the L2a region the flow into the ionization region is nearly viscous and the ^{3}He contribution takes the form $\{(3/4)^{0.5}1.66 \times 10^{-4}[4]\}$ where the additional factor accounts for the faster pumping speed for ^{3}He out of the ionization region.

In L1 and L2a the H_3^+ correction is primarily from H2 with an additional small contribution from CH_4. The magnitude of the correction for this contribution shown in Figure 4 is established through analysis of a calibration run carried out on the FU in 1985 where a hydrogen and helium mixture was introduced to the mass spectrometer over the full range of pressures expected during the descent. In this run the ^{3}He contribution to [3] was negligible and a correction to [3] of the form $[3]\text{H}_3^+ = ([3] - 1.05 \times 10^{11}[2]^2)$ gave a constant $[3]\text{H}_3^+/[2]$ ratio over much of the low pressure side of the L2a region. A correction of this form applied to the FU [3] counts from which the ^{3}He contribution has been subtracted gives the ratio in Figure 4 labeled H_3^+ corrections.

Final corrections to the [3]/[2] ratio to reflect the HD/H$_2$ abundance arise from instrumental discrimination between HD and H$_2$. This effect is approximately 10 in L1 as determined from introduction of a known mixture of HD and H$_2$ into the EU L1. It is even higher in L2 as illustrated in Figure 2 where the gas mixture is beaming directly into the ionization region. Our present best estimate of the D/H ratio is derived from an average of points near the minimum of these two curves. Points above the minimum for each leak are thought to arise from additional contributions

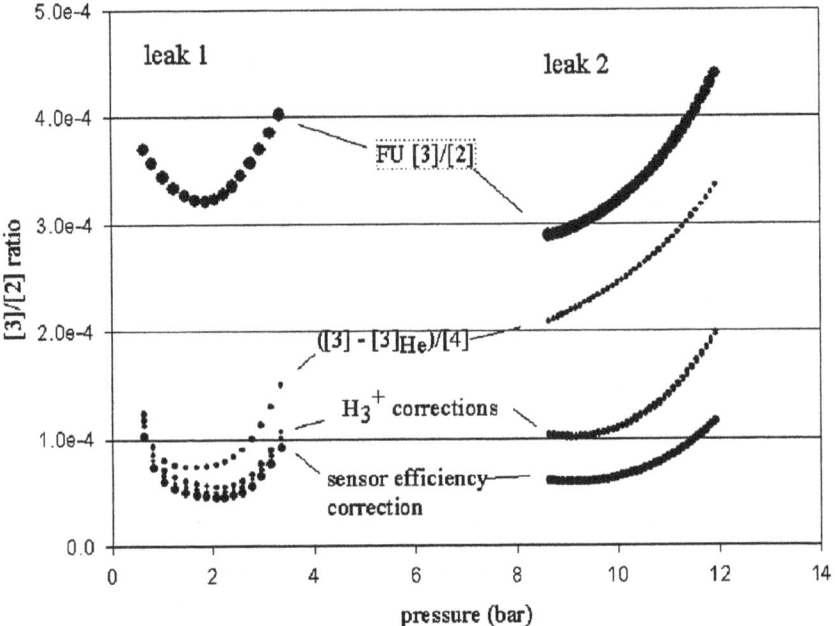

Figure 4. The effect of sequential removal of contributions to the 3 amu signal on the [3]/[2] ratio. The minimum in the bottom curves gives the HD/H$_2$ ratio derived from each of the two direct leak experiments. This value is twice the D/H ratio. The points shown represent values selected at regular intervals along polynomial fits to the [3] and the [2] data.

to H$_3^+$. The ratio derived from this analysis is

$$D/H = (2.6 \pm 0.7) \times 10^{-5}.$$

The errors for both this ratio and the ^{3}He/^{4}He ratio are 1σ limits derived from GPMS count variations and then increased to reflect additional systematic and instrumental uncertainties. Investigations of the dependence of [3] on pressure are continuing with the EU with the goal of increasing the precision of the analysis.

3. Comparison with Other Determinations of D/H

A summary of several Jovian D/H determinations by remote sensing is given in Table I. This table also gives protosolar D/H inferred from a comparison of (^{3}He/^{4}He) in the solar wind (SW) and in meteorites. The latter value of D/H must be identical to the Jovian value if the assumptions described above concerning Jupiter's formation are valid.

Table I

type	D/H value	authors (label in Fig. 5)	measurement
CH_3D	$(2.1 \pm 0.6) \times 10^{-5}$	Gautier and Owen (1989) (GO)	Voyager 7.7 μm, Bjoraker et al. (1986) CH_3D (5 μm)
CH_3D	$(1.2 \pm 0.5) \times 10^{-5}$	Bjoraker et al. (1986) (B)	5μm airborne
CH_3D	$(3.6 \pm 0.5) \times 10^{-5}$	Carlson et al. (1993) (C)	Voyager 7.7μm (reanalysis)
HD	$(1.0 \pm 2.9) \times 10^{-5}$	Smith et al. (1989) (S)	ground based visible
HD	$(1.8^{+1.1}_{-0.5}) \times 10^{-5}$	Lellouch et al. (1996) Encrenaz (1998) (LE)	ISO 37.7 μm
SW	$(2.6 \pm 1.0) \times 10^{-5}$	Geiss (1993)	Apollo, ISSE-3 solar wind measurements
SW	$(3.01 \pm 0.17) \times 10^{-5}$	Gautier and Morel (1997) (GM)	Apollo, ISSEE-3, Ulysses, solar wind measurements, earlier GPMS ^{3}He/^{4}He
SW	$(2.1 \pm 0.5) \times 10^{-5}$	Geiss and Gloeckler (1998) (GG)	Apollo, ISSEE-3, Ulysses, solar wind measurements
in situ	$(2.6 \pm 0.7) \times 10^{-5}$	this work	GPMS value [superceding previous GPMS report of $(5 \pm 2) \times 10^{-5}$ of Niemann et al. (1996)]

3.1. SPECTROSCOPIC MEASUREMENT OF D/H

One method to measure D/H in Jupiter's atmosphere is by optical spectroscopy. For example, a determination of the CH_3D/CH_4 ratio and an analysis which then takes into account the fractionation that occurs in the production of methane gives an inferred D/H as it would be measured in H_2 (Lecluse et al. 1996). Three such determinations giving the values labeled "CH_3D" are shown in Table I.

The first, by Gautier and Owen (1989), represents the result of an analysis of Voyager 1 data at 7.7μm combined with Bjoraker 5μm observations to get CH_3D/H_2. The second, by Bjoraker et al. (1986), is from airborne spectra at 5μm. The third is the result of a reanalysis of Voyager IR data (Carlson et al. 1993). Each of these values changes when adjusted for a C/H ratio 2.9 times solar (solar C/H $= 3.62 \times 10^{-4}$, Anders and Grevesse, 1989) and a ^{4}He/H_2 ratio of 0.157 (Niemann et al., 1996; von Zahn and Hunten, 1996). In particular, the Bjoraker et al. (1986) value increases to greater than 2×10^{-5} and the Carlson et al. (1993) value increases to greater than 4×10^{-5}.

The measurements labeled "HD" in Table I give a preliminary value of HD/H_2 as determined from recent IR observations of the HD R(2) line at 37.7μm by the short wavelength IR spectrometer on ISO (Encrenaz et al., 1996 and reanalyzed by Lellouch et al., 1996; Encrenaz, 1998). The ISO result is provisional, as analysis is still continuing. Ground based measurements of HD/H_2 in the visible spectrum have also been carried out by Smith et al. (1989).

On normalization to the GPMS derived CH_4/H_2 and He/H_2 ratios, the original Voyager measurements, the ISO observations, the airborne IR observations, the ground based visible observations, and the GPMS results are all consistent within the measurement errors. The analysis of the Voyager 1 CH_3D measurements by Carlson *et al.* (1993) gives a somewhat higher value.

3.2. D/H DERIVED FROM SOLAR WIND MEASUREMENTS OF $^3HE/^4HE$

$(D/H)_{ps}$ for the protosun can also be determined by measuring $(^3He/^4He)_{sw}$ in the solar wind. Deuterium in the young sun was converted very efficiently to ^3He. The helium has subsequently remained virtually unprocessed, as is implied by the continuing presence in the Sun of the more reactive ^9Be (Geiss and Reeves, 1972). Thus

$$(D/H)_{ps} = (^4He/H)_{ps}\{(^3He/^4He)_s - (^3He/^4He)_{ps}\} \qquad (3)$$

(Geiss and Reeves, 1972) where $(^3He/^4He)_{ps}$ is the $^3He/^4He$ ratio in the protosolar cloud and $(^3He/^4He)_{sw}$ in the solar wind today is corrected for modifications occurring in the corona and chromosphere as well as for changes that have occurred over time in the solar interior (Gautier and Morel, 1977; Geiss and Gloeckler, 1998). For $(^3He/^4He)_{ps}$ either the meteoritic value

$$(^3He/^4He)_{ps} \equiv (^3He/^4He)_m = (1.5 \pm 0.3) \times 10^{-4} \qquad (4)$$

(Black, 1972; Eberhardt, 1974; Geiss and Reeves, 1972; Geiss, 1993) or the GPMS value $(1.66 \pm 0.05) \times 10^{-4}$, which agrees with it may be used. The result as given by Geiss and Gloeckler (1998) is $(D/H)_{ps} = (2.1 \pm 0.5) \times 10^{-5}$. Gautier and Morel (1997) obtained $(D/H)_{ps} = (3.01 \pm 0.17) \times 10^{-5}$. This was in part the consequence of their using the earlier smaller GPMS value for $^3He/^4He$ (1.1×10^{-4}) given by Niemann *et al.* (1996) and in part from use of a higher $(^3He/^4He)_{sw}$ than that of Geiss and Gloeckler (1998). These results for $(D/H)_{ps}$ agree with the in situ GPMS measurement as well as most of the spectroscopic measurements.

The cited values for Jovian D/H or $(DH)_{ps}$ are shown in Figure 5. We note in passing that simply substituting the GPMS value for $(^3He/^4He)_{ps}$ of Equation (3) together with the initial Ulysses value of $(^3He/^4He)_{sw}$ (Bodmer *et al.*, 1995) gives $(D/H)_{ps} = (2.7 \pm 0.3) \times 10^{-5}$, in excellent agreement with the GPMS result

4. Solar System and Local Interstellar Medium D/H and $^3He/H$

In the nearby Local Interstellar Medium (LISM) Linsky *et al.* (1996) have measured an isotopic ratio $D/H = (1.6 \pm 0.12) \times 10^{-5}$. This is the present day value of D/H, in this region of the galaxy (which is not where the solar nebula was formed). The GPMS and the other three low values of D/H measured in the Jovian atmosphere indicate a modest 45% consumption of deuterium in the galactic neighborhood

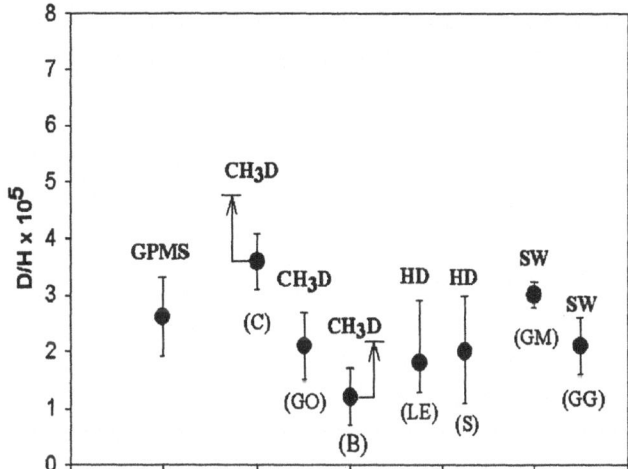

Figure 5. The D/H ratios derived through the measurements summarized in Table I. The arrow from the (C) value points to a value obtained by Lecluse *et al.* (1996) following a normalization to a methane mole fraction of 1.8×10^{-3} and using an isotopic enrichment factor of 1.25. The (B) value would similarly be increased with use of a more recent methane mole fraction. In the case of (GM) their value of 3.01×10^{-5} would be reduced with use of our new ^{3}He/^{4}He result.

during the past 4.55 Gy. In this regard it is interesting to examine the behavior of (^{3}He + D)/H. An isotopic analysis of helium ions entering the solar system from the surrounding interstellar cloud has been performed from measurements of pickup ions by the ion composition spectrometer on Ulysses (Gloeckler and Geiss, 1998). This measurement gives ^{3}He/H = $(2.48^{+0.85}_{-0.79}) \times 10^{-5}$ assuming ^{4}He/H to be 0.1. Thus in the LISM (^{3}He + D)/H = $(4.08 \pm 0.97) \times 10^{-5}$. The protosolar value of (D + ^{3}He)/H would be $(4.3 \pm 0.7) \times 10^{-5}$ if ^{4}He/H = 0.1 and the GPMS measurement of D/H and ^{3}He/^{4}He in the Jovian atmosphere represent conditions in the protosolar cloud. This, as well as those obtained by combining most other Jovian D/H measurements with either the meteoritic or GPMS values for ^{3}He/^{4}He, are indistinguishable from the LISM value. Thus they are consistent with no change in the ratio during the past 4.55 Gy and thus no significant consumption of ^{3}He.

5. Discussion and Conclusions

The ^{3}He/^{4}He ratio measured by the GPMS in Jupiter's atmosphere is within the error bars of the ratio in the "planetary" component of meteorites. The higher meteoritic value may reflect fractionation of the solar nebular gas in favor of the heavier isotope. The value of D/H that we have derived in Jupiter's hydrogen is consistent with several previous remote sensing determinations but excludes others. Further calibration work with the EU may allow us to reduce the error bars on D/H somewhat. We have established the anticipated result that Jupiter indeed represents a repository of solar nebula material unmodified by nuclear reactions for the last 4.55 Gy. The small change in the 'local' value of D/H that occurred during the last 4.55 Gy according to the GPMS, SW, ISO, and LISM measurements is consistent with values of D/H in the range $2 - 5 \times 10^{-5}$ for intergalactic clouds adsorbing Lyman (from quasars at very large redshifts (Tytler *et al.*, 1997, Burles and Tytler, 1998). The GPMS measurements are more difficult to reconcile with primordial values of D/H as high as 10^{-4} (Rugers and Hogan 1996; Songaila *et al.*, 1997; Webb *et al.*, 1997). In principle, one can test models for galactic evolution by comparing values of D/H and ^{3}He/^{4}He measured at different times. If there is no major consumption of ^{3}He, the ratio $(D+^3He)/H$ should remain constant in time. However, present uncertainties on the LISM value for ^{3}He/^{4}He prevent this test from being rigorous. The lowest value of LISM ^{3}He/^{4}He$= 1.69 \times 10^{-4}$, overlaps the high end of the GPMS determination: 1.7×10^{-4}. This alone suggests negligible consumption of ^{3}He during the last 4.55 Gy, and is consistent with the results for $(D+^3He)/H$. Future higher precision studies of the Interstellar ^{3}He/^{4}He would make this approach more useful.

Acknowledgements

This work was supported by NASA. One of us (TMD) thanks Johannes Geiss and his colleagues at the International Space Science Institute for providing the opportunity to work on this paper at ISSI.

References

Anders, E. and Grevesse, N.: 1989, 'Abundances of the Elements: Meteoritic and Solar', *Geochim. Cosmochim. Acta* **53**, 197–214, 1989.

Balsiger, H., *et al.*: 1995, 'D/H and 18O/16O Ratios in the Hydronium Ion and in Neutral Water from in situ Ion Measurements in Comet Halley', *J. Geophys. Res.* **100**, 5827–5834.

Bjoraker, G.L., *et al.*: 1986a, 'The Gas Composition of Jupiter Derived from $5 - \mu$m airborne Spectroscopic Observations', *Icarus* **66**, 579–609.

Black, D.C.: 1972, 'On the Origins of Trapped Helium, Neon, and Argon Isotopic Variations in Meteorites -I', *Cosmochim. Acta* **36**, 347.

Bocklee-Morvan, D. *et al.*: 1998, 'Deuterated Water in Comet C/199 (Hyakutake) and its Implications for the Structure of the Primitive Solar Nebula', *Icarus*, in press.

Bodmer, R. *et al.*: 1995, 'Solar Wind Helium Isotopic Composition from SWICS/Ulysses', *Space Sci. Rev.* **72**, 61–64.

Burles, S. and Tytler, D.: 1997, 'Cosmological Deuterium Abundance and The Baryon Density of the Universe'. *Science*, submitted.

Carlson, B. E., *et al.*: 1993, 'Tropospheric Gas Composition and Cloud Structure of The Jovian North Equatorial Belt', *J. Geophys. Res.* **98**, 5251–5290.

Eberhardt, P., Reber, M., Krankowsky, D. and Hodges, R.R.: 1995, 'The D/H and 18O/16O Ratios in Water from Comet P/Halley', *Astron. Astrophys.* **302**, 301–318.

Eberhardt, P.: 1974, 'A Neon-E-Rich Phase in the Orgueil Carbonaceous Chrondrite', *Earth Planet.Sci.Lett.* **24**, 182.

Encrenaz, Th. *et al.*: 1996, 'First Results of ISO-SWS Observations from Jupiter', *Astron. and Astrophys.* **315**, L397–402.

Encrenaz, Th.: 1998, personal communication.

Gautier, D. and Morel, P.: 1997, 'A Reestimate of the Protosolar $(^2H/^1H)_p$ Ratio from $(^3He/^4He)_{sw}$ Solar Wind Measurements', *Astron. and Astrophys.* **323**, L9–L12.

Gautier, D. and Owen, T.: 1989, 'The Composition of Outer Planet Atmospheres', in 'Origin and Evolution of Planetary and Satellite Atmospheres', Atreya, S.K. *et al.* (eds), University of Arizona Press, Tucson, pp. 487–512.

Geiss, J.: 1993, 'Primordial abundance of hydrogen and helium isotopes', in 'Origin and Evolution of the Elements', eds. N. Prantzos, E. Vangioni-Flam, and M. Cassé, Cambridge University Press, Cambridge, pp. 89–106.

Geiss, J. and Gloeckler, G.: 1998, 'Abundances of Deuterium and Helium-3 in the Protosolar Cloud', *Space Sci. Rev.*, this volume.

Geiss, J. and Reeves, H.: 1972, 'Cosmic and Solar System Abundances of Deuterium and Helium-3', *Astron. Astrophys.* **18**, 126.

Gloeckler, G.: 1996, 'The Abundance of Atomic ^1H, ^4He and ^3He in the Local Interstellar Cloud Piom Pickup Ion Observations with Swics on Ulysses', *Space Sci. Rev.* **78**, 335–346.

Gloeckler, G. and Geiss, J.: 1996, 'Measurements of the Abundance of ^3He in the Local Interstellar Cloud', *Nature* **381**, 210.

Gloeckler, G. and Geiss, J.: 1998, 'Measurement of the Abundance of Helium-3 in the Sun and in the Local Interstellar Cloud With SWICS on Ulysses', *Space Sci. Rev.*, this volume.

Lecluse, C., *et al.*: 1996, 'Deuterium Enrichment in Giant Planets', *Planet. Space Sci.* **44**, 1579–1592.

Lellouch, E., *et al.*: 1996, 'Delimination of the D/H Ratio on Jupiter from ISO/SWS Observations' presented at the 28th DPS meeting, Tucson, 23–26 October.

Linsky, J. L.: 1996, 'GHRS Observations of thc LISM', *Space Sci. Rev.* **78**, 157-164.

Meier, R. *et al.*: 1998, 'A Determination of the HDO/H2O Ratio in Comet C/1995 O1 (Hale-Bopp)', *Science* **279**, 842–844.

Niemann, H.B. *et al.*: 1996, 'The Galileo Probe Mass Spectrometer: Composition of Jupiter's Atmosphere', *Science* **272**, 846–849.

Niemann, H.B. *et al.*: 1992, 'Galileo Mass Spectrometer Experiment', *Space Sci. Rev.* **60**, 111–142.

Niemann, H.B. *et al.*: 1998, 'The Composition of the Jovian Atmosphere as Determined by the Galileo Probe Mass Spectrometer', *J. Geophys. Res.*, in press.

Rugers, M. and Hogan, C. J.: 1996, 'Confirmation of high deuterium abundance in quasar absorbers', *Astrophys. J.* **469**, L1–L4.

Smith, W.H., Schempp, W.V., Baines K.H.: 1989, *Astrophys. J.* **336**, 967.

Songaila, A., *et al.*: 1997, 'A High D Abundance in the Early Universe', *Nature* **385**, 137–139.

Tytler, D., *et al.*: 1996, 'Cosmological Baryon Density Derived from the Deuterium Abundance at Redshift $z = 3.57$', *Nature* **381**, 207–209.

Tytler, D., *et al.*: 1997, 'New Keck Spectra of Q0014+813: Annulling the Case for High Deuterium Abundance', Astro-ph/9612121 (preprint).

von Zahn, U. and Hunten, D. M.: 1996, 'The Helium Mass Fraction in Jupiter's Atmosphere', *Science* **272**, 849–851.

Webb, J.K. *et al.*: 1997, 'A High Deuterium Abundancc at Redshift $z = 0.7$', *Nature* **388**, 250–252.

EVIDENCE OF ELEMENT DIFFUSION INSIDE THE SUN AND THE STARS AND ITS CONSEQUENCES ON THE LITHIUM PRIMORDIAL ABUNDANCE

S. VAUCLAIR

Laboratoire d'Astrophysique de Toulouse, CNRS UMR 5572, 14, av. E.Belin, 31400 Toulouse, France

Abstract. The process of element segregation in stars (also called "microscopic diffusion") has to be introduced in all computations of stellar structure to obtain consistent models. Although recognized by the pioneers of stellar physics, this process has long been forgotten, except for white dwarfs and for the so-called "chemically peculiar stars". More recently helioseismology has given evidence that this process occurs in the Sun, and leads to helium and heavier element depletion by about 20 percent. Some macroscopic motions (mild mixing) must also occur below the convection zone in order to account for the lithium depletion. These motions do not prevent the segregation : they only slightly smooth the abundance gradients. These results are presented here and the connexion with the ^{3}He abundance is discussed. The importance of these processes for Pop II stars is also developped.

Key words: Stellar structure, Element diffusion, Helioseismology, Lithium

1. Introduction

Element diffusion inside the stars represents a basic physical process which cannot be ignored in the computations of stellar structure. When stars form out of gas clouds, they build density, pressure and temperature gradients which force the various chemical species present in the stellar gas to move with respect to one another. As a consequence, the abundances observed at the surfaces of stars are not always representative of their protosolar values, even if they did not suffer any nuclear processing.

The resulting abundance variations depend on several effects. What we use to call "microscopic" diffusion of the chemical elements represents in fact a competition between two kinds of processes. First the atoms move under the influence of external forces (due to gravity, radiation, etc.), second they collide with other atoms and share the acquired momentum with them in a random way, which slows down their motion. This competition leads to a process of element segregation with a time scale decreasing with increasing density.

In cool stars, this segregation occurs below the outer convection zones. The observational consequences are smaller than for hotter stars as the convection zones are deeper and the density larger at the bottom. For the same reason, the effects should be more important in Pop II stars than in Pop I stars of the same effective temperature, as the convection zones are shallower for lower metallicities.

Although recognized by the pioneers of the study of stellar structure, this fundamental physical process was long forgotten in the computations of stellar models,

Space Science Reviews **84**: 265–271, 1998.

except for white dwarfs (Schatzman, 1945). Only with the discovery of large abundance anomalies in main-sequence type stars (the so-called Ap and Am stars), which present characteristic variations with the effective temperature, was element diffusion brought into light fifty years later (Michaud, 1970, see other references in Vauclair and Vauclair, 1982).

At that time, the effects of element diffusion were supposed to be important only when the diffusion time scale was smaller than the stellar age. In the Vauclair and Vauclair (1982) review paper, Fig. 1 shows the regions in the HR diagram where diffusion could lead to "observable" abundance variations. The Sun was excluded, as diffusion could not lead in it to abundance variations larger than some ten percent. At that time evidences of abundance variations could be obtained through spectroscopic observations only, and there was no hope to be able to detect differences of order 20 percent.

In the present days, due to helioseismology, we know the internal structure of the Sun with a high precision, and evidences for the occurence of element diffusion are obtained with a high degree of accuracy. Abundance variations of the order of a few percent now become indirectly detectable, by comparisons of the theoretical computations with the results of the inversion of pulsating modes. The confirmation by helioseismology of the predictions concerning element diffusion in the Sun represents a great success for the theory of stellar structure. We have entered a new area in this respect.

2. Evidence of element diffusion from helioseismology

The Sun is the most well known of all the stars and it can be used as a precise test for theoretical studies. Its mass, radius, luminosity and age are known with a high degree of precision (see values and references in Richard et al., 1996). The photospheric abundances have been also precisely determined (Grevesse, 1991).

For the light elements, the abundance determinations show that lithium has been depleted by a factor of about 140 while beryllium is generally believed to be depleted by a factor 2 compared to the meteoritic value. These values have widely been used to constraint the solar models. However, while the lithium depletion factor seems well established, the beryllium value is still subject to caution. Balachandran (1997) argues that the beryllium depletion is not real because of insufficient inclusion of continuous opacity in the abundance determination. Her new treatment leads to a solar value identical to the meteoritic value.

Observations of the ^{3}He/^{4}He ratio in the solar wind and in the lunar rocks (Geiss et al., 1972; Geiss, 1993; Gloecker and Geiss, 1997) show that this ratio may not have increased by more than $\cong$ 10% since 3 Gyr in the Sun. While the occurence of some mild mixing below the solar convection zone is needed to explain the lithium depletion and, as we will see below, is consistent with helioseismology, the ^{3}He/^{4}He observations put a strict constraint on its efficiency.

The study of the internal structure of the Sun entered a new age with helio-seismology. Several ground based networks and the SoHO mission continuously observe the solar oscillations. In particular GONG, the "Global Oscillation Network Project", gathers six sites around the world with six identical Doppler instruments. These instruments observe the phase shift of the Ni 676.8nm line with 3 images every minute, 1.8 sites observing simultaneously.

Millions of solar p-modes have been detected. The inversion of the measured frequencies yields accurate and detailed information about the sound velocity in the Sun's interior, which in turn leads to constraints on the equation of state, opacities, chemical composition. Precise informations on the differential rotation inside the Sun have also been obtained. (see the special Science issue on GONG, vol. 272, 31 May 1996).

Many authors have computed the gravitational and thermal diffusion of helium and heavier elements in the Sun with various approximations (Cox, Guzik and Kidman, 1989; Bahcall and Pinsonneault, 1992; Proffitt, 1994; Thoul, Bahcall and Loeb, 1994; Richard *et al.*, 1996; Basu, 1997). Most models are quite consistent with each other, which is encouraging.

In Richard *et al.* (1996), computations of the solar internal structure have been done with the Toulouse code, a version of the Geneva stellar evolution code in which we have precisely included microscopic diffusion processes and possibilities of testing the effects of macroscopic motions like turbulence and stellar winds. Microscopic diffusion was treated as in Charbonnel, Vauclair and Zahn (1992), and the diffusion coefficients were computed using the Paquette *et al.* (1986) approximation. The abundance variations were followed separately for helium and 14 heavy elements, and they were iterated so that the final abundances correspond to those given by Grevesse (1991).

These models have been compared to the results of helioseismology in col-laboration with the Warsaw group (Dziembowski *et al.*, 1994). The values of the function $u = P/\rho$ as obtained from the models are compared to those of the "seismic Sun" (Fig. 1). The best solar model is obtained by taking into account both the element segregation induced by diffusion and a mild mixing necessary to account for the lithium depletion. Then the comparison between the models and the "seismic Sun" is very good below the convection zone. However in this model the ^3He increase with time in the convection zone is too large compared to the observations.

More recent solar models have been computed with a smaller mildly mixed zone below the convection zone in order to account for the lithium depletion and the constraint on the ^3He enhancement (Vauclair and Richard, 1997a). In these models the critical μ-gradient able to stabilize the mixing processes was varied as a parameter, between 5 and 1.5×10^{-13}. The variation of the $^3He/^4He$ ratio was found small enough during these last 3 Gyrs as soon as $(\nabla \ln \mu)_{crit} \leq 2 \times 10^{-13}$.

If the constraint on the beryllium depletion is relaxed, it allows a still thinner mixed zone. It is interesting in this case to compute the minimum enhancement

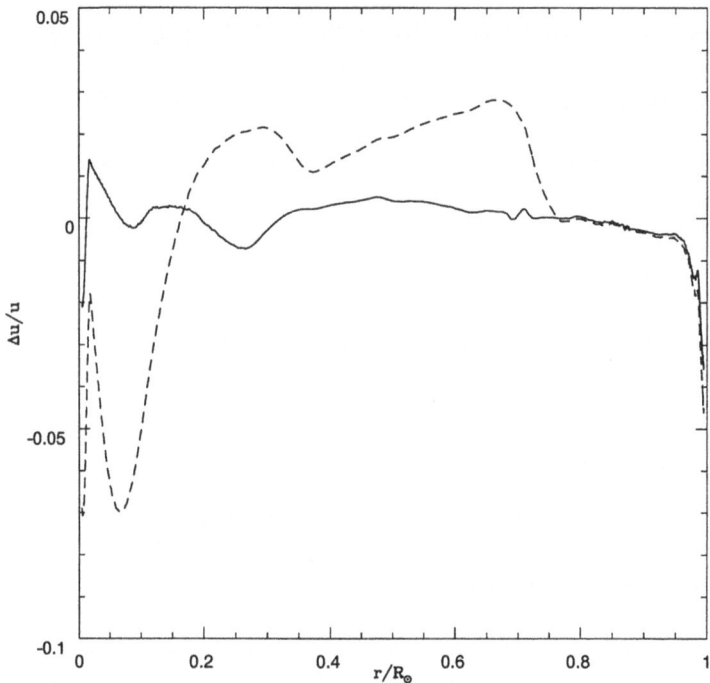

Figure 1. Differences between the $(u = \frac{P}{\rho})$ function in the seismic Sun and the best model of Richard *et al.* (1996) (solid line) or model with core mixing (dashed line). Mixing near the solar core leads to a modification of the solar structure which is incompatible with the seismic Sun (after Vauclair and Richard, 1997a).

of the ^{3}He/^{4}He implied by the lithium observed depletion. Vauclair and Richard (1997b) show that it is possible to deplete lithium by a factor larger than 100 as observed and not increase ^{3}He/^{4}He by more than 5 percent since the solar origin. In any case the evaluations of the ^{3}He evolution with time must take all these effects into account. Introducing microscopic diffusion alone (Gautier and Morel, 1996) is not consistent with the lithium observations.

The comparison between the models and the "seismic Sun" (Figs. 1 and 2) is also able to rule out the core mixing processes which have been invoked to account for the solar neutrino deficiency (e.g. Morel and Schatzman, 1996). Although such a core mixing can indeed reduce the neutrino fluxes, it leads to a solar model inconsistent with helioseismology (Richard and Vauclair, 1996).

3. Application to Population II Stars

Microscopic diffusion in stars is now considered as a "standard" process, which has to be included in all computations of stellar models. It cannot be ignored when deriving the lithium primordial abundance from the observations of Pop II stars.

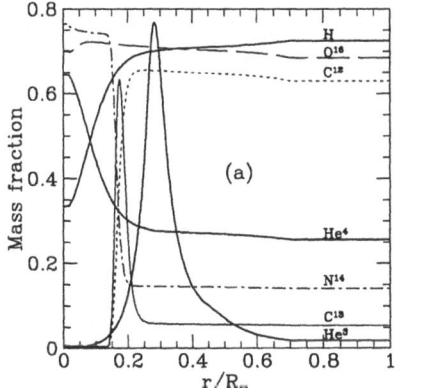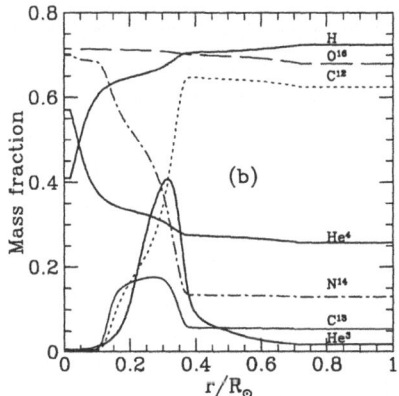

Figure 2. Chemical composition inside solar models (the conventions are the same as in Richard *et al.*, 1996). Graph (a) represents the case of the best solar model in Richard *et al.*, with no core mixing. The influence of the mixing layer below the convection zone on the ^{3}He abundance is clearly seen on the right side of the ^{3}He peak. Graph (b) represents the model with a partially mixed core. In this model the neutrino fluxes are decreased, but it is not compatible with helioseismology.

The lithium abundances observed in the oldest stars of our Galaxy lie around 2×10^{-9} in number compared to hydrogen, as soon as their effective temperature is larger than 5500 K (the "Spite plateau", Spite and Spite, 1982). While cooler stars present a large dispersion with strong evidences of lithium depletion, the abundance scatter in the "plateau" is very small (Thorburn, 1994; Ryan *et al.*, 1996; Spite *et al.*, 1996; Bonifacio and Molaro, 1997). Furthermore the observations of the ^{6}Li isotope in a few halo stars (Smith *et al.*, 1992; Hobbs and Thorburn, 1994), if confirmed, represent an evidence that lithium has suffered no, or only very small nuclear destruction in these stars.

A good understanding of Pop II stars together with a realistic derivation of the primordial lithium must rely on a consistent study of the internal structure of these stars, taking real physics into account. Models were calculated with the Toulouse-Geneva stellar evolutionary code with improved microphysics as in Richard *et al.* (1996). The radiative opacities are from Iglesias and Rogers (1996), completed with the atomic and molecular opacities by Alexander and Ferguson (1994). The equation of state is described with a set of MHD tables (Mihalas, Hummer and Däppen, 1988) specifically calculated for the mass and metallicity domain of Pop II stars (Charbonnel *et al.*, 1997). Stellar models of 0.65, 0.70, 0.75 and $0.80 M_\odot$ for $Z = 10^{-4}$ have been computed from the pre-main sequence up to the turn-off.

In the parameter-free models in which no mixing neither mass loss are introduced, lithium is depleted at the surface due to microscopic diffusion. Its value increases inside the star up to a maximum and then decreases due to nuclear destruction. None of the stars keep the original lithium abundance in any part of its internal structure (Vauclair and Charbonnel, 1995, 1997). If the stars did not suffer any macroscopic motions during their whole lifetime except for ordinary convection, they should show a lithium depletion increasing for hotter stars. The fact that the observations do not show this trend is an evidence that stars are indeed subject to mixing and/or mass loss during their lifetime (as we know is the case for the Sun itself). The maximum lithium value inside the stars slightly decreases with stellar mass for a given age. This is due to the fact that in more massive stars the convection zone is shallower so that the distance between its bottom layer and the nuclear destruction zone is larger. Meanwhile the lithium depletion in the convection zone is more important as the density at the bottom is smaller.

Vauclair and Charbonnel (1995) discussed this problem in terms of stellar winds. They showed that the lithium behavior in Pop II stars could be explained with the assumption that they suffered mass loss during all their lifetime with an average rate about 10 to 30 times larger than the present solar wind. In this case the primordial abundance is given by the upper envelope of the observations.

Contrary to Ryan *et al.* (1996), Bonifacio and Molaro (1997) find no intrinsic dispersion for the lithium abundances in the plateau stars. If this result is confirmed, it suggests that the macroscopic process which competes with microscopic diffusion forces the lithium abundance close to its maximum possible value (Vauclair and Charbonnel, 1997).

The best fit of the maximum lithium value with these observations is obtained for an initial abundance of 2.30 to 2.45 dex. This leads for the primordial lithium value: $\log \mathrm{Li}_p = 2.4 \pm 0.1$. When compared to BBN computations (e.g. Schramm, 1998) these results lead to two possible ranges for the baryonic number, between approximately 1 to 1.7×10^{-10} on one hand, and 3.5 to 6×10^{-10} on the second hand. For $H_0 = 50$, these values correspond respectively to $0.015 < {}_b < 0.03$ and $0.05 < {}_b < 0.09$.

References

Alexander, D.R., Fergusson, J.W.: 94, *ApJ* **437**, 879.

Bahcall, J.N., Pinsonneault, M.H.: 1992, *Reviews of Modern Physics* **64**, 885.

Balachandran, S.: 1997, preprint

Basu, S.: 1997, preprint.

Bonifacio, P., Molaro, P.: 1997, *MNRAS* **285**, 847.

Charbonnel, C., Vauclair, S., Zahn, J.P.: 1992, A&A **255**, 191 (CVZ).

Charbonnel, C., Däppen, W., Bernasconi, P.A., Maeder, A., Meynet, G., Schaerer, D., Mowlavi, N.: 1997, *A&A*, submitted.

Cox, A.N., Guzik, J.A., Kidman, R.B.: 1989, *ApJ* **342**, 1187.

Dziembowski, W.Q., Goode, P.R., Pamyatnikh, A.A., Sienkiewicz, R.: 1994, *ApJ* **432**, 417.

Gautier and Morel: 1997 *A&A* **323**, L9.

Geiss, J., Buhler,F., Cerutti, H., Eberhardt, P., Filleux, Ch.: 1972, *Apollo Preliminary Science Report, NASA SP-315, section 14.*

Geiss, J.: 1993, *Origin and Evolution of the Elements*, ed. Prantzos, Vangioni-Flam, and Cassé (Cambridge Univ. Press), 90.

Gloecker, G., Geiss, J.: 1997, preprint.

Grevesse, N.: 1991, *A&A* **242**, 488.

Hobbs, L., Thorburn, J.A.: 1994, *ApJ* **428**, L25.

Iglesias, C.A., Rogers, F.J.: 1996, *ApJ* **464**, 943.

Michaud, G.: 1970, *ApJ* **160**, 641.
Mihalas, D., Hummer, D.G., Däppen, W.: 1988, *ApJ* **331**, 815.
Morel, P., Schatzman, E.: 1996, *A&A* **310**, 982.
Paquette, C., Pelletier, C., Fontaine, G., Michaud, G.: 1986, *ApJS* **61**, 177.
Proffitt, C.R.: 1994, *ApJ* **425**, 849.
Richard, O., Vauclair, S., Charbonnel, C., Dziembowski, W.A.: 1996, *A&A* **312**, 1000.
Richard, O., Vauclair, S.: 1997, *A&A* **322**, 671.
Ryan, S.G., Beers, T.C., Deliyannis, C.P., Thorburn, J.: 1996, *ApJ* **458**, 543.
Schatzman, E.: 1945, *Ann. d'Astr.* **8**, 143.
Schramm, D.: 1998, *Space Sci. Rev.*, this volume.
Smith, V.V., Lambert, D.L., Nissen, P.E.: 1992, *ApJ* **408**, 262.
Spite, M., Spite, F.: 1982, *A&A* **115**, 357.
Spite, M., François, P., Nissen, P.E., Spite, F.: 1996, *A&A* **307**, 172.
Thorburn, J.A.: 1994, *ApJ* **421**, 318.
Thoul, A.A., Bahcall, J.N., Loeb, A.: 1994, *ApJ* **421**, 828.
Vauclair, S., Vauclair, G.: 1982, *ARA&A* **20**, 37.
Vauclair, S., Charbonnel, C.: 1995, *A&A* **295**, 715.
Vauclair, S., Charbonnel, C.: 1997, preprint.
Vauclair, S., Richard, O.: 1997a, Proceedings of the Los Alamos meeting *A half century of stellar pulsation interpretations: a tribute to Arthur N. Cox"*, June 16–20 , 1997.
Vauclair, S., Richard, O.: 1997b, preprint.

Address for correspondence: Sylvie.Vauclair@obs-mip.fr

VI: LOCAL INTERSTELLAR MEDIUM

MEASUREMENT OF THE ABUNDANCE OF HELIUM-3 IN THE SUN AND IN THE LOCAL INTERSTELLAR CLOUD WITH SWICS ON ULYSSES

GEORGE GLOECKLER

Department of Physics and IPST, University of Maryland, College Park, MD 20742, USA,
and
Department of Atmospheric, Oceanic and Space Sciences, University of Michigan, Ann Arbor, MI 48109, USA.

JOHANNES GEISS

International Space Science Institute, Hallerstrasse 6, CH–3012 Bern, Switzerland.

Abstract. The abundance of ^{3}He in the present day local interstellar cloud (LIC) and in the sun has important implications for the study of galactic evolution and for estimating the production of light nuclei in the early universe. Data from the Solar Wind Ion Composition Spectrometer (SWICS) on Ulysses is used to measure the isotopic ratio of helium (^{3}He/^{4}He $= \Gamma$) both in the solar wind and the local interstellar cloud. For the solar wind, the unique high-latitude orbit of Ulysses allows us to study this ratio in the slow and highly dynamic wind in the ecliptic plane as well as the steady high-latitude wind of the polar coronal holes. The ^{3}He$^+$/^{4}He$^+$ ratio in the local cloud is derived from the isotopic ratio of pickup helium measured in the high-speed solar wind. In the LIC the ratio is found to be $(2.48^{+0.68}_{-0.62}) \times 10^{-4}$ with the 1-σ uncertainty resulting almost entirely from statistical error. In the solar wind, Γ is determined with great statistical accuracy but shows systematic differences between fast and slow solar wind streams. The slow wind ratio is variable. Its weighted average value $(4.08 \pm 0.25) \times 10^{-4}$ is, within uncertainties, in agreement with the Apollo SWC results. The high wind ratio is less variable but smaller. The average Γ in the fast wind is $(3.3 \pm 0.3) \times 10^{-4}$.

1. Introduction

Knowledge of the isotopic ratio of helium in the local interstellar cloud (LIC) and in the outer convective zone (OCZ) of the sun has importance for cosmology and stellar evolution (e.g. Geiss and Gloeckler, 1998). We obtain the present day ^{3}He abundance from the isotopic analysis of pickup helium, using a longer averaging period than in our previous work (Gloeckler and Geiss, 1996). The helium isotopic ratio in the solar wind, $\Gamma_{SW} = {}^3$He^{++}/^{4}He^{++}, provides a good and reliable sample of $(^3$He $+ {}^2$H) in the protosolar cloud some 4.6 Gy ago (Geiss and Gloeckler, 1998). Here we determine Γ_{SW} for various solar wind flow conditions both in the fast, high-latitude, and in the slow, in-ecliptic wind using SWICS/Ulysses data from 1991 to mid-1997. Previous studies (Geiss *et al.*, 1970; Geiss *et al.*, 1972; Coplan *et al.*, 1984; Bochsler, 1984) were based on data taken in the slow solar wind during several years of relatively high solar activity. Using the 6.5 years of SWICS data we determine the dependence of Γ_{SW} on solar activity and solar wind characteristics such as its speed, and FIP strength based on the Fe/O ratio. From this analysis we find the base-line value of the solar wind ^{3}He^{++}/^{4}He^{++} ratio with good precision.

Space Science Reviews **84**: 275–284, 1998.
© 1998 *Kluwer Academic Publishers.*

2. The ^{3}He/^{4}He Ratio in the Local Interstellar Cloud

The orbit of Ulysses and the low background capabilities of the SWICS instrument (Gloeckler *et al.*, 1992) made it possible to discover and study many of the interstellar pickup ions (Gloeckler *et al.*, 1993; Geiss *et al.*, 1994). Interstellar pickup ions are created deep inside the heliosphere through ionization of interstellar atoms (e.g. Gloeckler *et al.*, 1997a; Gloeckler and Geiss, 1998), and provide new information on the elemental and isotopic abundance in the local interstellar cloud.

The local interstellar cloud ^{3}He/^{4}He ratio, Γ_{LIC}, is derived from the measured velocity distributions of interstellar pickup helium isotopes (Gloeckler and Geiss, 1996) as illustrated in Fig. 1. Because virtually all physical and instrument parameters that are used to determine the pickup ion distributions for ^{3}He$^+$ are the same as for ^{4}He$^+$, Γ_{LIC} reduces to the ratio of counts divided by the ratio of probabilities of detecting the two helium isotopes respectively. The only possible other systematic correction comes from the ratio of the anisotropy factors. However, because the distribution function anisotropy of pickup ^{4}He^{++} was found to be the same (within errors) as that of ^{4}He$^+$ (Gloeckler and Geiss, 1998) it is reasonable to assume that the ^{3}He$^+$ anisotropy (which has a rigidity between that of ^{4}He^{++} and ^{4}He$^+$) is also equal to that of ^{4}He$^+$.

Because pickup ^{3}He$^+$ has very low abundance, long accumulation times and the most stringent coincidence conditions were necessary to positively identify these ions. The mass/charge (m/q) histogram of ions of masses between 2 and 6 amu and W (ion speed/solar wind speed) from 1.6 to 2 is shown in the top panel of Fig. 2. Since singly-charged helium ions are absent in the solar wind, especially in this speed range, the identification of these ions as pickup ions is virtually certain. These triple coincidence data were accumulated during a 40 month period in the high speed solar wind (>700 km/s) in order to increase counting efficiency of ^{3}He$^+$. Despite the low count rate the peak at m/q of about 3 (due to ^{3}He$^+$) is well separated from the two neighboring peaks (pickup ^{4}He^{++} and ^{4}He$^+$) and well above the residual small background of less than one count in three years.

The distribution function of ^{3}He$^+$ (divided by 2.4×10^{-4}) is compared to that of ^{4}He$^+$ in the bottom panel of Fig. 2. These spectra were averaged during all times when Ulysses was in the high-speed solar wind of both the south and north polar coronal holes. Within the statistical errors the five point ^{3}He$^+$ is the same as the ^{4}He$^+$ distribution (solid curve) averaged over the same time period reinforcing the case for the correct identification of interstellar ^{3}He$^+$. The interstellar ^{3}He$^+$/^{4}He$^+$ ratio obtained by least-squares (χ^2) fit from the ratio of the two speed distributions is 2.48×10^{-4}. The 1-σ statistical uncertainties $(\chi^2_{min} + 1)$ are $+0.63 \times 10^{-4}$ and -0.57×10^{-4} and systematic errors are estimated to be $\pm 0.25 \times 10^{-4}$. Our present value, derived from a more extended data set than was used in Gloeckler and Geiss (1996), is slightly higher than our previous ratio (Gloeckler and Geiss, 1996), but well within the statistical errors of both measurements.

Interstellar number density of atoms with mass m, ρ_m

- $\rho_m = F_{m/q}(W, R, \Theta) \Big/ \int_{\Omega_{\text{inst}}} d\Omega \, f_{m/q}(w, \theta, \phi)$

W = ion speed / solar wind speed V_{SW}
R, Θ are spacecraft position coordinates; $\Theta = 0°$ is along motion of solar system
Integration over instrument view angles

Measured velocity distribution of pickup ions with mass per charge m/q

- $F_{m/q}(W, R, \Theta) = C_{\text{inst}}\{r_{m/q}(R, \Theta)/\eta_{m/q}(W)\}W^{-4}$

C_{inst} = instrument factor
$r_{m/q}$ = count rate of pickup ion
$\eta_{m/q}$ = count efficiency of pickup ion

Velocity distribution of pickup ions in the solar wind frame as a function of
$\mathbf{w} = \mathbf{W} - \mathbf{V}_{\text{SW}}/|V_{\text{SW}}|$

- $f_{m/q}(w, \theta, \phi) = (3/8\pi)w^{-3/2}\{(\beta_{m/q,\text{prod}})(R_0^2/R)\} \times$
  $\{N(w, R, \Theta, \beta_{m,\text{loss}}/V_0)\}\{G(\lambda_{m/q}, \theta, \phi)\}$

$\beta_{m/q,\text{prod}}$ = rate of pickup ion production by photoionization and charge exchange
     of atom of mass m
$\beta_{m,\text{loss}}$ = ionization loss rate of interstellar atoms of mass m
$R_0^2 = (1 \text{ AU})^2 = (1.5 \times 10^{13} \text{ cm})^2$
N = normalized ($N = 1$ in LIC) spatial distribution of interstellar atoms in
     heliosphere
V_0 = relative speed of the interstellar cloud and the Sun
G = anisotropy function of the pickup ion velocity distribution
$\lambda_{m/q}$ = pitch-angle scattering mean free path

Interstellar ^{3}He/^{4}He density ratio = $\Gamma_{\text{LIC}} \approx$

- $F_{^3\text{He}^+}(W)/F_{^4\text{He}^+}(W) \approx \{r_3(W)/r_4(W)\}\{\eta_4(W)/\eta_3(W)\}\{G(\lambda_4)/G(\lambda_3)\}$

Figure 1. Computational steps relating the number density ratio of helium isotopes in the local interstellar cloud to the phase space density ratio of the corresponding helium pickup ion in the heliosphere. Because the distribution functions of ^{4}He$^+$ (especially ^{3}He$^+$) can only be measured in a limited portion of phase space, model velocity distributions in the solar wind frame, $f(w, \theta, \phi)$, are used to fill in the missing portions of phase space.

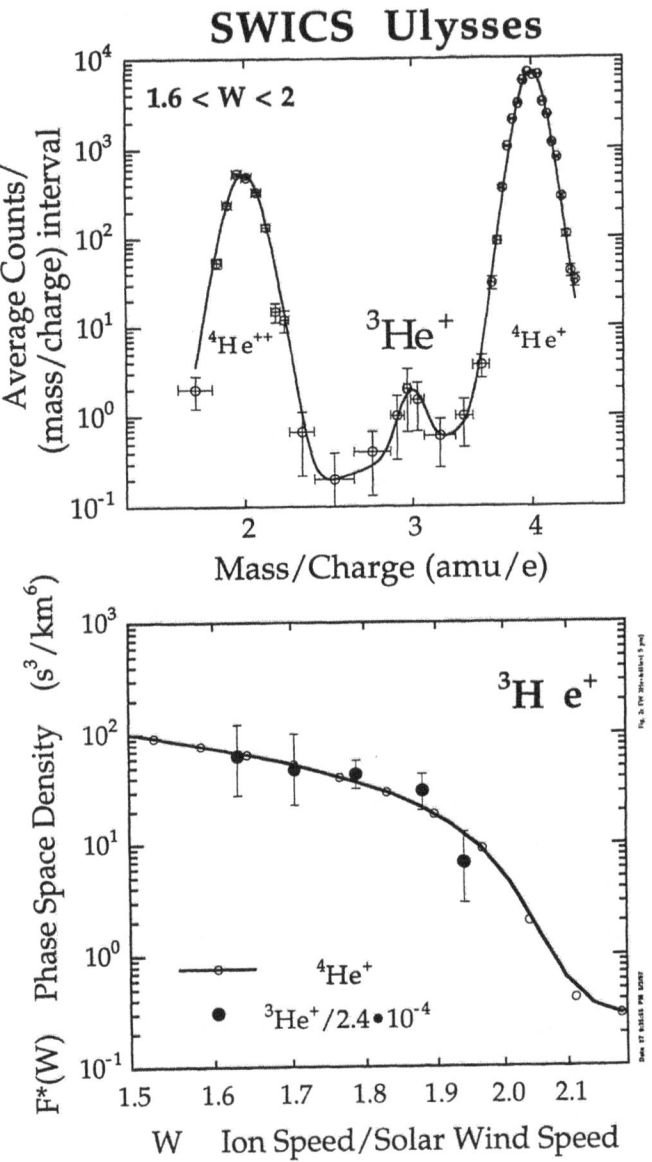

Figure 2. Top: Average triple-coincidence count rate density of ions selected to have masses between 2 and 6 amu and speeds W between 1.6 and 2 versus mass/charge (m/q). The ^{3}He$^+$ peak at $m/q = 3$ is well separated from the two adjacent much larger peaks at $m/q = 2$ and 4 from pickup ^{4}He^{++} and ^{4}He$^+$ respectively. Triple coincidence is required to measure both mass and mass/charge of the ion (Gloeckler *et al.*, 1992) and to reduce background sufficiently to see the ^{3}He$^+$ peak.

Bottom: Phase space density, scaled by $(438/V_{SW})^4$, of ^{4}He$^+$ (open circles) and ^{3}He$^+$ divided by 2.4×10^{-4}, (filled circles) versus W. The similarity of the two spectra is evident despite the large statistical errors for ^{3}He$^+$. Below $W = 1.6$ the triple coincidence efficiency for ^{3}He$^+$ becomes extremely small, and above $W = 2$ no ^{3}He$^+$ was detected, consistent with the sharp cutoff of pickup ion spectra at $W = 2$ (e.g. Gloeckler and Geiss, 1998).

3. The ^{3}He/^{4}He Ratios in the Solar Wind

Unlike pickup ion distributions which are broad and have a cutoff at $W = 2$ (e.g. Gloeckler and Geiss, 1998), the solar wind velocity distribution is narrow and well represented by a kappa function in the solar wind frame.

$$f(w) = f_0[1 + (w/\theta)^2/\kappa]^{-(\kappa+1)} \tag{1}$$

$$V_{th} = (1 - 1.5/\kappa)^{-1/2}\theta \tag{2}$$

$$w = |(\mathbf{V}_{ion} - \mathbf{V}_{SW})|/|\mathbf{V}_{SW}| \tag{3}$$

The isotopic helium density ratio in the solar wind, $\Gamma_{SW} = {}^3\mathrm{He}^{++}/{}^4\mathrm{He}^{++}$ is derived from the respective integrated phase space densities of these isotopes as indicated in Fig. 1. However, because the full distribution functions are measured (unlike for pickup ions), knowledge of the shape of the distribution functions is not required and Γ_{SW} is simply equal to $(N_{^3\mathrm{He}^{++}}/N_{^4\mathrm{He}^{++}}) \times \langle(\eta_{^4\mathrm{He}^{++}}/\eta_{^3\mathrm{He}^{++}})\rangle$. Here $N_{^3\mathrm{He}^{++}}$ and $N_{^4\mathrm{He}^{++}}$ are the total counts in a given time period of the two isotopes respectively and $\langle(\eta_{^4\mathrm{He}^{++}}/\eta_{^3\mathrm{He}^{++}})\rangle$ is the average ratio of efficiencies for these ions at the average measured solar wind speed for that period.

3.1. ANALYSIS

The solar wind Γ_{SW} may be obtained from the SWICS data in a number of different ways (Gloeckler et al., 1992; Bodmer, 1996). However, each method used requires knowledge of the corresponding efficiency ratio at the appropriate speed for these helium isotopes. Unfortunately, preflight calibration with $^3\mathrm{He}^{++}$ could not be obtained and thus $^3\mathrm{He}^{++}$ efficiencies for the SWICS/Ulysses instrument were not determined. In this analysis we have therefore used a method that is least sensitive to our lack of measured $^3\mathrm{He}^{++}$ efficiencies. We use data requiring only double coincidence efficiencies which (a) vary little with solar wind speed, (b) can be determined (for $^4\mathrm{He}^{++}$) from flight data, and (c) are nearly equal for $^3\mathrm{He}^{++}$ and $^4\mathrm{He}^{++}$ at the same speed (as indicated by measurements with the SWICS spare instrument as well as model predictions).

The $^3\mathrm{He}^{++}$ counts (averaged over a given time interval) are obtained directly from the SWICS pulse-height (PHA) data corresponding to double-coincidence (i.e. time-of-flight) selected to have m/q between 1.42 and 1.60 amu/e and mass (m) less than 5 amu. When plotted against $W_{^3\mathrm{He}^{++}}$, the speed of $^3\mathrm{He}^{++}$ divided by the solar wind speed, a distinct peak is always seen in the counts spectrum at $W_{^3\mathrm{He}^{++}} = 1$, indicating measurements of solar wind $^3\mathrm{He}^{++}$. Because the maximum count rate corresponding to $^3\mathrm{He}^{++}$ was always low, no corrections for PHA event saturation were required. That is, all $^3\mathrm{He}^{++}$ detected were registered as pulse-height events. However, a background correction resulting from spill-over from the five orders of magnitude more abundant solar wind protons was necessary. This correction reduced $^3\mathrm{He}^{++}$ counts primarily below $W_{^3\mathrm{He}^{++}} \sim 0.95$. The total reduction in the integrated counts, however, was smaller than 10% in all cases.

A selection similar to that used for $^3\text{He}^{++}$ could not be used for $^4\text{He}^{++}$ because its high count rate resulted in substantial pulse-height saturation. To correct for this, we used the unsaturated triple-coincidence $^4\text{He}^{++}$ rate MR1 (Gloeckler *et al.*, 1992) counts and multiplied these by the ratio of double-coincidence to triple-coincidence PHA counts selected for $^4\text{He}^{++}$. The PHA selection used for double-coincidence events was: m/q between 1.76 and 2.53 amu/e, and m between 2.6 and 5.6 amu as well as mass-0 (no energy) events. The triple-coincidence PHA data had the same m/q and m selection applied to it as rate MR1 ($1.571 < m/q < 2.521$ and $3.56 < m < 5.13$).

3.2. RESULTS

Using the method described above we have analyzed twelve time periods varying in length from 1.5 days to 310 days both in the slow in-ecliptic as well as the fast coronal hole solar wind. The count rate densities of $^3\text{He}^{++}$ and $^4\text{He}^{++}$ for two such periods are given in Fig. 3. In the top panel the long-term average in the steady, fast (779 km/s) solar wind of the north polar coronal hole shows identical shapes for the count density distributions of the two helium isotopes. The average solar wind Γ_{SW} for this high-speed period, computed from the ratio of the integrated counts from the kappa fits (eq. 1) to the distributions, was $(2.88 \pm 0.03) \times 10^{-4}$, the lowest value observed in the twelve periods analyzed. Essentially the same result ($\Gamma_{SW} = (2.85 \pm 0.04) \times 10^{-4}$) was obtained from the ratio of the sum of counts in the four W bins (0.95 to 1.05) with the highest counts.

The bottom panel of Fig. 3 is a plot of the count rate density distributions of the solar wind helium isotopes during a relatively unperturbed period in the slow solar wind (437 km/s) near the ecliptic. The distributions of both isotopes in the slow wind are narrower than in the fast wind and the $^4\text{He}^{++}$ spectrum is slightly broader than that of $^3\text{He}^{++}$. The isotopic ratio computed from the kappa fits (same value of kappa is used for both distributions) is $\Gamma_{SW} = (3.92 \pm 0.07) \times 10^{-4}$. In addition to the statistical errors a systematic error (estimated to be $\pm 6\%$) due to a combination of uncertainties in the ratio of efficiencies, $\langle (\eta_{^4\text{He}^{++}}/\eta_{^3\text{He}^{++}}) \rangle$, and background correction (for $^3\text{He}^{++}$) is assumed for all solar wind helium isotopic ratios given here.

4. Discussion and Conclusions

Our measurements of the solar wind helium isotopic ratio, Γ_{SW}, during eleven of the twelve time periods analyzed here are summarized in Fig. 4. During the 5-day interval of the twelfth period (days 168 to 173 of 1992) $\Gamma_{SW} = (1.11 \pm 0.26) \times 10^{-3}$ was more than a factor of two higher than the slow-wind average, possibly because the solar wind $^4\text{He}^{++}$ (and proton) densities were unusually low (30 times below the ambient values). During the in-ecliptic portion of Ulysses' trajectory (days 145

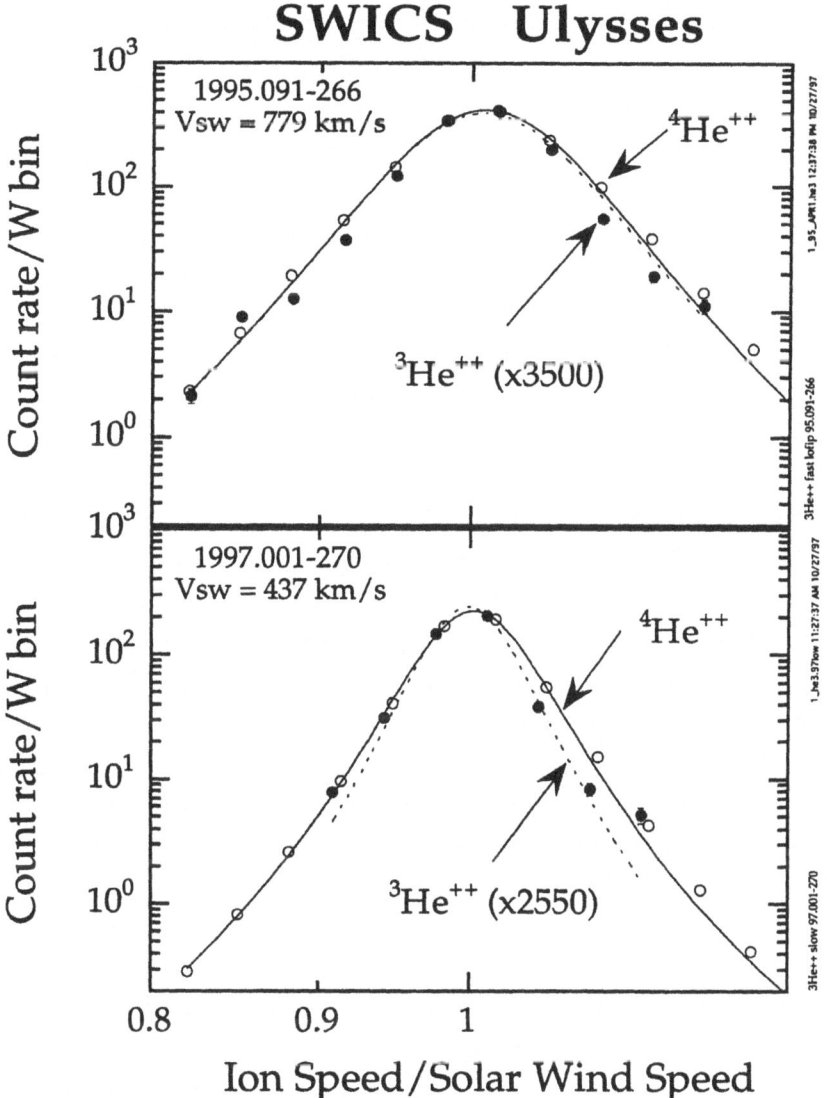

Figure 3. Average count rate density versus ion speed divided by solar wind speed (W) for ${}^4\text{He}^{++}$ and ${}^3\text{He}^{++}$ (multiplied by the factor in parentheses) in a high-speed (779 km/s) solar wind from the north coronal hole (top panel) and a low-speed wind (437 km/s) in the ecliptic plane. $\Gamma_{\text{sw}} = {}^3\text{He}^{++}/{}^4\text{He}^{++}$ is computed directly from the ratio of total counts obtained by integrating the kappa function fits to the respective distributions. Γ_{sw} measured in the coronal hole wind is lower than in the slow in-ecliptic wind.

to 545 since 1991) SWICS sampled the highly turbulent slow wind over radial distances from 3.2 to 5.4 AU. The means of the ratios in this slow wind are variable but tend to cluster around 4.15×10^{-4}, a value close to the average ratio (indicated

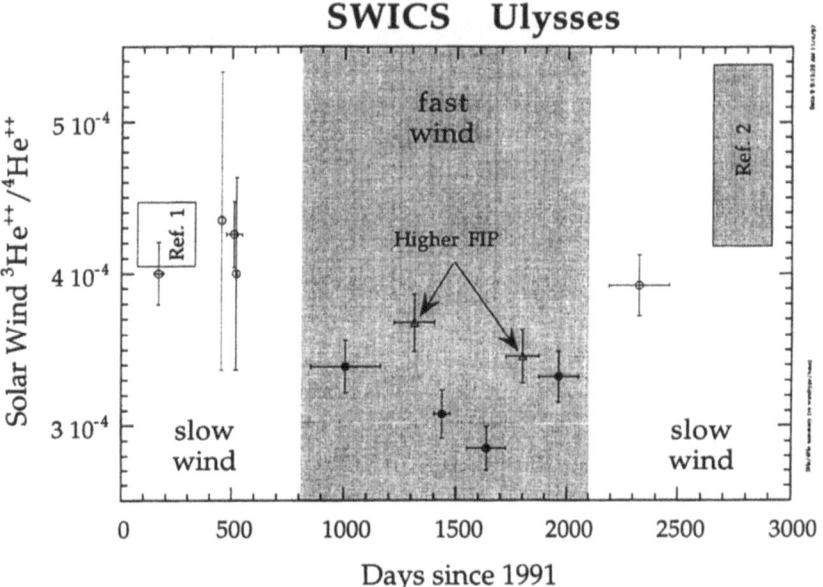

Figure 4. Solar wind isotopic helium ratio, $\Gamma_{SW} = {}^3\text{He}^{++}/{}^4\text{He}^{++}$, measured during selected time periods with SWICS on Ulysses. In the slow, in-ecliptic wind time periods Γ_{SW} is variable with a weighted slow-wind average of $(4.08\pm0.25) \times 10^{-4}$. In the high-latitude fast wind of the polar coronal holes (shaded region) Γ_{SW} has its lowest values. Sorting the coronal hole data by the measured (with SWICS) value of the solar wind Fe/O ratio suggests a dependence of Γ_{SW} on Fe/O. Time periods when Fe/O is at its lowest level (filled circles) have the lowest Γ_{SW}. Except for the shortest time periods, errors shown are due primarily to uncertainties in the ratio of double-coincidence efficiencies of the helium isotopes. Ref. 1 is the ${}^3\text{He}^{++}/{}^4\text{He}^{++}$ value reported by Geiss *et al.* (1970; 1972) and Ref. 2 the average ratio given by Bochsler (1984).

by the box labeled Ref. 1) obtained by Geiss *et al.* (1970; 1972) using foil collection techniques on the moon. Our other slow-wind average ratio was measured during the 270 day period in 1997 near 5 AU when Ulysses was again at low latitudes (see also bottom panel of Fig. 3). The solar wind was relatively quiet at this near-minimum phase of the solar activity cycle. Our value of $\Gamma_{SW} = (3.92\pm0.25) \times 10^{-4}$ in the slow wind in 1997 (day ~ 2300 since 1991) is well within the limits of the Geiss *et al.* (1972) average, but outside the error limits of the ratios reported by Bochsler (1984), shown as the box labeled Ref. 2, and especially by Coplan *et al.* (1984), both results obtained with the ICE instrument on ISEE-3. Variability of the ${}^3\text{He}^{++}/{}^4\text{He}^{++}$ ratio in the slow wind has been noted previously (e.g. Geiss *et al.*, 1970; 1972), and recent results (Gloeckler *et al.*, 1997b) indicate unusually high ratios ($\Gamma_{SW} \sim 6 \times 10^{-3}$) in Coronal Mass Ejections (CMEs) near solar minimum. The time-weighted average of all the slow wind data presented here gives a value for Γ_{SW} of $(4.08 \pm 0.25) \times 10^{-4}$.

In the fast wind of the polar coronal holes (shaded region) we obtained low Γ_{SW} ratios. In four of the six time periods (filled circles) we measured the lowest solar wind Fe/O (i.e. lowest FIP ratio) of the entire Ulysses observation period. The other two time periods, had a distinctly higher FIP (Fe/O) ratio although not nearly as high as in the slow wind. The average $^3He^{++}/^4He^{++}$ ratio of the four lowest FIP periods was $(3.16\pm0.25) \times 10^{-4}$ and of the six fast wind periods $(3.3\pm0.3) \times 10^{-4}$, in reasonable agreement with the lowest value obtained by Bodmer (1996) using SWICS data in the polar holes. Combining our lowest solar wind value with the (model-dependent) upper limit for the photospheric $^3He/^1H = 2.3 \times 10^{-5}$ deduced from gamma-ray spectroscopy of solar flares (Murphy *et al.*, 1997) we obtain a photospheric $^4He/^1H = 0.080 \pm 0.006$. This is in remarkable agreement with the definitive helium to hydrogen ratio of 0.084 obtained from helioseismology.

While it is clear that the high-speed coronal hole wind has a lower Γ_{SW} than the in-ecliptic slow wind, the reason for this difference is not obvious. The solar wind $^3He^{++}/^4He^{++}$ may decrease with solar activity (the fast wind periods were near solar minimum of the present cycle). The results shown in Fig. 4 also suggest a correlation with Fe/O which is an indicator of strength of the FIP bias of solar wind elemental abundances. Lower values of Γ_{SW} in the fast solar wind are also observed by Bodmer (1996), although the variation with speed is complicated. It is clearly important to study more thoroughly the variabilities in the solar wind helium isotopic ratio and to find the causes for it.

Based on six years of SWICS/Ulysses data, we have established that the $^3He^{++}/^4He^{++}$ ratio in the fast wind is definitely lower than the average slow wind ratio. Such a systematic difference was not found in the earlier work based on ICI/ISEE-3 (Coplan *et al.*, 1984; Bochsler, 1984) and SWICS/Ulysses data (Bodmer, 1996). The difference is not entirely unexpected because of basic differences in the acceleration dynamics of the slow wind and the fast streams (cf. Geiss and Gloeckler, 1998). However, the difference is surprisingly high, and this may be due to unrecognized admixtures into slow wind averages of plasma parcels with anomalously high $^3He^{++}/^4He^{++}$ ratio as those found by us during days 168–173 or in CME plasmas. Thus, simple averages of $^3He^{++}/^4He^{++}$ ratios in the slow wind have to be corrected for such a bias when deriving the OCZ value.

The OCZ value of $^3He/^4He$ of $(3.8 \pm 0.5) \times 10^{-4}$ given by Geiss and Gloeckler (1998) is based on the ratio measured in both the slow wind and the fast streams. If at a later time, an improved theoretical understanding of the dynamics in the solar wind source region confirms that helium isotopes are much less fractionated in the fast streams than in the slow wind, the $^3He^{++}/^4He^{++}$ ratio in the OCZ could best be derived from the fast wind data alone, and this would lower the OCZ value. For example, if further study confirms that the fast stream plasma with the lowest Fe/O ratio is the most representative of the OCZ, the $^3He/^4He$ ratio of $(3.8 \pm 0.5) \times 10^{-4}$ in the OCZ would be lowered by 10 to 15%.

Acknowledgements

We are very grateful to the many individuals at the University of Maryland, the University of Bern, the Max-Planck-Institute für Aeronomie and the Technical University of Braunschweig for their many outstanding contributions to the success of the SWICS experiment on Ulysses during the many years of its development phase. We thank Christine Gloeckler for her help with data reduction. This work was supported in part by NASA/JPL contract 955460 (GG) and by the International Space Science Institute and the Swiss National Science Foundation (JG).

References

Bochsler, P.: 1984, *Helium and Oxygen in the solar wind: Dynamic properties and abundances of elements and Helium isotopes as observed with the ISEE-3 plasma composition experiment*, Habilitationsschrift, Univ. of Bern.

Bodmer, R.: 1996, *The Helium Isotopic Ratio as a Test for Mi nor Ion Fractionation in the Solar Wind Acceleration Process: SWICS/ULYSSES Data Compared with Results from a Multifluid Models*, Ph.D. Thesis, Univ. of Bern.

Coplan, M. A., Ogilvie, K. W., Bochsler, P. and Geiss, J.: 1984,*Solar Phys.* **93**, 415.

Geiss, J., Eberhardt, P., Bühler, F., Meister, J. and Signer, P.: 1970, *J Geophys. Res.* **75**, 5972.

Geiss, J., Bühler, F., Cerutti, H., Eberhardt, P. and Filleux, C.: 1972, in *Apollo 16 Preliminary Science Report, NASA SP-315*, section 14.

Geiss, J. and Gloeckler, G.: 1998, *Space Sci. Rev.*, this issue.

Geiss, J., Gloeckler, G., Mall, U., von Steiger, R., Galvin, A. B., and Ogilvie, K. W.: 1994, "Interstellar oxygen, nitrogen, and neon in the heliosphere", *Astr. Astrophys.* **282**, 924–933.

Gloeckler, G., *et al.*: 1992, Geiss, J., Balsiger, H., Bedini, P., Cain, J. C., Fischer, J., Fisk, L. A., Galvin, A. B., Gliem, F., Hamilton, D. C., Hollweg, J. V., Ipavich, F. M., Joss, R., Livi, S., Lundgren, R., Mall, U., McKenzie, J. F., Ogilvie, K. W., Ottens, F., Rieck, W., Tums, E. O., von Steiger, R., Weiss, W., and Wilken, B.: 1992, "The solar wind ion composition spectrometer", *Astr. Astrophys. Suppl. Ser.* **92**, 267–289.

Gloeckler, G., *et al.*: 1993, Geiss, J., Balsiger, H., Fisk, L. A., Galvin, A. B., Ipavich, F. M., Ogilvie, K. W., von Steiger, R., and Wilken, B.: 1993, *Science* **261**, 70.

Gloeckler, G. and Geiss, J.: 1996, "Abundance of ^{3}He in the local interstellar cloud", *Nature* **381**, 210.

Gloeckler, G. and Geiss, J.: 1998, "Interstellar and Inner Source Pickup Ions Observed with SWICS on Ulysses", *Space Sci. Rev.*, in press.

Gloeckler, G., Fisk, L. A. and Geiss, J: 1997a, "Anomalously small magnetic field in the local interstellar cloud", *Nature* **386**, 374–377.

Gloeckler, G., Galvin, A. B., Hamilton, D. C., Ipavich, F. M., Fisk, L. A., Geiss, J., Bochsler, P., and Wilken, B.: 1997b, "The Unusual Solar Wind in the High-Density Pulse of the Magnetic Cloud Associated With the Halo CME of January, 1997", Spring AGU meeting, Baltimore, MD, *EOS Trans. American Geophys. Union* **47**.

Murphy, R. J.,. Share, G. H,. Grove, J. E, Johnson, W. N., Kinzer, R. L., Kurfess, J. D., Strickman, M. S. and Jung, G. V.: 1997, "Accelerated Particle Composition and Energetics and Ambient Abundances from Gamma-Ray Spectroscopy of the 1991 June 4 Solar Flare, *Astrophys J.*, submitted.

DEUTERIUM ABUNDANCE IN THE LOCAL ISM AND POSSIBLE SPATIAL VARIATIONS

JEFFREY L. LINSKY

JILA, University of Colorado and NIST, Boulder, CO 80309-0440 USA

Abstract. Excellent HST/GHRS spectra of interstellar hydrogen and deuterium Lyman-α absorption toward nearby stars allow us to identify systematic errors that have plagued earlier work and to measure accurate values of the D/H ratio in local interstellar gas. Analysis of 12 sightlines through the Local Interstellar Cloud leads to a mean value of D/H $= (1.50 \pm 0.10) \times 10^{-5}$ with all data points lying within $\pm 1\sigma$ of the mean. Whether or not the D/H ratio has different values elsewhere in the Galaxy and beyond is a very important open question that will be one of the major objectives of the Far Ultraviolet Spectroscopic Explorer (FUSE) mission.

Key words: LISM, interstellar medium, deuterium

1. Introduction

An accurate measurement of $(D/H)_{LISM}$, the D/H abundance ratio in the local interstellar medium (LISM), and an assessment of spatial variations of D/H in the Galaxy are required to address two critically important questions in contemporary astrophysics. First, the largest credible D/H ratio in our Galaxy will provide a lower limit to the primordial D/H ratio, $(D/H)_{prim}$, which constrains the critical density, Ω_B, of baryons present in both luminous and "dark" forms. Second, $(D/H)_{LISM}$ is the end result of an incompletely understood complex set of chemical evolution processes in the Galaxy. Comparison of the $(D/H)_{LISM}$ ratio with D/H ratios characteristic of the protosolar nebula and in interstellar gas located elsewhere in the Galactic disk and halo will test our understanding of stellar evolution, stellar mass loss, interstellar physics, and the rate of infall and chemical composition of halo gas. Testing Galactic chemical evolution codes against both D/H and metal abundances in different environments with different histories will lead to a more detailed understanding of the evolution of our Galaxy and will test for the first time the mixing time scales for interstellar matter.

Recently, the precise value of $(D/H)_{LISM}$ has acquired greater importance as the previously announced high value of $(D/H) = 2 \times 10^{-4}$ in the absorption spectrum toward Q0014+813 is apparently spurious (Tytler *et al.* 1996). Thus $(D/H)_{LISM}$ is likely much closer to $(D/H)_{prim}$ than some authors had thought, and our understanding of Galactic chemical evolution will be tested by a measurement of the small difference (perhaps only a factor of 2) between $(D/H)_{prim}$ and $(D/H)_{LISM}$. Accurate measurements of D/H are clearly required for such tests of the chemical evolution codes and their underlying assumptions.

Lyman line absorption is generally recognized as the most reliable technique for inferring the present value of D/H in our local environment. This is a consequence

Space Science Reviews **84**: 285–296, 1998.

of the Sun being surrounded by a cloud of warm, partially ionized gas (e.g., Lallement *et al.*, 1995; Wood and Linsky, 1997a) with very few molecules. At a gas temperature of about 7,000 K (Linsky *et al.*, 1993), the ionization and adsorption on to grains is nearly the same for H and D. With a line separation of 81 km s^{-1}, Lyman-α absorption by H and D can be observed with high S/N in HST/GHRS spectra toward nearby stars, provided the hydrogen column density is not so large as to obliterate the D line ($N_{HI} < 10^{18.7}$ cm^{-2}). The launch of the Far Ultraviolet Spectroscopic Explorer (FUSE) in 1998 will allow us to observe the less opaque higher Lyman lines to extend this method to more distant lines of sight (LOS). Other techniques for measuring the D/H ratio, including studies of deuterated molecules in cold molecular clouds and the search for the deuterium analog of the hydrogen 21 cm line, have major difficulties, leaving the Lyman lines as the most useful D/H diagnostics.

Ferlet *et al.* (1996) reviewed D/H measurements obtained primarily with the Copernicus and IUE spectrographs. These pre-HST studies of Lyman line absorption toward both hot and cool stars left a confused picture in which the large range of permitted values of D/H for each LOS left open the possibility that D/H spatial variations of a factor of 2 or larger could be present on very short spatial scales. The flood of beautiful new GHRS spectra has changed this picture dramatically. The first clear indication of this paradigm shift was the measurement of D/H $= (1.60^{+0.14}_{-0.19}) \times 10^{-5}$ for the Capella LOS (Linsky *et al.*, 1993; 1995). Since this result lies outside of the published error bars for all of the previous results for this LOS, I believe that the older results at least for the late-type stars are unreliable because of systematic errors that were not considered when these lower quality data were analyzed and thus not included in the published error bars. The GHRS spectra have allowed us to develop analysis techniques that minimize these systematic errors, leading to far more reliable values of D/H. Table I summarizes the results of these analyses of GHRS spectra.

2. Minimizing Random and Systematic Errors in the Analysis of the Lyman Lines

2.1. THE EFFECTS OF BETTER QUALITY DATA

GHRS echelle spectra have far higher S/N and resolution (3.6 km s^{-1}) than Copernicus (15 km s^{-1}) and IUE (25–30 km s^{-1}) spectra. Since the core of the Lyman-α line is very saturated (optical depths of $10^5 - 10^6$) and is located on the flat part of the curve of growth, high S/N and spectral resolution are critical for inferring H column densities from the steep outer edges of the core absorption profile. Even with the best available GHRS spectra, however, the H column densities are usually more uncertain than the D column densities. Accurate fits to the outer edges of the core absorption alone may explain much of the previous scatter in the D/H values.

Table I

Summary of GHRS Observations of the LISM.

Star	d (pc)	l (°)	b (°)	Grating[†]	Clouds in LOS	D/H (10^{-5}) (in LIC)	(others)
α Cen A* [1,2]	1.3	316	−01	EA, EB	G		1.2 ± 0.7
α Cen B [2]	1.3	316	−01	EA, EB	G		1.2 ± 0.7
Sirius [1,3,4]	2.7	227	−09	EB, M	L+1	(1.65)	(1.65)
ε Eri [11]	3.3	196	−48	EA, EB	L	1.4 ± 0.4	
ε Ind [5]	3.4	336	−48	EA, EB	G/L	1.6 ± 0.4	
Procyon [6]	3.5	214	+13	EB, M	L+1	1.6 ± 0.4	
α Aql [1]	5.0	48	−09	EB	L+2		
α PsA [1]	6.7	21	−65	EB			
Vega [1]	7.5	68	+19	EB	L+2		
β Gem [11]	10.6	192	+23	EA, EB	L+1	1.4 ± 0.4	1.6 ± 0.4
β Leo [1]	12.2	251	+71	EB	L+2		
Capella* [6,7]	12.5	163	+05	EA, EB, M	L	$1.60^{+0.14}_{-0.19}$	
β Cas [8]	14	118	−03	M	L	1.6 ± 0.4	
[11]						1.7 ± 0.3	
β Cet [8]	16	111	−81	EB, M	2	2.2 ± 1.1	
β Pic [1]	16.5	258	−31	EB	L		
α Tri [11]	18	139	−31	EA, EB	L	1.6 ± 0.6	1.0 ± 0.6
λ And [5]	24	110	−15	EA	L	1.7 ± 0.5	
δ Cas [1]	27	127	−02	EB	L		
HR1099* [8]	33	185	−41	EA, EB, M	L+2	1.46 ± 0.09	
G191-B2B [9]	48	156	+07	EB, M	L+2	$1.4^{+0.1}_{-0.3}$	(1.),(1.5)
σ Gem* [11]	56	191	+23	EA, EB	L+1	1.4 ± 0.4	1.3 ± 0.4
HZ 43 [12]	63	054	+84	EA, EB	NGP		(1.6)
31 Com [8]	80	115	+89	EB, M	NGP	1.5 ± 0.4	
[11]						2.0 ± 0.4	
ε CMa [10]	187	240	−11	EB, M	L+5	(1.65)	

Quantities in parenthesis are assumed D/H values that lead to good profile fits.
* These stars were observed twice. Capella, HR 1099 and σ Gem were observed near opposite quadratures.
[†] Gratings: EA = Echelle-A, EB = Echelle-B, M = G140M or G160M.
References: [1] Lallement et al. (1995), [2] Linsky and Wood (1996), [3] Lallement et al. (1994), [4] Bertin et al. (1995), [5] Wood et al. (1996), [6] Linsky et al. (1995), [7] Linsky et al. (1993), [8] Piskunov et al. (1997), [9] Lemoine et al. (1996), [10] Gry et al. (1995), [11] Dring et al. (1997), [12] Landsman et al. (1996).

2.2. SYSTEMATIC ERRORS INTRODUCED BY COMPLEX VELOCITY STRUCTURE AND UNCERTAIN HYDROGEN PREDICTORS

Another critical issue is the presence of absorption components at many velocities in a stellar spectrum. For example, ultra-high resolution spectra of the Na I and Ca II resonance lines typically show many closely spaced narrow velocity components

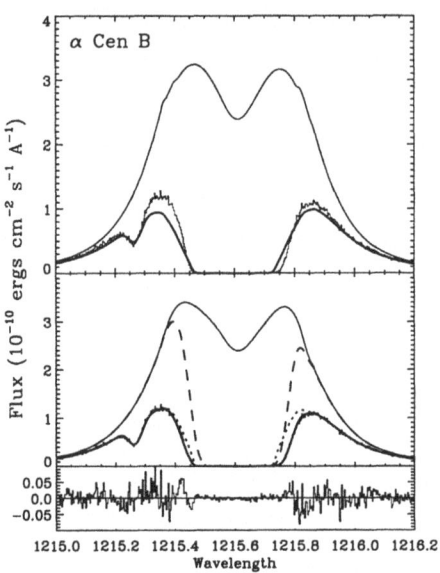

Figure 1. Upper panel: comparison of the observed Echelle-A spectrum (noisy line) of α Cen B with the assumed intrinsic stellar spectrum (smooth thin line) and the best constrained one-component model fit (thick solid line). Middle panel: best two-component model with the absorption due to the ISM component only (dotted line), absorption due to the H wall component only (dashed line), and the total absorption (thick solid line). Lower panel: residuals between the observed profile and the two-component fit (from Linsky and Wood, 1996).

even for short LOS (e.g., Welty *et al.*, 1996). Since thermal broadening is much larger in the low mass H and D lines than in the metal lines, it is difficult to isolate the absorption due to individual velocity components in the H and D lines. One can include these velocity components in the analysis of the H and D lines, but the relative column densities observed in the metal lines may not be good predictors of the relative column densities in the H and D lines. It is commonly assumed that N I and O I are good predictors of H I and D I column densities because all four species have nearly the same ionization potentials and thus should have the same ionization equilibria. However, Vidal-Madjar (these proceedings) showed that for the three velocity components along the LOS to the hot white dwarf G191-B2B, the D/N column density ratios differ by a factor of 3 and the D/O column density ratios differ by a factor of 10. Since there is no apparent explanation for these large discrepancies, one must conclude that even N I and O I are not infallible predictors of H column densities. Thus it is difficult to infer D/H for individual velocity components along a complex LOS, and the inferred mean value of D/H along a LOS could be biased by saturation.

2.3. SYSTEMATIC ERRORS INTRODUCED BY HYDROGEN WALLS

Velocity components with column densities orders of magnitude smaller than the main interstellar absorption component pose an especially difficult problem, since these components are not detectable in metal lines, but they could be optically thick in the H Lyman-α line. Linsky and Wood's (1996) analysis of the very short (1.3 pc) LOS to α Cen A and α Cen B shows that by not including such velocity components when they are present the inferred H column densities can be a factor of 2 too large and the D/H ratios can thus be a factor of 2 too small.

The upper panel of Figure 1 shows the observed Lyman-α profile toward α Cen B and Linsky and Wood's best fit model in which the velocity and temperature of the interstellar H I are constrained to be the same as the values obtained by fitting the D I, Mg II, and Fe II lines. Clearly there is missing opacity near zero observed flux on the red side of the interstellar absorption, indicating the need to include additional absorption by gas that (i) is redshifted compared to the interstellar flow velocity of 18.0 km s^{-1}, (ii) is hotter than the interstellar gas (required to fit the gentler slope of the red side of the absorption compared to the blue side), and (iii) has a relatively low column density (the additional absorption has no Voigt wings). Because there is missing opacity at zero flux, no sensible change in the assumed intrinsic stellar profile can explain the discrepancy. The middle panel of Figure 1 shows their least-squares fit to the observed profile by a two-component model. The first component of this model has the same interstellar velocity and broadening parameters as those derived from the D I, Mg II, and Fe II lines, but a smaller hydrogen column because the second component will explain the deep absorption on the red side of the interstellar absorption feature. The gas in the second component turns out to be hot ($T = 29,000 \pm 5,000$ K), has a low column density ($\log N_{\text{H I}} = 14.74 \pm 0.24$), and is redshifted by 2–4 km s^{-1} relative to the main component of the interstellar gas. Given the highly saturated nature of the main component, the inclusion of this second component, even though it has only 1/1000 the H column density of the first component, lowers the hydrogen column of the first component by a factor of 2 and thereby raises the inferred D/H ratio from $\approx 6 \times 10^{-6}$ to $(1.2 \pm 0.7) \times 10^{-5}$. The uncertainty in D/H is large because there are many parameters to be determined.

When Linsky and Wood derived these results they were unaware of the location of the second absorption component of H I along the LOS to α Cen. The location became clear later at the 1995 July 12-13 meeting of the IUGG in Boulder, Colorado, where Baranov, Zank, and Williams presented their calculations of the interaction between the solar wind and the incoming interstellar flow. Their models (Baranov *et al.*, 1995a; 1995b; Pauls *et al.*, 1995), which include charge exchange between the outflowing solar wind protons and the inflowing H I atoms, show that on both sides of the heliopause there is a region of decelerated, hot hydrogen with higher density than in the LIC (see Fig. 2). This H I pileup region, which is located about 200 AU in the upstream direction (depending on the proton density in the

LIC), has been called the "hydrogen wall." Because the computed column density, temperature, and flow velocity of H I in the hydrogen wall agree with the parameters derived independently for the second component toward α Cen, Linsky and Wood (1996) concluded that the second component originates in the wall around the heliosphere. Before this time the hydrogen wall was an interesting theoretical concept with no observational confirmation, although Lyman-α backscattering observations (e.g., Quémerais et al., 1995) indicated that $n_{H\ I}$ increases outward toward the heliopause.

Do hydrogen walls exist around other stars? A stellar hydrogen wall would be seen as a second absorption component shifted to shorter wavelengths compared to the interstellar gas flowing toward the star, because in order to detect the stellar wall it would have to be viewed from the upwind direction. Wood, Alexander, and Linsky (1996) found that a one-component model for the interstellar absorption toward ϵ Ind could not explain the absorption on the blue side of the interstellar Lyman-α line. They concluded that a second component blueshifted by 18 ± 6 km s^{-1} with respect to the interstellar flow was needed with $\log N_{H\ I} = 14.2 \pm 0.2$ and $T = 100,000 \pm 20,000$ K. The high temperature and large blueshift are consistent with the higher inflow velocity of 64.0 km s^{-1} toward this rapidly moving star. They also found evidence for a hydrogen wall around λ And with a smaller blueshift and temperature. Dring et al. (1997) showed that a H wall is also present around ϵ Eri, and Wood and Linsky (1997b) found evidence for hydrogen walls around the high velocity stars 61 Cyg A and 40 Eri A. Thus the inclusion of solar and stellar hydrogen wall absorption when the viewing angles are appropriate is required for obtaining accurate D/H values.

2.4. SYSTEMATIC ERRORS INTRODUCED BY THE UNKNOWN EMISSION LINE PROFILE

Another source of systematic errors is the unknown stellar Lyman-α emission line which serves as the "continuum" against which one measures interstellar absorption. One method for minimizing this problem is to analyze spectroscopic binary systems observed at opposite quadratures (when the orbital radial velocities are a maximum) so that the combined stellar emission line profile has a different shape in the two observations, whereas the interstellar absorption is unchanged. The additional information provided by the two observations allows one to infer the intrinsic stellar emission line profiles and the interstellar absorption nearly independent of each other. Linsky et al. (1995) and Piskunov et al. (1997) have used this approach to analyze the LOS toward Capella and HR 1099, and Dring et al. (1997) have used a similar approach to study the LOS toward σ Gem.

High radial velocity stars provide an opportunity for deriving $N_{H\ I}$ more accurately from the optically thin line wings than from the saturated core. For the star ϵ Ind, Wood et al. (1996) inferred the interstellar Lyman-α wing absorption (see

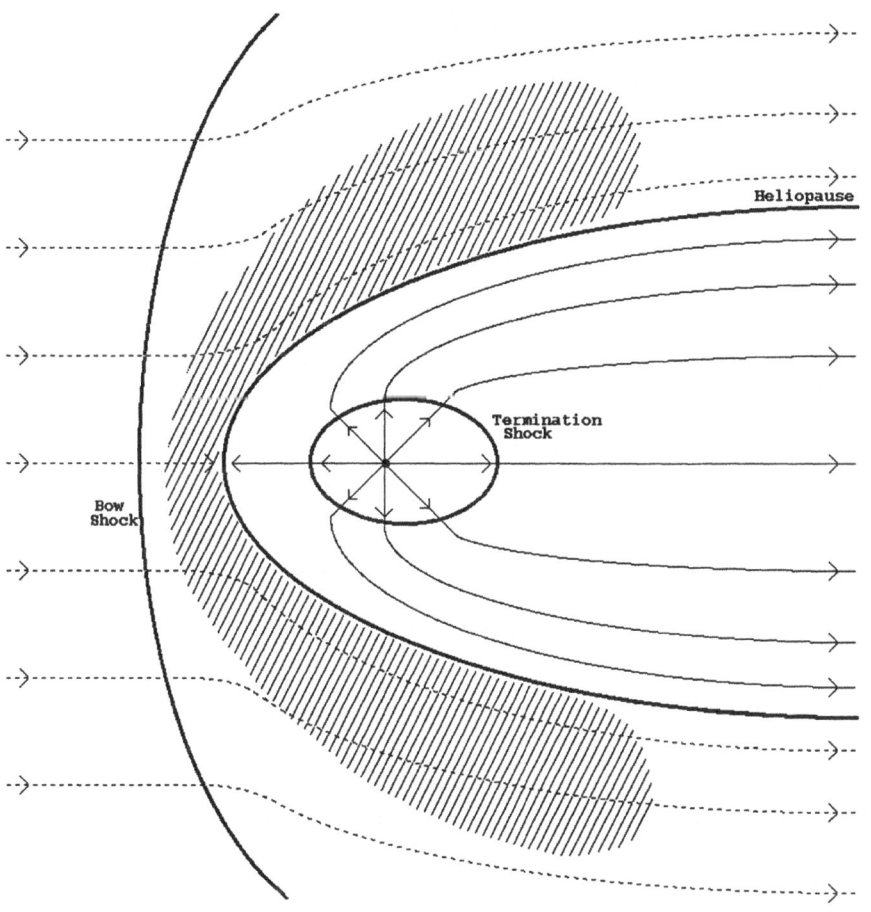

Figure 2. A schematic picture of the heliosphere showing the bow shock (where the incoming interstellar gas flow becomes subsonic), the termination shock (where the solar wind becomes subsonic), the heliopause surface dividing the two plasma flows, and the general location of the hydrogen wall (shaded region) where the inflowing neutral hydrogen is decelerated, compressed and heated by charge exchange reactions with solar wind protons. The hydrogen wall actually extends into the heliopause toward the termination shock.

Fig. 3) by reconstructing the stellar Lyman-α emission line wings assuming only that they are symmetric about the stellar radial velocity (-38.9 km s^{-1}).

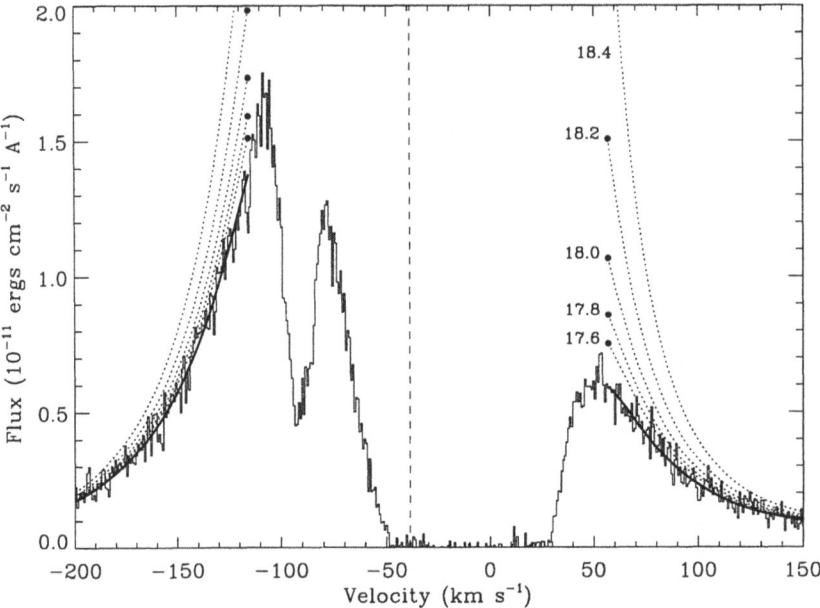

Figure 3. Reconstructions of the wings of the Lyman-α line profile of ϵ Ind assuming different values of log $N_{\mathrm{H\,I}}$ (numbers next to the dotted lines). The thick solid lines are polynomial fits to the wings of the observed Lyman-α profile. The dotted lines are estimates of the intrinsic Lyman-α line wings for different assumed values of log $N_{\mathrm{H\,I}}$. The selected best value for log $N_{\mathrm{H\,I}}$ is that which produces a stellar emission line centered on the stellar radial velocity (dashed line) (from Wood *et al.*, 1996).

3. Kinematic and Physical Properties of the LISM

Lallement and Bertin (1992) proposed that the Sun lies inside a cloud, which they called the Local Interstellar Cloud (LIC), because the LOS velocities toward 6 nearby stars observed in ground-based Ca II spectra and the velocity of interstellar He I flowing into the solar system are consistent with a single flow vector. GHRS spectra of the Mg II and Fe II resonance lines (2796, 2803, and 2600 Å) formed in the LOS toward other nearby stars (Lallement *et al.*, 1995) confirmed this picture with the flow vector magnitude 26 ± 1 km s^{-1} directed from Galactic coordinates $l = 186° \pm 3°$ and $b = -16° \pm 3°$ in the heliocentric rest frame. In the local standard of rest (defined by the motion of nearby stars), the LIC flow is from Galactic coordinates $l = 331.9°$ and $b = +4.6°$. The direction of this flow suggests that it originates from the expansion of a large superbubble created by supernovae and stellar winds from the Scorpius-Centaurus OB Association (Crutcher, 1982; Frisch, 1995).

GHRS spectra are confirming that the kinematical structure of the LISM is indeed very complex. Most lines of sight show at least one velocity component in addition to the LIC, even for stars as close as Sirius (2.7 pc) and Procyon (3.5 pc), indicating that additional clouds lie outside of the LIC at short distances. Table I lists the specific clouds now identified on the LOS toward the stars observed with the GHRS, where L refers to the LIC. In the Galactic Center direction the Sun lies close to the edge of the G cloud as the α Cen stars show interstellar absorption only at the velocity of this cloud. The Sun also likely lies very close to the edge of the LIC toward the North Galactic Pole as 31 Comae (Piskunov *et al.*, 1997) shows only one velocity component that is inconsistent with the LIC vector.

The temperature and nonthermal broadening of interstellar gas can be measured by comparing line widths of low mass elements (H and D) and high mass elements (e.g., Mg and Fe). For the LOS to Capella, Linsky *et al.* (1995) derived $T = 7000 \pm 500 \pm 400\,\mathrm{K}$ and $\xi = 1.6 \pm 0.4 \pm 0.2\,\mathrm{km\ s^{-1}}$, where the second uncertainty refers to the likely systematic errors due primarily to the uncertain intrinsic stellar emission lines. They also found $T = 6900 \pm 80 \pm 300$ K and $\xi = 1.21 \pm 0.27$ for the Procyon LOS. Temperatures and turbulent velocities measured for other LOS through the LIC are consistent with these values. For example, Lallement *et al.* (1994) found that $T = 7600 \pm 3000$ K and $\xi = 1.4^{+0.6}_{-1.4}\,\mathrm{km\ s^{-1}}$ for the LIC component toward Sirius, and Gry *et al.* (1995) found that $T = 7200 \pm 2000$ K and $\xi = 2.0 \pm 0.3\,\mathrm{km\ s^{-1}}$ for the LIC component toward ϵ CMa. Recent analyses (Piskunov *et al.*, 1997) of the LIC components toward HR 1099, 31 Com, β Cet, and β Cas yield similar results, and *in situ* measurements of the LISM H I and He I atoms flowing through the heliosphere (cf. Lallement *et al.*, 1994) yield consistent values for the temperature. I therefore conclude that $T \approx 7000$ K and $\xi \approx 1.2\,\mathrm{km\ s^{-1}}$ in the LIC.

Other clouds have different parameters. The G cloud, for example, is cooler with $T = 5400 \pm 500$ K and $\xi = 1.20 \pm 0.25\,\mathrm{km\ s^{-1}}$ along the LOS to α Cen. Component 2 toward ϵ CMa is also cooler with $T = 3600 \pm 1500$ K and $\xi = 1.85 \pm 0.3\,\mathrm{km\ s^{-1}}$ (Gry *et al.*, 1995). Hotter gas is inferred for several clouds toward ϵ CMa and for some of the gas toward Sirius (Bertin *et al.*, 1995).

4. A Summary of What is Now Known about D/H in the LISM

Table I lists all of the published D/H values derived for different clouds along the LOS toward nearby stars with GHRS observations. Quantities in parenthesis are assumed rather than derived and are not included in the subsequent analysis. Figure 4 shows the derived D/H ratios for 12 stars with interstellar radial velocities indicating that their LOS pass through the Local Interstellar Cloud (LIC) and 7 stars with LOS that pass through other nearby warm clouds. The mean value for the LIC is (D/H) = $(1.50 \pm 0.10) \times 10^{-5}$ and the $\pm 1\sigma$ error bars for all 12 data points are consistent with the mean value. The horizontal and dashed lines in Figure 4 show the mean relation for all data points, (D/H) = $(1.47 \pm 0.18) \times 10^{-5}$. This figure shows that there is no trend in the D/H ratios with distance. Figure 5, in

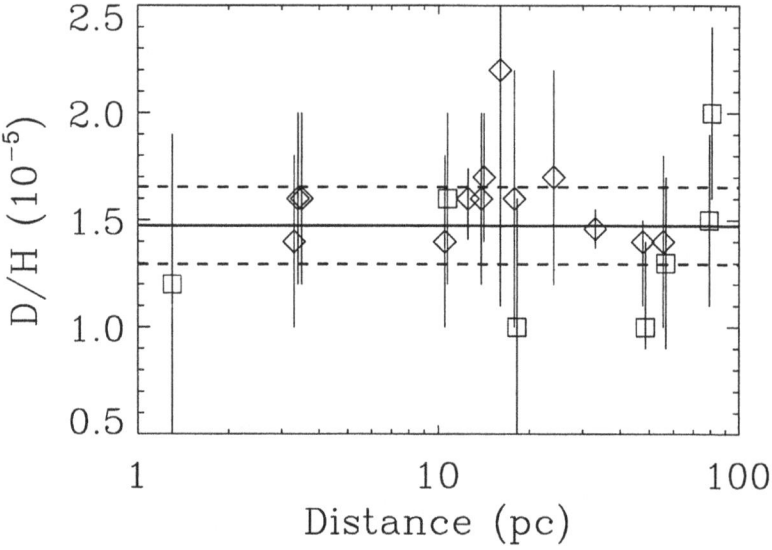

Figure 4. D/H ratios for interstellar gas toward all nearby stars observed with the GHRS. Diamond symbols are for gas in the LIC and square symbols are for other warm clouds. The solid and dashed horizontal lines represent the mean value of D/H and the $\pm 1\sigma$ error of the mean for all data points.

which the same data are plotted with respect to Galactic longitude, shows that there is no trend with Galactic longitude either.

The LIC data lead me to conclude that the value of D/H in the tiny region of the Galaxy occupied by the LIC is constant and reasonably well known, but we are just beginning to sample more distant lines of sight which may show different D/H ratios. The data for the other clouds are more scattered with $(D/H) = (1.28 \pm 0.36) \times 10^{-5}$. Whether or not this scatter represents measurement errors or indicates real D/H differences between the LIC and the other clouds can only be answered with more high quality observations obtained with FUSE and STIS. FUSE will observe the less opaque high Lyman lines and thus permit us to study D/H further out in the Galactic disk, in the Galactic halo, and along some LOS toward active galactic nuclei. These spectra will be a challenge to analyze compared to the GHRS spectra because the resolution (10–12 km s^{-1}) and S/N will be lower and the LOS toward the more distant targets will be complex. Nevertheless, we look forward to analyzing the FUSE data after the launch of the satellite scheduled for October 1998.

This work is supported by NASA through grant S-56460-D. I thank Dr. Brian Wood for the many discussions and collaborations upon which this review is based. I also thank ISSI for its hospitality and support.

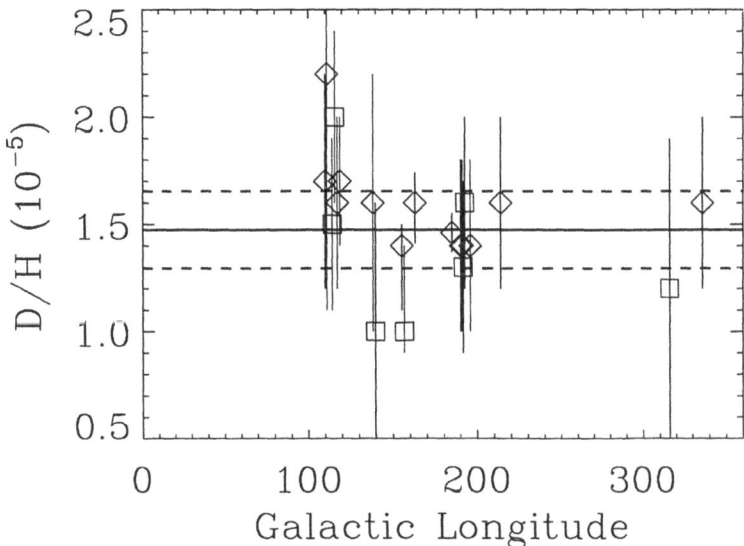

Figure 5. Same as Figure 4, except that the D/H ratios are plotted with respect to Galactic longitude.

References

Baranov, V. B., and Malama, Y. G.: 1995a, *JGR* **98**, 15157.
Baranov, V. B., and Malama, Y. G.: 1995b, *JGR* **100**, A8, 14755.
Bertin, P., Vidal-Madjar, A., Lallement, R., Ferlet, R., and Lemoine, M.: 1995, *A&A* **302**, 889.
Crutcher, R.M.: 1982, *ApJ* **254**, 82.
Dring, A.R., Linsky, J., Murthy, J., Henry, R.C., Moos, W., Vidal-Madjar, A., Audouze, J., and Landsman, W.: 1997, *Ap.J.*, in press.
Ferlet, R. *et al.* in *Science with the HST-II*, ed. B. Benvenuti *et al.* (Space Telescope Science Institute, Baltimore, 1996), p. 450.
Frisch, P. C.: 1995, *Science* **265**, 1423.
Gry, C., Lemonon, L., Vidal-Madjar, A., Lemoine, M., and Ferlet, R, 1995, *A&A* **302**, 497.
Lallement, R., and Bertin, P.: 1992, *A&A* **266**, 479.
Lallement, R., Bertin, P., Ferlet, R., Vidal-Madjar, A., and Bertaux, J. L.: 1994, *A&A* **286**, 898.
Lallement, R., Ferlet, R., Lagrange, A. M., Lemoine, M., and Vidal-Madjar, A.: 1995, *A&A* **304**, 461.
Landsman, W., Sofia, U.J., and Bergeron, P.: 1996, in *Science with the HST-II*, ed. B. Benvenuti *et al.* (Space Telescope Science Institute, Baltimore), p. 454.
Lemoine, M., Vidal-Madjar, A., Bertin, P., Ferlet, R., Gry, C., and Lallement, R.: 1996, *A&A* **308**, 601.
Linsky, J. L., Brown, A., Gayley, K., Diplas, A., Savage, B. D., Ayres, T. R., Landsman, W., Shore, S. N., and Heap, S. R.: 1993, *Ap.J.* **402**, 694.
Linsky, J. L., Diplas, A., Wood, B. E., Brown, A., Ayres, T. R., and Savage, B. D.: 1995, *Ap.J.* **451**, 335.
Linsky, J. L. and Wood, B. E.: 1996, *Ap.J.* **463**, 254.
Pauls, H. L., Zank, G. P., and Williams, L. L.: 1995, *JGR* **100**, 21595.
Piskunov, N., Wood, B., Linsky, J. L., Dempsey, R. C., and Ayres, T. R.: 1997, *Ap.J.* **474**, 315.
Quémerais, E., Sandel, B. R., Lallement, R., and Bertaux, J.-L.: 1995, *A&A* **299**, 249.

Tytler, D., Burles, S. and Kirkman, D.: 1997, *Ap.J.*, in press.
Welty, D. *et al.*, *Ap.J. Suppl.* **106**, 533 (1996).
Wood, B.E. and Linsky, J.L.: 1997a, *Ap.J* **474**, L39.
Wood, B.E. and Linsky, J.L.: 1997b, *Ap.J*, in press.
Wood, B. E., Alexander, W. R., and Linsky, J. L.: 1996, *Ap.J.* **470**, 1157.

DEUTERIUM OBSERVATIONS IN THE GALAXY

A. VIDAL-MADJAR and R. FERLET

Institut d'Astrophysique de Paris, CNRS,
98bis Bd. Arago, 75014 Paris, France.

M. LEMOINE

Department of Astronomy and Astrophysics, Enrico Fermi Institute,
The University of Chicago, Chicago, IL 60637-1433

Abstract. An accurate measurement of the primordial value of D/H would provide one of the best tests of nucleosynthesis models for the early Universe and the baryon density. Such evaluations have been traditionally made using present estimations of the deuterium abundance in the interstellar medium, extrapolated backwards in time with the use of galactic evolution models. Direct estimations of the primordial deuterium abundance have been carried out only recently in QSOs absorbers at high redshift.

We will summarize galactic observations of deuterium and suggest that, perhaps, a single D/H value for the interstellar medium is not representative. These evaluations mainly came from observations completed in the far UV with first the Copernicus satellite over the Lyman lines series followed then by H and D Lyman-alpha lines observations with both the IUE and the GHRS on the Hubble Space Telescope. We discuss different known systematics and show that the situation is not yet clear. It is not possible today to claim that we know "the" D/H value in the interstellar medium, if any.

Overall and in the context of additional D observations made in the solar system, we conclude that the actual evolution of deuterium from Big-Bang nucleosynthesis to now is not yet understood. More observations, recently made with IMAPS (the Interstellar Medium Absorption Profile Spectrograph) and hopefully to be made with FUSE (the Far Ultraviolet Spectroscopic Explorer to be launched in the fall of 1998), at higher spectral resolution or in many different galactic sites are certainly needed to help us reach a better global view of the evolution of that key element, and thus better constrain any evaluation of its primordial abundance.

Key words: Deuterium, Interstellar Abundance, Cosmology

1. Introduction

Deuterium is only produced in significant amount during primordial Big Bang nucleosynthesis (BBN) and thoroughly destroyed in stellar interiors. Hence, any abundance of deuterium measured at any metallicity should provide a lower limit to the primordial deuterium abundance. Deuterium is thus a key element in cosmology and in galactic chemical evolution (see e.g. Audouze and Tinsley, 1976; Gautier and Owen, 1983; Vidal-Madjar and Gry, 1984; Boesgaard and Steigman, 1985; Olive *et al.*, 1990; Pagel, 1992; Vangioni-Flam and Cassé, 1994; Prantzos, 1996). Indeed, its primordial abundance is the best tracer of the baryonic density parameter of the Universe $_B$, and the decrease of its abundance along the galactic evolution should trace the amount of star formation (among others).

The first, although indirect, measurement of the deuterium abundance of astrophysical significance was carried out through ^{3}He evaluation in the solar wind, leading to D/H$\simeq 2.5 \pm 1.0 \times 10^{-5}$ (Geiss and Reeves, 1972), a value representative

Space Science Reviews **84**: 297–308, 1998.
© 1998 *Kluwer Academic Publishers.*

of 4.5 Gyrs ago. The first measurements of the interstellar D/H ratio, representative of the present epoch, were reported shortly thereafter (Rogerson and York, 1973). Their value of D/H$\simeq 1.4 \pm 0.2 \times 10^{-5}$ has since then nearly not changed, whatever the availability of adequate instrumentation was. For more than a decade, these interstellar abundances have been used to constrain BBN in a direct way.

However, abundances measured at lower metallicities are less contaminated by the effect of galactic evolution. This is the reason why the deuterium abundance is now being chased in high redshift absorbers on quasar lines of sight: the composition of these clouds of very low metallicity should reflect the actual primordial deuterium value.

In the following, we discuss the different measurements of the deuterium abundance: in the ISM, in the pre-solar nebula, and in high redshift absorbers. We conclude eventually that many aspects of the evolution of deuterium in the Universe are yet unknown.

2. Interstellar observations

There are several methods to measure the interstellar abundance of deuterium (see Vidal-Madjar, 1991; Ferlet, 1992). One of them is to observe deuterated molecules such as HD, DCN, etc... and to form the ratio of the deuterated molecule column density to its non-deuterated counterpart (H_2, HCN, etc....). More than twenty different deuterated species have been identified in the ISM, with abundances relative to the non-deuterated counterpart ranging from 10^{-2} to 10^{-6}. Conversely, this means that fractionation effects are important, and that, as a consequence, this method cannot provide a precise estimate of the true interstellar D/H ratio; rather, this method is used in conjunction with estimates of the interstellar D/H ratio to gather information on the chemistry of the ISM.

Another way to derive the D/H ratio comes through radio observations of the hyperfine line of DI at 92cm. The detection of this line is however extremely difficult, and no firm detection has ever been reported. The detection of this line would allow to probe more distant interstellar media than the local medium discussed below; however, because a large column density of D is necessary to provide even a weak spin-flip transition, these observations aim at molecular complexes. As a result, the upper limit derived toward Cas A (Heiles *et al.*, 1993): D/H$\leq 2.1 \times 10^{-6}$ may as well result from a large differential fraction of D and H being in molecular form in these clouds, as from the fact that one expects the D/H ratio to be lower closer to the galactic center (since D is destroyed in stellar processing).

Finally, the only way to derive a reliable estimate of the interstellar D/H ratio is to observe the atomic transitions of D and H of the Lyman series in the far-UV, in absorption in the local ISM against the background continuum of cool or hot stars. These observations have been performed using the Copernicus and the IUE satellites, and now the Hubble Space Telescope. Both types of target stars present pros and cons.

2.1. COOL STARS

The main advantage of observing cool stars is that they can be selected in the vicinity of the Sun. This results in low HI column densities, and "trivial" to nearly "trivial" lines of sight. In effect, due to the low atomic weight of HI and DI, to the DI–HI -82 km/s isotopic shift, and to the abundance of HI in the local medium, the DI line cannot be detected at Lyman α in the wing of the HI line for HI column densities larger than 10^{19} cm^{-2}. Also, the presence of several interstellar components with different b-values may imply a large error on the HI column density if these components are unresolved. For this reason, deriving the HI column density has always been the limiting factor of accurate D/H ratios measurements. Note that the spectral resolutions of Copernicus and IUE were respectively 15 and 30 km s^{-1}, and, as a consequence, a non-trivial line of sight, even in the local ISM, would generally go unresolved. Even though HST–GHRS now offers a spectral resolution of 3.5 km s^{-1}, the thermal width of the DI line in the local ISM is $\simeq 8$ km s^{-1}, so that one has to observe lines of heavier species (thinner lines) to fully use the resolving power of HST, and build up a coherent line of sight velocity structure. This is one of the first difficulties inherent to the "cool stars" approach: the detailed structure of the line of sight could be found only through the observation of the FeII and the MgII ions, which are unfortunately present in both HI and HII regions and thus may not trace properly the HI gas. In particular, species like NI and OI could not be observed (see below).

Moreover, the chromospheric Lyman α emission line has to be modeled to set the continuum for the interstellar absorption. Such a procedure necessarily introduces systematic errors.

Nevertheless, this method has provided the most precise measurement of the local D/H ratio in the direction of Capella, using HST–GHRS:
$(D/H)_{\alpha Aur} = 1.60 \pm 0.09^{+0.05}_{-0.10} \times 10^{-5}$ (Linsky $et\ al.$, 1993; 1995).

Several more cool stars have been observed with HST since then (Linsky $et\ al.$, 1995: Capella, Procyon; Linsky and Wood, 1996: α Cen A, α Cen B; Piskunov $et\ al.$, 1997: HR 1099, 31 Com, β Cet, β Cas; Dring $et\ al.$, 1997: α Tri, ϵ Eri, σ Gem, β Gem). Although all compatible with the Capella evaluation, none of these results is precise enough to place any new constraints on this evaluation.

2.2. HOT STARS

Hot stars are unfortunately located further away from the Sun, so that one always has to face a high HI column density and often a non-trivial line of sight structure. In these cases, DI could not be detected at Lyα, and one has to observe higher order lines, $e.g.$ Lyγ, Lyδ, Lyϵ; hence these measurements have primarily come through

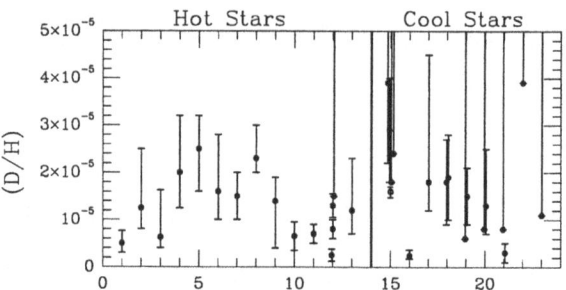

Figure 1. Measurements of the D/H ratio in the local ISM. The left hand-side box collects data obtained toward hot stars, while the right hand-side one shows cool stars observations. The x-axis has no physical significance, and merely labels the different stars. Data points next to each other, within less than 1 x-axis unit, correspond to the same target star. The open circle represents the Linsky *et al.* (1993; 1995) measurement using HST toward Capella. All other data points come from Copernicus or IUE observations.

Copernicus observations. The stellar continuum is however smooth at the location of the interstellar absorption and, moreover, the NI triplet at 1200 Å as well as other NI lines are available to probe the velocity structure of the line of sight. In particular, NI and OI were shown to be excellent tracers of HI in the ISM (Ferlet, 1981; York *et al.*, 1983). The interstellar void identified in the direction to hot stars in the Canis Majoris tunnel (Gry *et al.*, 1995) has allowed HST observations of DI at Lyα, but no further constraints have resulted so far.

2.3. PRESENT STATUS

All published D/H ratios are collected in Figure 1, distinguishing hot stars from cool stars observations (all references except Allen *et al.*, 1992 can be found in Vidal-Madjar, 1991; Ferlet, 1992; note that the most recent cool stars estimations from HST are not included). The D/H ratios range from $\sim 5 \times 10^{-6}$ to $\sim 4 \times 10^{-5}$. A large scatter is clearly detected in Figure 1 and represents differences of the D/H ratio in the local ISM, that may be as large as a factor $\simeq 4$ over scales as small as a few parsecs. The essential question is: do these variations really exist?

Unfortunately, no one can answer this question for the moment. To progress, one may have either to re-analyze all these data in a consistent way, looking for possible undetected systematics, or to complete new observations in the direction of a great variety of targets. In effect each type of target will generate its own type of problems. Both approaches are underway.

As an example, one should recall that time variations of the D/H ratio have already been reported toward ϵ Per (Gry *et al.*, 1983), which were interpreted as due to the ejection of high velocity hydrogen atoms from the star. But since this perturbation can only enhance the D/H ratio, it is worth noting that in at least four cases the D/H ratio was found to be really low: $0.7 \pm 0.2 \times 10^{-5}$ and $0.65 \pm 0.3 \times 10^{-5}$ toward δ and ϵ Ori (Laurent *et al.*, 1979); $0.8 \pm 0.2 \times 10^{-5}$ toward

λ Sco (York, 1983) and $0.5 \pm 0.3 \times 10^{-5}$ toward θ Car (Allen *et al.*, 1992). In each cases, authors discussed in details possible systematics but concluded that none of the identified ones could explain such values. These low values may thus be real. If so, explanations to these possible fluctuations of the D/H ratio have been put forward as early as Vidal-Madjar *et al.* (1978), Bruston *et al.* (1981) as possibly due to a selective radiation pressure effect.

To try to make some progress, we have inaugurated the use of a new type of targets that should solve many of the intrinsic difficulties of the problem, namely nearby white dwarfs.

2.4. ACTUAL PROSPECTS

Observing white dwarfs has many advantages. Such targets can be chosen near to the Sun, circumventing the main disadvantage of hot stars, and they can also be chosen in the high temperature range, so as to provide a smooth stellar profile at Lyα. At the same time, the NI triplet at 1200 Å as well as the OI line at 1302 Å would be available, allowing thus an accurate sampling of the line of sight. Such observations have now been conducted using HST toward two white dwarfs: G191–B2B (Lemoine *et al.*, 1996; Vidal-Madjar *et al.*, 1997b) and Hz43 (Landsman *et al.*, 1996).

In the case of G191–B2B, data were obtained in Cycles 1 and 5 at high resolution ($\simeq$ 3.3 km s^{-1}) for Lyα, NI 1200 Å, OI 1302 Å, SiIII 1304 Å, MgII 2800 Å, FeII 2343 Å and SiIII 1206 Å. The line of sight velocity structure coherent in all these lines (about 15, including those observed at lower resolution) comprises one HI region – the Local Cloud observed toward Capella – together with two HII regions; we refer to these components as blue, white and red, according to their positions along the wavelength scale. Both HII regions are clearly seen in SiIII, but their presence is also felt in strong lines such as HI Lyα and OI 1302 Å. The analysis in terms of column densities is still underway, but it seems already clear that the column density ratio of the blue to the red component varies from ion to ion. In particular, if the D/H ratio for the red component common to the G191–B2B and Capella sight-lines (these stars are separated by only $8°$ on the sky) is forced to be that found by Linsky *et al.* (1993; 1995), then the D/H ratio for the blue component appears significantly lower.

In Figure 2 are superposed the G191–B2B spectra obtained at the highest resolution of Ech-A of the GHRS, all plotted in velocity space to allow an easy comparison of the different absorption features. In both NI and SiII lines, at least two components are obviously present. However, only one of them (at about $+9$ km s^{-1}, dotted line) is detected in the SiIII line (the feature near $+29$ km s^{-1} is stellar), while a third component close to $+13$ km s^{-1} is needed to fit the SiIII line. The existence of this third component is further confirmed in the OI line in order to deepen enough the OI absorption feature between both NI and SiII components. Freezing then the velocity separations and b-values, it is possible to fit the DI line.

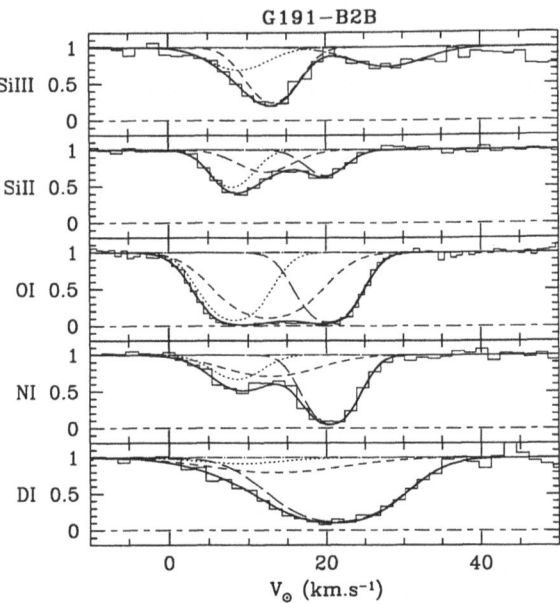

Figure 2. Are superposed in velocity space the absorption features seen in the different species indicated on the left hand side of the figure. Two obvious components are seen in SiII and NI. However, a third component in between the two others is imposed by the SiIII and OI lines. The two components at the shortest wavelengths, seen in SiIII, are HII regions, while the component at ~ 20 km s^{-1} is the Local Cloud, an HI region. It is clear that the D/H ratio cannot be the same in the red and the blue component (see text). Also, it is obvious that deducing lines of sight velocity structure for D/H evaluations from ionized species could be extremely misleading.

The result is that if one assumes that D/H is 1.6×10^{-5} in the red component (which corresponds to the Capella one), then the D/H ratio in the blue component has to be of the order of 5.5×10^{-6}!

It is interesting to note that a D/H value clearly below 10^{-5} seems to be required by the observations, suggesting again that D/H variations in the local ISM may be real.

3. The D/H evaluations in the solar system

The pre-solar value for the deuterium abundance has traditionally been derived from meteorites and lunar soil: D/H= $2.6 \pm 1.0 \times 10^{-5}$ (Geiss, 1993). The giant planets Jupiter and Saturn are considered to be undisturbed deuterium reservoirs, free from production or losses processes, preserving the abundance of their light elements since the formation of the solar system 4.5 billion years ago (Owen *et al.*, 1986). The first measurements of the D/H ratio in the Jovian atmosphere have been performed through methane and its deuterated counterpart CH$_3$D yielding D/H= $5.1 \pm 2.2 \times 10^{-5}$ (Beer and Taylor, 1973). Abundant molecules like H$_2$

were also used yielding lower values: D/H=2.1 ± 0.4 × 10^{-5} (Trauger *et al.*, 1973). More recent molecular measurements, also model-dependent, gives D/H= 2.5 ± 0.7 × 10^{-5} (Gautier and Owen, 1983).

Recently, new important measurements of the D/H ratio implying very different methods were carried out. One is based on the far infrared ISO observation of the HD molecule in Jupiter (Encrenaz *et al.*, 1996) and leads to D/H=2.2 ± 0.5 × 10^{-5}, although some systematics are suspected. Another is based on the direct observation with HST–GHRS of both HI and DI Lyman α emission at the limb of Jupiter for the first time (Ben Jaffel *et al.*, 1994; 1997). The third one is an in situ measurement with a mass spectrometer onboard the Galileo probe (Niemann *et al.*, 1996). These two latter yield similar ratios: 5.9 ± 1.4 × 10^{-5} and 5.0 ± 2.0 × 10^{-5} respectively (this last value has been however revised toward the lower part of the range, i.e. 3.6 ± 0.6 × 10^{-5}).

These last values are nevertheless clearly higher than the meteoritic one (Geiss, 1993) and the other Jovian ones (see e.g. Bjoraker *et al.*, 1986) by ∼ 2σ. That may call to a reconsideration of the pre-solar abundance of deuterium since systematics may be inherent to each observational approach (see e.g. Lecluse *et al.*, 1996).

4. Lines of sight toward QSOs

A very promising approach to evaluate the D/H ratio was initiated a few years ago by observing directly with large ground based telescopes the Lyman lines redshifted in the visible part of the spectrum (see e.g. Carswell *et al.*, 1987; Webb *et al.*, 1991). These lines are seen in absorption in QSOs spectra produced by the absorption of HI gas present in the intergalactic medium in front of the QSOs.

The very interesting point of this approach is that one may probe directly a primordial cloud (*i.e.* a very low metallicity system), and thus have direct access to the primordial D/H value. On the other hand, the difficulty is twofold: *i)* the relative faintness of the sources; *ii)* the large number of absorbers per unit redshift, so that there is always the possibility that the observed deuterium feature would be in fact a mimicking low-column density HI absorber. The advent of very large telescopes can at least partly overcome the first point. Space-based UV telescopes, such as HST, offer the possibility of observing absorbers at low redshift, where the probability of contamination is greatly reduced. Note as well that there is a large scatter of metallicity with redshift, so that there are many low-redshift absorber candidates that are truly metal-poor (see Timmes *et al.*, 1997). Thus, one can hope that measurements of the primordial deuterium abundance will be possible at low redshift.

As for today, the different reported estimations are presented in Table I. For high redshift QSOs absorption systems, they are obtained from observations with the 10m Keck1 or the 4m Kitt Peak telescopes; for the lower redshift system, data are from HST. Actually, only four detections of deuterium rather than upper limits are

Table I

D/H measurements in absorption systems toward QSOs.

QSO	z_{abs}	D/H	Ref.
Q0014+8118	3.320	$\leq 2.5 \times 10^{-4}$	Songaila et al., 1994; Carswell et al., 1994
Q0014+8118	3.320	$1.9 \pm 0.5 \times 10^{-4}$	Rugers and Hogan, 1996
Q1009+2956	2.504	$2.5 \pm 0.5 \times 10^{-5}$	Burles and Tytler, 1996
Q1937−1009	3.572	$2.3 \pm 0.3 \times 10^{-5}$	Tytler et al., 1996
Q0420−3851	3.086	$> 2. \times 10^{-5}$	Carswell et al., 1996
Q1202−0725	4.672	$\leq 1.5 \times 10^{-4}$	Wampler et al., 1997
Q1937−1009	3.572	$> 4. \times 10^{-5}$	Songaila et al., 1997
Q1718+4807	0.701	$2.0 \pm 0.5 \times 10^{-4}$	Webb et al., 1997a
Q0454−2203	0.482	$< 1.3 \times 10^{-5}$	Webb et al., 1997b

claimed, in complete disagreement with each other (1σ random error in Table I), giving either a high ($\sim 2 \times 10^{-4}$, Rugers and Hogan, 1996a; 1996b; Webb et al., 1997a) or a low ($\sim 2.5 \times 10^{-5}$, Burles and Tytler, 1996; Tytler et al., 1996) D/H ratio.

The generic method employed in these studies is as follows. Metal lines such as CIV, SiIV, MgII are used to determine the velocity structure in the redshift region where DI is being chased. Taking this velocity structure into account, all the Lyman series up to the Lyman limit and the limit itself are then fitted for estimating the HI column density and b(HI). In general, the high order lines yield an estimate of b, while the Lyman limit yields an estimate of N(HI). But one has to recall that again the usual limiting factor in determining the D/H ratio is the evaluation of N(HI). Whenever D is detected at Lyα, the estimate of D/H is probably reasonable if all the above steps have been successfully completed.

Although there is always the possibility that the DI line be in fact an HI interloper, the actual DI detections were claimed on the basis of the following arguments: the b-value for the DI line was found to be consistent with b(DI)/b(HI)$\sim 1/\sqrt{2}$ which corresponds to a pure thermal case, and the velocity shift between the DI and the HI line was observed to be very similar to the isotopic shift.

However, three caveats can at least be identified in this type of analysis, the magnitude of which are yet unknown. First, as it was shown in the case of the white dwarf G191-B2B, ionized metal species do not trace correctly the HI gas. Hence, one should imagine that the redshift of HI absorbers do not match those derived from metal lines. Furthermore, there could be substantial substructure in these HI clouds, with no hope of tracing them from either HI or ionized metal lines. An interesting study would be to observe these clouds at higher resolution (< 8 km s^{-1}) in OI (NI is not expected to be detectable in these very metal-poor clouds). Second, the estimate of N(HI) from the Lyman limit is not reliable if the residual flux below the limit is uncertain, or is contaminated by instrumental noise (Songaila et al., 1997). Third, the assumption of Voigt profiles in the analysis could be extremely misleading (although not always necessary as shown by Jenkins, 1996) as suggested by Levshakov *et al.* (1996) and Levshakov and Takahara (1996), particularly in the case of QSOs for which the number of available absorption metallic lines is greatly reduced.

To increase the general confusion toward what should be considered as **the** primordial D/H value, a 2σ upper limit D/H< $1. \times 10^{-5}$ has been set from HST observations of a $z_{abs} = 0.4823$ absorber (Webb *et al.*, 1997b; Table I; see an illustration of these observations in Vidal-Madjar *et al.*, 1997a). These data were gathered in Cycles 5 and 6 at $\lambda \sim 16$ km s^{-1}, and signal-to-noise ratios $\sim 5 - 8$. They cover the Lyman limit, Lyϵ, Lyδ, Lyγ, Lyα, SiIII; further ground-based observations of MgII were collected at the Cerro Tololo International Observatory. The generic method outlined above was followed to derive D/H, although no MgII is seen in the system. Deuterium is not detected at Lyα. But the position of the HI absorber at Lyα is accurately known from the other lines, and precludes any deuterium absorption, whence the strong limit: D/H< 1.3×10^{-5}. This result could nevertheless be modified if substructure were present (this study is underway).

5. Who's wrong, who's right?

If the result of Webb *et al.* (1997b) are observationally confirmed toward some other QSOs, then there will be definitely something rotten in some remote kingdom... No model of chemical evolution is able to account for an increase of the deuterium abundance with metallicity. In that situation, we will thus have to assume either the distribution of cosmic deuterium is strongly inhomogeneous, or such a low value is the result of a strong contamination by deuterium-poor gas. However, it is not so easy to deplete deuterium without producing metals, although a particular configuration of the line of sight aiming through a stellar wind contaminated region could do the job. A remote possibility is to produce significant amount of deuterium in some still unknown astrophysical site... Keeping these remarks in mind, let us discard this result from the following discussion, because too perturbing!

We now assume that Tytler *et al.* (1996) is right, *i.e.* the primordial deuterium abundance is low, $\sim 2.5 \times 10^{-5}$ as preferred by Hata *et al.* (1996). It is then easy to argue that the high deuterium abundances sometimes measured are due to interlopers, eventhough their expected small probability of occurrence, eventhough the claimed detected signature of DI b-value. In order to match a low primordial deuterium abundance with other light elements, one would have to accomodate either large systematics in the determination of the primordial ^{4}He mass fraction, or, equivalently, further light degrees of freedom (equivalent to neutrino flavors) at BBN. The reader is referred to Copi *et al.* (1996a) for a discussion of this point. The ensuing chemical evolution of deuterium would be rather straightforward, possibly involving infall to avoid too much deuterium destruction.

Let now assume that Tytler *et al.* (1996) is wrong, *i.e.* the high values of D/H are the correct primordial ones. Although in excellent agreement with the primordial abundances of ^{7}Li and ^{4}He (Cardall and Fuller, 1996), such high values would lead to several other problems. The implied low baryonic density parameter $_B$ would strengthen the so-called "baryon cluster catastrophy". The chemical evolution of deuterium would be quite difficult to account for, as it would require an astration factor $\sim$ 15 over $\sim$ 10 Gyrs when standard models destroy D by no more than a factor $\sim$ 3. Moreover, most models already overproduce ^{3}He after 10 Gyrs of evolution. Since D is converted to ^{3}He in stellar interiors, a high D/H ratio would make this situation worse. However, Scully *et al.* (1996; 1997) have developed a promising new galactic evolution model which, as well as the so called "standard" galactic evolution models is able to take into account all the galactic observables, but could at the same time solve all these problems. In such a case the D/H observations may have triggered a new vision of the galactic evolution.

Finally, let put everyone right and happy. The deuterium abundance seen in one absorber or the other, or both, could have been strongly affected by some unknown mechanism (e.g. Timmes *et al.*, 1997; Jedamzik and Fuller, 1996). However there is, up to now, no compelling mechanism able to reconcile the observed values. As well and after all, there might not be a unique primordial deuterium abundance. For instance, the presence of isocurvature baryon fluctuations at BBN would affect the yields differently in different regions. However, it seems that Copi *et al.* (1996b) and Jedamzik and Fuller (1996) do not quite agree on whether or not such fluctuations of an amplitude and a scale large enough to explain the variations of D/H at high redshifts can be made compatible with the high degree of isotropy of the Cosmic Microwave Background (CMB).

Last, the remaining possibility is that no one is right (Levshakov and Takahara, 1996). For sure, that would be fun and will ask for many more observations!

6. Conclusion

The previous discussion may be summarized in the following manner. The different – and discordant – D/H evaluations are shown as a function of time, with no a priori bias to select one over another. Some of the differences are too large to be accounted for. It is clear that all measurements cannot be correct and some still unknown systematics should be identified. Certainly each one will find its preferred value in each domain, but our impression is that for the moment we are far from having understood the whole story.

As a matter of fact, if the variations of the D/H ratio in the local interstellar medium are illusory, then one could quote as an average of the published values: $(D/H)_{ISM} \simeq 1.3 \pm 0.4 \times 10^{-5}$. The rather large error bar arises from a subjective although conservative viewpoint. On the contrary, if D/H does vary in the ISM, one has to understand why; until then, *no measurement of the D/H ratio in the ISM or the IGM should be quoted as reliable.* Moreover, one should expect these variations to be larger in reality. The actual value might in fact be very different from what is observed if these variations are systematic, *i.e.* act in one way only; this in turn would heavily bear on the chemical evolution of deuterium. It also appears that the upper bound on $_B$ is obtained from BBN predictions through the interstellar abundance of deuterium: this bound would have to be removed until the variations and their cause are properly understood.

There is a hope that the IMAPS and the FUSE missions will solve these problems. IMAPS allows new observations of the ISM in the direction of bright stars already studied with the Copernicus satellite, but at a much higher resolution of $R \sim 120000$, while FUSE will probe the ISM much further away, within the whole galaxy and up to extragalactic low-redshift objects. Both these instruments should give access to more precise D/H evaluations and possibly may reveal gradients of the deuterium abundance with galactocentric distance and with galactic height in the halo. These dedicated studies should greatly clarify the problem of the chemical evolution of deuterium, and hence much better constrain our understanding of what should be **THE** primordial D/H value.

Acknowledgements

We thank J. Audouze, M. Cassé, C. Copi, D. Schramm, J. Truran, M. Turner, E. Vangioni-Flam and D. York for useful discussions. M.L. acknowledges support by the NASA, DoE, and NSF at the University of Chicago.

References

Audouze, J., and Tinsley, B.M.: 1976, *Ann. Rev. Astron. Astrophys.* **14**, 43.
Allen, M., Jenkins, E.B., and Snow T.P.: 1992, *Astrophys. J. Suppl.* **83**, 261.
Beer, R., and Taylor, F.: 1973, *Astrophys. J.* **179**, 309.
Ben Jaffel, L., *et al.*: 1994, *Bull. Am. Astron. Soc.* **26**, 1100.
Ben Jaffel, L., *et al.*: 1997, *The Scientific return of GIIRS*, in press.
Bjoraker, G., Larson, H., and Kunde, V.: 1986, *Icarus* **66**, 579.
Boesgaard, A.M., and Steigman, G.: 1985, *Ann. Rev. Astron. Astrophys.* **23**, 319.
Bruston, P., Audouze, J., Vidal-Madjar, A., and Laurent, C.: 1981, *Astrophys. J.* **243**, 161.
Burles, S., and Tytler, D.: 1996, *preprint* astro-ph/9603070.
Cardall, G.M., and Fuller, G.: 1996, *preprint* astro-ph/9603071.
Carswell, R.F., Webb, J.K., Baldwin, J.A., and Atwood, B.: 1987, *Astrophys. J.* **319**, 709.
Carswell, R.F., *et al.*: 1994, *M.N.R.A.S.* **268**, L1.
Carswell, R.F., *et al.*: 1996, *M.N.R.A.S.* **278**, 506.
Copi, C.J., Schramm, D.N., and Turner, M.S.: 1996a, *preprint* astro-ph/9606059.
Copi, C.J., Olive, K.A., and Schramm, D.N.: 1996b, *preprint* astro-ph/9606156.
Dring, A., Murthy, J., Henry, R.C., *et al.*: 1997, *Astrophys. J.*, in press.
Encrenaz, T., *et al.*: 1997, *Astron. Astrophys.*, in press.
Ferlet, R.: 1981, *Astron. Astrophys. Letters* **98**, L1.
Ferlet, R.: 1992, *IAU Symposium 150*, 85.
Gautier, D., and Owen, T.: 1983, *Nature* **304**, 691.

Geiss, J.: 1993, *Origin and Evolution of the Elements*, *CUP*, 89.
Geiss, J., and Reeves, H.: 1972, *Astron. Astrophys.* **18**, 126.
Gry, C., Laurent, C., and Vidal-Madjar, A.: 1983, *Astron. Astrophys.* **124**, 99.
Gry, C., *et al.*: 1995, *Astron. Astrophys.* **302**, 497.
Hata, N., Steigman, G., Bludman, S., and Langacker, P.: 1996, *preprint* astro-ph/9603087.
Heiles, C., McCullough, P., and Glassgold, A.: 1993, *Astrophys. J. Suppl.* **89**, 271.
Jedamzik, K., and Fuller, G.: 1995, *Astrophys. J.* **452**, 33.
Jedamzik, K., and Fuller, G.: 1996, *preprint* astro-ph/9609103.
Jenkins, E.B.: 1996, *preprint*.
Landsman, W., Sofia, U.J., and Bergeron, P.: 1996, *Science with the Hubble Space Telescope - II*, STScI, 454.
Laurent, C., Vidal-Madjar, A., and York, D.G.: 1979, *Astrophys. J.* **229**, 923.
Lecluse, C., Robert, F., Gautier, D., and Guiraud, M.: 1996, *Plan. Space Sci.*, in press.
Lemoine, M., *et al.*: 1996, *Astron. Astrophys.* **308**, 601.
Levshakov, S.A., and Takahara, F. 1996, *M.N.R.A.S.* **279**, 651.
Levshakov, S.A., Kegel, W.H., and Mazets, I.E.: 1996, *Yukawa Institut, Kyoto preprint YITP-96-23*.
Linsky, J., *et al.*: 1993, *Astrophys. J.* **402**, 694.
Linsky, J., *et al.*: 1995, *Astrophys. J.* **451**, 335.
Linsky, J., and Wood, B.E.: 1996, *Astrophys. J.* **463**, 254.
Niemann, H.B., *et al.*: 1996, *Science* **272**, 846.
Olive, K., Schramm, D., Steigman, G., and Walker, T.: 1990, *Phys. Rev. Letters*B236, 454.
Owen, T., Lutz, B., and De Bergh, C.: 1986, *Nature* **320**, 244.
Pagel, B., *et al.*: 1992, *M.N.R.A.S.* **255**, 325.
Piskunov, N., *et al.*: 1997, *Astrophys. J.*, in press.
Prantzos, N.: 1996, *Astron. Astrophys.* **310**, 106.
Rogerson, J., and York, D.: 1973, *Astrophys. J. Letters* **186**, L95.
Rugers M., and Hogan, C.J.: 1996a, *Astrophys. J. Letters* **549**, L1.
Rugers M., and Hogan, C.J.: 1996b, *Astron. J.* **111**, 2135.
Scully, S.T., Cassé, M., Olive. K.A., Schramm, D.N., Truran, J., and Vangioni-Flam, E.: 1996, *Astrophys. J.* **462**, 960.
Scully, S.T., Cassé, M., Olive. K.A., Vangioni-Flam, E.: 1997, *Astrophys. J.* **476**, 521.
Songaila, A., Cowie, L.L., Hogan, C.J., and Rugers, M.: 1994, *Nature* **368**, 599.
Songaila, A., Wampler, E.J., and Cowie, L.L.: 1997, *Nature*, in press.
Timmes, F.X., Truran, J.W., Lauroesch, J.T., and York, D.G.: 1997, *Astrophys. J.*, in press.
Trauger, J., Roesler, F., Carleton, N., and Traub, W.: 1973, *Astrophys. J. Letters* **184**, L137.
Tytler, D., Fan, X.-M., and Burles, S.: 1996, *Nature* **381**, 207.
Vangioni-Flam, E., and Cassé, M.: 1994, *Astrophys. J.* **427**, 618.
Vidal-Madjar, A., Laurent, C., Bruston, P., and Audouze, J.: 1978, *Astrophys. J.* **223**, 589.
Vidal-Madjar, A., and Gry, C.: 1984, *Astron. Astrophys.* **138**, 285.
Vidal-Madjar, A.: 1991, *Adv. Space Res.* **11**, 97.
Vidal-Madjar, A., Ferlet, R., Lemoine, M.: 1997a, *The Scientific return of GHRS*, in press.
Vidal-Madjar, A., *et al.*: 1997b, in preparation.
Wampler, *et al.*: 1997, *Astron. Astrophys.*, in press.
Webb, J.K., *et al.*: 1991, *M.N.R.A.S.* **250**, 657.
Webb, J.K., Carswell, R.F., Lanzetta, K.M., Ferlet, R., Lemoine, M., Vidal-Madjar, A., and Bowen, D.V.: 1997a, *Nature* **388**, 250.
Webb, J.K., *et al.*: 1997b, in preparation.
York, D.G.: 1983, *Astrophys. J.* **264**, 172.
York, D.G., *et al.*: 1983, *Astrophys. J. Letters* **266**, L55.

Address for correspondence: Institut d'Astrophysique de Paris, CNRS, 98bis Bd. Arago, 75014 Paris, France.

LOCAL INTERSTELLAR MEDIUM

Summary of Working Group VI

JEFFREY L. LINSKY

JILA, University of Colorado and NIST, Boulder, CO 80309-0440 USA

T. L. WILSON

Max-Planck-Institute für Radioastronomie, Bonn, Germany

R. T. ROOD

Dept. of Astronomy, University of Virginia, Charlottesvile, VA 22903 USA

Abstract. This report summarizes the issues discussed in Working Group VI concerning the accuracy of measurements of D/H and ^{3}He/H in the local interstellar medium, possible systematic errors, and emerging trends in the results.

Key words: LISM, interstellar medium, deuterium, ^{3}He, helium-3

1. Introduction

The light element nuclei D, ^{3}He, ^{4}He, and Li are relics of the period 100–1000 seconds after the Big Bang when the physical properties of the early universe first permitted these nuclei to be stable. After these nuclei "froze out" of the primordial fireball, their relative abundances compared to H evolved through nuclear reactions in stellar cores. This astrated material was then dispersed into the interstellar medium by supernova explosions and stellar winds, and became incorporated into subsequent generations of stars. Thus we must understand the physical processes that govern the evolution of the light elements, before we can infer the properties of the early universe, in particular the density of baryons in both luminous and "dark" forms. While the rates of the important nuclear reactions are reasonably well known, we have limited knowledge of stellar interior structure (including accurate descriptions of diffusion, convection, and internal dynamics), stellar mass loss rates, star formation rates, mixing of interstellar gas within the Galactic disk, and mixing of disk and halo gas (which may or may not have primordial abundances). The problems of Galactic chemical evolution and the determination of the properties of the early universe are thus inexorably intertwined. Accurate measurements of the light element abundances can illuminate these processes. The alternative path of measuring primordial abundances in extragalactic sources is fraught with a different set of problems that are discussed elsewhere in this volume (Burles and Tytler, 1998; Hogan, 1998).

Working Group VI discussed the available measurements of D/H and ^{3}He/H in the local interstellar medium, the tiny portion of the Galactic disk extending out to about 1 kpc from the Sun, and considered the information that these data provide concerning Galactic chemical evolution. D and ^{3}He play critical but different roles in the chemical evolution of the Galaxy. Epstein, Lattimer, and Schramm (1976) showed that since nuclear reactions destroy D more easily than the other light elements, the evolution of D is always toward lower D/H ratios with no known source of D since the Big Bang. Thus the evolution of D/H is monotonic and Big Bang nucleosynthesis calculations (e.g., Walker *et al.*, 1991) show that D/H has a steep slope with respect to the density parameter η_{10}, which measures the baryon/photon ratio. For these reasons and given the increasing number of accurate D/H measurements from absorption line spectra, the D/H ratio is thought to provide the tightest constraint on the baryonic density of the universe.

The Big Bang also created ^{3}He nuclei; subsequent evolution, however, has been more complex than for D. As described by Rood *et al.* (1998), D burning efficiently creates ^{3}He that is then converted more slowly into ^{4}He. Stellar evolution codes typically show a net creation of ^{3}He with time primarily in low mass stars, but this result must be tested empirically.

The observational material discussed by Working Group VI is contained in the papers published elsewhere in this volume concerning D (Linsky, 1998; Vidal-Madjar, Ferlet, and Lemoine, 1998) and ^{3}He (Rood *et al.*, 1998). The paper by Geiss and Gloeckler (1998) on D and ^{3}He in the protosolar cloud is also important for the discussion. We summarize the discussions in Working Group VI under three topics.

2. How Accurate are the Published Values of D/H and ^{3}He/H for Lines of Sight Through the Local Interstellar Medium?

Vidal-Madjar *et al.* (1998) summarize the arguments as to why the analysis of the H and D Lyman line absorption provides the most reliable D/H ratios. For hydrogen column densities $N_{HI} < 10^{19}$ cm^{-2}, Copernicus, IUE, and the Hubble Space Telescope (HST) provide high resolution Lyman-α spectra from which both the H and D lines can be resolved and their column densities measured. The high resolution spectra obtained with the Goddard High Resolution Spectrograph (GHRS) on HST represent an enormous advance over previous data sets by virtue of the high spectral resolution (3.5 km s^{-1}) and signal/noise that exceeds 100:1. For larger hydrogen column densities, one must study the less opaque Lyman-β and higher Lyman lines, which are not observable by HST for Galactic sources, so that the broad H absorption does not obliterate the D line. Until recently only Copernicus spectra were available, but now the IMAPS rocket and soon the Far Ultraviolet Spectrograph Explorer (FUSE) spacecraft will obtain high resolution spectra of the higher Lyman lines. The fundamental issue in the analysis of these data is the role that systematic errors can play. Both Linsky (1998) and Vidal-Madjar *et al.* (1998) discuss this vital topic in detail for lines of sight toward nearby cool stars and hot white dwarfs and the more distant OB stars. They consider the following sources of systematic error:

Complex lines of sight: GHRS spectra of metal lines and ultra high resolution ground-based spectra of the Na I and Ca II resonance lines (e.g., Welty *et al.*, 1996) show that even very short lines of sight toward the nearby cool stars contain several velocity components, which indicate the presence of separate interstellar clouds along the line of sight. Linsky (1998) lists the number of known velocity components toward the nearby stars, but the longer lines of sight toward hot stars must be more complex. This complexity can introduce errors in the derived D/H ratios because the H Lyman lines are very saturated and the metal lines (e.g., Mg II and Fe II) are not good predictors of the H I and D I column densities. Vidal-Madjar *et al.* (1998) argue that O I and N I lines are better predictors of the H I and D I column densities and thus should be used to determine how to apportion the H I and D I column densities among the different clouds for a complex line of sight. Their application of this technique to the analysis of the line of sight to the hot white dwarf G191-B2B with three velocity components results in significant differences in the D/H ratios for two of these interstellar clouds. Disentangling the saturated H I absorption along complex lines of sight should be viewed skeptically and should be tested by analyzing the higher Lyman lines. More opaque lines of sight toward more distant OB stars may be simpler to analyze as one can infer the total H column density from the Lyman line wings rather than the saturated line cores, but it will still be difficult to apportion the H and D column densities among the various clouds in the line of sight. As a result, the inferred D/H ratio will be a line-of-sight average and thus not indicate the possible variations in D/H in the Galactic disk.

Previously unknown weak absorbers: Since line center optical depths of the strongest metal resonance lines are 10^{-3} to 10^{-4} that of H I Lyman-α, there can be clouds in the line of sight toward nearby stars that are invisible in the metal lines but optically thick in Lyman-α. An important example is that "hydrogen walls" surrounding the Sun and stars are produced when the incoming interstellar gas interacts with the stellar wind (c.f. Linsky, 1998). These redshifted and blueshifted absorption features, if not recognized and taken into account, can increase the inferred H column density and lower the inferred D/H ratio by a factor of 2.

Shape of the background continuum: This is not an important issue for the OB stars and hot white dwarfs as the background continuum is flat or nearly so. For cool stars the background continuum is the stellar Lyman-α emission line whose shape must be guessed. Since the H I absorption in the Lyman lines is saturated, uncertainties in the shape of the Lyman-α emission line can lead to errors in the inferred H column density. The errors in the inferred

D column density are much smaller because the D line is not saturated and its narrow width allows the background "continuum" to be estimated more reliably. Linsky (1998) describes two techniques for minimizing the systematic errors associated with the background continuum — inference of the H I column density from the line wings of stars with large radial velocities, and analysis of spectroscopic binary systems observed at opposite quadratures such that the stellar emission line changes shape while the interstellar absorption is constant. These two techniques lead to more accurate D/H ratios, but these systematic errors may not yet be fully understood.

Unknown unknowns: There may be additional systematic errors that have not yet been recognized. There is a good chance that the analysis of high resolution spectra of the higher Lyman lines by FUSE will identify possible new systematic errors.

The best hope for high precision D/H measurements is to analyze simple lines of sight for which the known systematic errors can be identified and included in the analysis. For the highest signal/noise spectra of cool stars with simple lines of sight and for stars with high or variable radial velocity, the errors in D/H appear to lie in the range 10–20% (Linsky, 1998). The best D/H values for lines of sight toward hot stars may be about 30%.

Analyses of the ^{3}He data provide a different set of challenges and systematic errors. Rood et al. (1998) describe these problems in detail. We summarize these problems and comment on how they may change our picture of the evolution of the light element abundances. Outside of the solar system the only technique for measuring the ^{3}He/H ratio is by analyzing observations of the 3.46 cm hyperfine transition of ^{3}He$^+$. This line is extremely weak, and the line-to-continuum ratio is small, usually $\sim 10^{-3}$ or less. These factors make any such measurement difficult, the most important factor being reflections in the radio telescope that produce baseline ripples. In order to convert these weak signals into ^{3}He/H ratios, one must model the H II regions or planetary nebulae (PNe) to estimate the density structure (clumping) and the ionization fractions of He and H. Important inputs for models of such structure are provided by VLA images. Non-LTE excitation of He$^+$ does *not* contribute to the line formation process. An important part of the analysis consists of comparisons with H and He radio recombination lines. The radio recombination lines, from large Rydberg atoms, however, *are* affected by pressure broadening and non-LTE excitation effects. While all of these issues have been addressed as best one can, the many complex steps required to estimate ^{3}He/H ratios point to the need for additional observational and theoretical work. Further measurements with new or newly refurbished radio telescopes are definitely needed.

3. Is there a Consensus on Representative Values of D/H and ^{3}He/H in the Local Interstellar Medium?

The degree of agreement concerning representative values for D/H decreases rapidly with distance to the location of the interstellar gas. The Sun resides inside the Local Interstellar Cloud (LIC), which is a warm, partially ionized cloud apparently moving as a rigid parcel of gas through space and extending only a very short distance from the Sun toward α Cen and perhaps 10 pc in other directions. All lines of sight with projected velocity components in agreement with the LIC flow vector have D/H values derived from GHRS spectra consistent with the LIC mean value D/H = $(1.50\pm0.10) \times 10^{-5}$ (Linsky, 1998). Since this result is based on the analysis of 12 lines of sight using different analysis techniques and the mean value of D/H is consistent with the $\pm 1\sigma$ error bars of all of these measurements, this result for the LIC is generally accepted. However, additional observations especially in the less opaque higher Lyman lines are needed to test whether systematic errors are under control.

For interstellar gas in other nearby clouds and further away in the Galactic disk, the situation is much less clear. In the GHRS data set presented by Linsky (1998), D/H ratios for other nearby clouds are more widely scattered with a tendency for somewhat lower values than in the LIC. For 6 lines of sight through other clouds, the mean value is D/H = $(1.28 \pm 0.36) \times 10^{-5}$ (Linsky, 1998). This is not inconsistent with the previous result for the LIC, but it hints at variations on a distance scale of tens of parsecs. Vidal-Madjar *et al.* (1998) presented their analysis of GHRS spectra of the hot white dwarf G191-B2B, which shows three interstellar absorption components. One of these has a velocity and D/H ratio consistent with the LIC, but their analysis of another component (with a radial velocity of $+13$ km s^{-1}) shows D/H < 1×10^{-5} and this ratio could be of order 5.5×10^{-6}. While the complexity of the line of sight and the many free parameters in the analysis point to the need for independent confirmation of this result, their analysis argues for D/H values much lower than the LIC values on spatial scales smaller than the 48 pc distance to G191-B2B. A further indication that D/H values may be significantly lower away from the LIC is that the published D/H ratios for lines of sight toward four OB stars lie in the range $(0.5 \pm 0.3) \times 10^{-5}$ to $(0.8 \pm 0.2) \times 10^{-5}$ based on analyses of Copernicus spectra of the high Lyman lines (cf. Vidal-Madjar *et al.*, 1998).

If confirmed by the analysis of upcoming FUSE observations, the following model may be appropriate. The OB stars with low interstellar D/H ratios are located in regions of star formation in Orion, Scorpius, and Carina where the interstellar gas is likely to have been highly processed by one or more generations of stars. A large degree of astration would lead to low values of D/H but also relatively high values of the metal abundances. In this picture the LIC and other clouds near the Sun may consist of less astrated material than new star forming regions. This model should be tested by determining whether the interstellar medium along these lines of sight is metal rich compared to the LIC. On the other hand, the complex lines of sight toward the OB stars may produce systematic errors in the D/H ratios that have not yet been considered. FUSE spectra should help address this question.

We note that the published low values of D/H in the presumably more astrated young clouds of interstellar gas near the young OB stars are consistent with the derived higher value of D/H $= (2.1 \pm 0.5) \times 10^{-5}$ for the protosolar nebula (Geiss and Gloeckler, 1998), which likely refers to gas that was less astrated than the present day LIC. However, the roughly 0.6×10^{-5} decrease in D/H between the protosolar cloud (PSC) and the LIC may not be consistent with the roughly 1.3×10^{-5} increase in ^{3}He/H over the 4.6×10^9 yr interval between the PSC and the LIC (Geiss and Gloeckler, 1998), given the plausible assumption that essentially all of the D in the Sun has been converted to ^{3}He and the less certain assumption that the light element abundances depend only on galactocentric distance. However, the analysis of FUSE observations may well reveal that interstellar gas in the Galactic disk contains a rich diversity of D/H values even at the same galactocentric distance representing the complex history of star formation, mass loss, and interstellar gas mixing over the lifetime of the Galaxy.

The ^{3}He data for HII regions represents the present day ISM. Rood et al. (1998) proposed that from their analysis of all studied lines of sight, these data are consistent with a value ^{3}He/H $= 1.5^{+1.0}_{-0.5} \times 10^{-5}$, where the uncertainty has been increased a factor of 2 compared to simple random errors of measurement as a rough estimate of systematic errors. Two unexpected results from the study of HII regions are that: (1) the ^{3}He/H ratio does not appear to depend on galactocentric distance, and (2) the average ratio is not very different from the protosolar value. The ^{3}He data for the low-mass progenitor PNe refer to nucleosynthesis production in solar mass stars. The one definite detection of ^{3}He in NGC 3242 gives a ^{3}He/H ratio of $\sim 10^{-3}$; there are two other PNe with similar ratios, but with larger uncertainties. If these PNe are typical, then the ISM should be much richer in ^{3}He. More study is needed, but on the basis of these data there must be substantial revisions of either present models of the evolution of low mass stars or of contemporary Galactic chemical evolution theories.

4. Prospects for the Future

The study of the light element abundances is clearly a data-driven field. Advances in our understanding have come from sophisticated analyses of new data characterized by higher signal/noise, spectral resolution, and sensitivity. The Far Ultraviolet Spectroscopic Explorer, which is scheduled for launch in October 1998, will obtain spectra with resolution $\lambda/\Delta\lambda \approx 30,000$ in the 900–1180 Å spectral range containing the higher Lyman lines with sensitivity in excess of 10,000 times than of Copernicus. FUSE should be able to study hot stars throughout the Galactic disk and far into the halo, as well as nearby cool stars and white dwarfs beyond 100 pc. Also the IMAPS spectrometer used on sounding rockets and short duration space missions is obtaining very high resolution spectra of bright sources in the 900–1180 Å spectral region. Thus we anticipate that important new measurements of D/H in diverse regions of the Galaxy will become available within the next few years.

We also look forward to a dramatic increase in the accuracy and number of ^{3}He/H values in the Galaxy with the new radio telescopes, upgrades of existing telescopes, and improvements in theoretical modeling. However, as Rood *et al.* (1998) have clearly pointed out, this window of opportunity must be supported by the funding agencies.

References

Burles, S., and Tytler, D.: 1998, *Space Sci. Rev.*, this volume.
Epstein, R, Lattimer, J., and Schramm, D. N.: 1976, *Nature* **263**, 198.
Geiss, J., and Gloeckler, G.: 1998, *Space Sci. Rev.*, this volume.
Hogan, C. J.: 1998, *Space Sci. Rev.*, this volume.
Linsky, J. L.: 1998, *Space Sci. Rev.*, this volume.
Rood, R. T., Bania, T. M., Balser, D. S., and Wilson, T. L.: 1998, *Space Sci. Rev.*, this volume.
Vidal-Madjar, A., Ferlet, R., and Lemoine, M.: 1998, *Space Sci. Rev.*, this volume.
Walker, T., Steigman, G., Schramm, D. N., Olive, K., and Kang, H. S.: 1991, *Ap.J.* **376**, 51.
Welty, D., Morton, D. C., and Hobbs, L. M.: 1996, *Ap.J. Suppl.* **106**, 533.

EPILOGUE

CONCLUDING REMARKS

H. REEVES
Service d'Astrophysique, CEA, Saclay

This ISSI conference on the Light Elements pursued a double aim. First, to extrapolate backwards in time the abundances observed at given periods in order to recover their values at BBN. One can thereby determine, with increased accuracy, the nucleonic density of the universe, one of the key parameters of cosmological models. Second, to use these values as monitors constraining models of stellar and galactic phenomena. One useful property of these primordial abundances is their expected spatial uniformity, against which later spatial variations can be gauged and interpreted. Most of the references quoted here are found in the papers of this volume.

1. Deuterium

Since D is not produced importantly by stellar or galactic phenomena, we expect a gradual depletion of its primordial abundance as a function of time, as more and more of the galactic gas is incorporated in stars and later ejected as D-free gas. This depletion is plausibly tempered by the continuous infall of extragalactic gas on the galactic disk, with presumably pristine D. We have three sets of historical data, pertaining respectively to the present interstellar medium (ISM), the protosolar nebula and the early universe. Linsky (1998), as well as Vidal-Madjar *et al.* (1998) have presented and discussed the ISM D/H values in front of bright stars. Linsky estimates a value of D/H $= (1.5 \pm 0.1) \times 10^{-5}$ and shows (his fig. 4 and 5) that this value seems independant of distance within 100 pc from the Sun; this conclusion is contested by Vidal-Madjar *et al.* (1998).

The protosolar value was estimated by Geiss and Reeves (1971) by subtracting the ^{3}He/^{4}He solar wind value from the meteoritic value, on the assumption that the solar wind ^{3}He is the sum of the protosolar D burned in ^{3}He plus the protosolar ^{3}He. Over the years Geiss has progressively refined this estimate by introducing various minor corrections and by using the Jupiter ^{3}He/^{4}He value (Mahaffy *et al.*, 1998) instead of the meteoritic value. The present estimate is $(2.1 \pm 0.5) \times 10^{-5}$ (Geiss and Gloeckler, 1998). Jupiter values of D/H $= (2.6 \pm 0.7) \times 10^{-5}$ have been reported by Mahaffy *et al.* (1998). Other measurements (Gautier and Morel, 1989; Bjoraker *et al.*, 1986; Carlson *et al.*, 1983; Encrenaz *et al.*, 1996; Ben Jaffel *et al.*, 1996) are in the range 2–5 $\times 10^{-5}$. Because of this large range of uncertainties, together with the possibility of physicochemical processes altering the ratio in the protosolar nebula or in the outer layers of the planet, the Jupiter data does not appear to be of immediate usefulness in our search for the protosolar value.

Space Science Reviews **84**: 319–324, 1998.
© 1998 Kl.

Deuterium measurements by line absorption in high redshift clouds in front of quasars have led to a period of confusion. A high value of D/H $\sim 10^{-4}$ was obtained by Songaila et al. (1997, discussed by Hogan, 1998). A low value of $\sim 3 \times 10^{-5}$ (Burles and Tytler, 1998) was also reported. An analysis of the Burles and Tytler data by Levshakov et al. (1998) give a somewhat higher (and somewhat more uncertain) value of 4.5×10^{-5}. There appears now to be a consensus for such low values against the $\sim 10^{-4}$ high values. The possibility of spatial variations of deuterium in these absorbing clouds is discussed by Vidal-Madjar et al. (1998). The observational situation should improve with the coming FUSE satellite.

Several authors (Tosi, 1998; Prantzos, 1996; 1998; Matteucci and François, 1989; Ferrini et al., 1995; Chiappini et al., 1997; Vangioni-Flam et al., 1994) have presented consistent galactic evolution models incorporating most relevant observables: abundance ratios, stellar, gaseous and total mass fractions as a function of age and radial distance from the center of the Galaxy. The physical inputs of these models are: 1) the rate of star formation (SFR), 2) the Initial Mass Function (IMF) and 3) the rate of infall of extragalactic matter. The galactic evolution models generally include up-to-date (but still uncertain) nucleosynthesis yields on the various elements and isotopes. In this respect the observations of Beers et al. (1998) on the composition of very metal poor stars (tracing the earliest nucleosynthesis products of massive stars) are of great interest. The general conclusion is that models which account satisfactorily for the ensemble of observations generate only a modest amount of D depletion (at most by a factor of three). Attempts at artificially increasing that depletion get into severe trouble, mostly by the fact that they require in general large early SFR, resulting in too many heavy elements and too many low mass stars (thus worsening the G-dwarf problem in the solar neighborhood). Scully et al. (1997) have raised the possibility of circumventing the former difficulty by assuming very strong early outflows of galactic matter ejected by supernovae and the latter by truncating the low-mass part of the early stellar IMF. The recent work on the end of the "dark age" (Rees, 1998) and the general ionization of the intergalactic media around redshift $z = 5$, together with the evolution of the metallicity in high z galaxies (Prantzos, 1998, fig. 5) leave little room for the postulated high SFR period. The observed SFR as a function of redshift, somewhat higher than predicted by evolutionary models of spiral galaxies, can probably be accounted in terms of the early evolution of elliptical galaxies. Fields et al. (1998) have presented a modelisation of the early bursts of star formation in the cosmos.

In summary, if the low D/H value (~ 3–4×10^{-5}) is close to the primordial one, the situation is satisfactory from the point of view of galactic chemical evolution. If the high D/H value ($\sim 10^{-4}$) is correct, we run into severe troubles with the galactic models including the surplus of ^{3}He expected in this case. A diagram of the history of the abundances of deuterium is presented in the paper by Geiss and Gloeckler (1998): D/H seems to have decreased between the solar birth and now. However, in view of the dispersion of the age-metallicity relation, this decrease should be seen with some caution.

2. Helium-3

The recent measurement of ^{3}He/^{4}He $= (2.48^{+0.68}_{-0.62}) \times 10^{-4}$ in the ISM near the solar system by Gloeckler and Geiss (1998) is one of the highlights of this symposium. Rood et al. (1998) have obtained ^{3}He/^{4}He values in HII regions with different oxygen abundances, situated at various galactocentric distances. They are in approximate agreement with the protosolar and present value (Geiss and Gloeckler, 1998) and also with the Jupiter value: ^{3}He/^{4}He $= (1.66 \pm 0.05) \times 10^{-4}$ (Mahaffy et al., 1998). With a dispersion of a factor ~ 2, there appears to be no systematic trend of ^{3}He abundance with either galactocentric distance or metallicity. From the lack of variation for metallicities in the range $1/4 \, Z_\odot$ to $Z_\odot$ and from the fact that the observed values are in the range of BBN yields (i.e. for baryonic densities estimated from D, ^{3}He, ^{4}He and ^{7}Li), Rood et al. (1998) introduce the idea of a "helium-3 plateau", analogous to the Spite "lithium-7 plateau". In other words, quite unlike the Pop I ^{7}Li case, the stellar contribution to ^{3}He would not yet have emerged from the BBN contribution. Geiss and Gloeckler (1998) mention the small increase between the birth of the solar system and the present gas, but again it is difficult to assess the possible effects of the age metallicity dispersion.

These results are in conflict with standard theory of stellar nucleosynthesis. In stars, ^{3}He is produced during the Main Sequence phase by the D+p reaction. In the hot center of stars, ^{3}He is further transformed in ^{4}He by the ^{3}He+^{3}He or ^{3}He+^{4}He reactions. In the outer layers, however, the temperature is high enough to generate ^{3}He but not to destroy it. Stellar models show an accumulation of ^{3}He in the intermediate layers, reaching values of ^{3}He/^{4}He $= 10^{-2}$ (Vauclair, 1998, fig. 2). The fate of this isotope is governed by stellar internal processes from the Main Sequence to the Giant Branch and the Planetary Nebula (PN) phases through the effects of various dredge-up and mixing processes. The conventional stellar models predict an important yield of ^{3}He for stars less than 2 $M_\odot$. Observations of Planetary Nebulae (Rood et al., 1998) support this prediction with some detections of ^{3}He/H $\sim 10^{-3}$.

The problem lies with the stellar contribution of these PN during the life of the galaxy. Because small stars evolve slowly, this contribution should indeed appear quite late, but according to the models we should already have reached values of ^{3}He/^{4}He $\sim 10^{-3}$ in the interstellar medium, quite in disagreement with the Geiss and Gloeckler or the Rood et al. measurements. Charbonnel (1998) has discussed the relation between this problem and the problems raised by the observations of ^{12}C/^{13}C ratio and of N and O abundances in PN. The discrepancy could be solved by the operation of a new mixing phase in the Red Giant Branch, which would account for the destruction of ^{3}He and also for the unexpectedly low ^{12}C/^{13}C observed in some PN (Palla et al., 1998). Phenomenologically, the galactic ^{3}He problem would be resolved if a large fraction ($\sim 80\%$) of the PN did not have a positive net yield of ^{3}He (Tosi, 1998). The physical and/or astronomical justifications of this assumption remain to be found. It is interesting, in this respect, to mention the recent observations (Allen *et al.*, 1997) showing a marked underabundance (more than a factor two over the number expected from white dwarfs statistics) of the low mass PN, precisely those which are expected to generate ^{3}He.

3. Helium-4

Several authors have discussed possible improvements on the extrapolation of the ^{4}He measurements from metal-poor extragalactic HII regions back to BBN. The crucial $\Delta Y/\Delta Z$ parameter has been discussed by Høg *et al.* (1998) and by Skillman *et al.* (1998), while Steigman *et al.* (1998a) have considered the uncertainties due to temperature variations in the HII regions (which according to Grazina Stazinska -private conversation- are not expected to alter the estimated values in an important way). We are left with a crucial question: are the uncertainties on the value of $Y_P \sim 0.23$–0.24 (Thuan and Izotov, 1998) large enough to accomodate the baryon density obtained from the low D/H ratio?

4. Lithium-7

Observations of ^{7}Li/H in Pop II stars show a remarkable constancy for [Fe/H] between -3.5 and -1.3. The flatness of the curve as a function of metallicity and temperature (for T$<$5500 K) is discussed by Cayrel (1998) and Spite *et al.* (1998). This flatness (with a dispersion smaller than $\sim 50\%$), together with the apparently confirmed presence of ^{6}Li at [Fe/H]~ -2, suggest quite naturally a rather small amount of surface stellar depletion, at best comparable to the dispersion. In this case the mean value of ^{7}Li/H $= (1.3 \pm 0.6) \times 10^{-10}$ (Spite *et al.*, 1998) should give a fair representation of the BBN yield. The flatness of the curve, however, has far reaching implications for the physics of stellar surface layers. Vauclair (1998) has stressed the fact that the effect of atomic (microscopic) diffusion (Michaud, 1970) below the convective zone should, by itself and unless neutralized by other phenomena, show some mass dependance on the observed abundances, larger than the dispersion of the flat curve. Several groups (Deliyannis, personal communication; Vauclair 1998) have studied the possible influence of various physical phenomena such as: mixing via differential rotation, gravity waves and stellar winds. In all cases the flatness of the curve imposes important constraints on the efficiency of these processes.

   The flatness of the lithium abundance curve as a function of metallicity (Cayrel, 1998, figs. 1 and 3; Spite *et al.*, 1998, fig. 1) argues against heavy stellar astration and hence important D depletion in the corresponding metallicity range. More information on the physics of Pop II stellar surfaces can be obtained from combined observations of Li, Be and B (Duncan 1998; Primas, 1998), since these elements burn at different temperature. The discovery of two Pop II stars with B and Be lower than the mean value at that metallicity but with Li on the flat curve seems to be a sign of local variations of the corresponding cosmic ray fluxes. The observed value of the ^{6}Li/ H can not easily be related to the Be and B abundances since, contrary to Be and B, Li is partly generated by the $\alpha + \alpha$ spallation reactions and hence its abundance depends on the unknown $(\alpha,p / C,N,O)$ ratio in the early cosmic rays.

5. Cosmology

In view of the uncertainties on the D/H and ^{4}He abundances, the ratio of nucleon to photon number η is still rather uncertain. Adopting the lower values of D/H we get η_{10} (= $10^{10}\eta$) = 5.5 ± 2. This corresponds to a nucleonic mass density (measured in units of the critical density) $\Omega_b = 0.04 \pm 0.015$ for $H_0 = 70$ km/sec/Mpc. The best estimates of the total mass fraction Ω_T are around 0.3 (Tammann, 1998). These numbers suggest strongly the existence of a non-nucleonic "exotic" mass component (Schramm, 1998; Walker, 1998). The search for the nature of this component is still pursued actively. Steigman *et al.* (1998b) have used a combination of cosmological parameters (the age of the universe, the Hubble constant, the nucleonic mass fraction of galaxy clusters and a "shape" parameter characteristic of the mass distribution of large structures) in order to evaluate the parameter η independantly of the element abundances. In view of the uncertainties on some of these parameters (in particular the last one), the results are still approximate but encouragingly close to the BBN calculations. This attempt underlines the fact that all these observables are related and should be instrumental in the elaboration of a satisfactory cosmological theory. The values of Ω_b, Ω_T and the cosmological constant Λ play a crucial role in the models of the early universe. They govern the small angular temperature variations of the cosmic microwave background (CMB) and also the shape distribution of large structures. The height of the first peak of the CMB (at $\sim 1°$) is related to Ω_b (in relation with the amplitude of the sound waves in the first collapsing clouds). Its precise angular position is related to Ω_T (in relation with the corresponding curvature of space). It is of interest here to mention that very idea of a CMB by Gamow and his collaborators (Alpher *et al.*, 1948) emerged from the hypothesized formation of helium in the Big Bang. The two steps of the argument were the following: 1) in order to have nuclear reactions, the temperature should be in the MeV range; 2) to transform some but not all protons in helium, the mean free path for neutron capture (and hence the cosmic density) should be in a given density range. Combining these numbers gave the first estimate of the ratio of nucleons to photons and predicted the existence and temperature of the CMB today. The CMB was later discovered, BBN was formulated and shown to give a satisfactory account of the abundances of the elements and hence an estimate of the nucleonic density. Today we use its spatial variations to obtain the nucleonic density independantly from the helium and other light element abundances.

References

Allen, C., Carrigi, L., and Peimbert, M.: 1997, preprint.

Alpher, R., Bethe, H. and Gamow, G.: 1948, *Phys. Rev.* **73**, 803.

Beers, T. C., *et al.*: 1998, *Space Sci. Rev.*, this volume.

Ben Jaffel, L., *et al.*: 1996, preprint.

Bjoraker, G., *et al.*: 1986, *Icarus* **66**, 579.

Burles, S. and Tytler, D.: 1998, *Space Sci. Rev.*, this volume.

Carlson, B., *et al.*: 1993, *J. Geophys. Res.* **98**, 5251.

Cayrel, R.: 1998, *Space Sci. Rev.*, this volume.

Charbonnel, C.: 1998, *Space Sci. Rev.*, this volume.

Chiappini, C., Matteucci, F., and Gratton, R.: 1997, *ApJ* **477**, 765.

Duncan, D.: 1998, *Space Sci. Rev.*, this volume.

Encrenaz, Th., *et al.*: 1996, *A&A* **315**, L397.

Ferrini, F., *et al.*: 1994, *ApJ* **427**, 745.

Fields, B. D., Mathews, G. J., and Schramm, D. N.: 1998, *Space Sci. Rev.*, this volume.

Gautier, D. and Morel, P.: 1997, *A&A* **323**, L9.

Geiss, J. and Reeves, H.: 1972, *A&A* **18**, 126.

Geiss, J. and Gloeckler, G.: 1998, *Space Sci. Rev.*, this volume.

Gloeckler, G. and Geiss, J.: 1998, *Space Sci. Rev.*, this volume.

Høg, E., Pagel, B. E. J., Portinari, L., Thejll, P. A., MacDonald, J., and Girardi, L.: 1998, *Space Sci. Rev.*, this volume.

Hogan, C.: 1998, *Space Sci. Rev.*, this volume.

Levshakov, S. A., Kegel, W. H., and Takahara, F.: 1998, *Space Sci. Rev.*, this volume.

Linsky, J.: 1998, *Space Sci. Rev.*, this volume.

Mahaffy, P. R., Donahue, T. M., Atreya, S. K., Owen, T. C., and Niemann, H. B.: 1998, *Space Sci. Rev.*, this volume.

Matteucci, F. and François, P.: 1989, *MNRAS* **239**, 885.

Michaud, G.: 1970, *ApJ* **160**, 41.

Palla, F., Galli, D., Bachiller, R., and Pérez-Gutiérrez, M.: 1998, *Space Sci. Rev.*, this volume.

Prantzos, N.: 1996, *A&A* **310**, 106.

Prantzos, N.: 1998, *Space Sci. Rev.*, this volume.

Primas, F.: 1998, *Space Sci. Rev.*, this volume.

Rees, M. J.: 1998, *Space Sci. Rev.*, this volume.

Rood, R. T., Bania, T. M., Balser, D. S., and Wilson, T. L..: 1998, *Space Sci. Rev.*, this volume.

Schramm, D. N.: 1998, *Space Sci. Rev.*, this volume.

Scully, S., *et al.*: 1997, *ApJ* **476**, 521.

Skillman, E. D., Terlevich, E., and Terlevich, R.: 1998, *Space Sci. Rev.*, this volume.

Songaila, A., *et al.*: 1997, *Nature* **385**, 137.

Spite, F., Spite, M., and Hill, V.: 1998, *Space Sci. Rev.*, this volume.

Steigman, G., Viegas, S. M., and Gruenwald, R.: 1998a, *Space Sci. Rev.*, this volume.

Steigman, G., Hata, N., and Felten, J. E.: 1998b, *Space Sci. Rev.*, this volume.

Tammann, G. A.: 1998, *Space Sci. Rev.*, this volume.

Thuan, T. X. and Izotov, Yu. I.: 1998, *Space Sci. Rev.*, this volume.

Tosi, M.: 1998, *Space Sci. Rev.*, this volume.

Vangioni-Flam, E., Olive, K., and Prantzos, N.: 1994, *ApJ* **148**, 3.

Vauclair, S.: 1998, *Space Sci. Rev.*, this volume.

Vidal-Madjar, A., Ferlet, R., and Lemoine, M.: 1998, *Space Sci. Rev.*, this volume.

Walker, T. P.: 1998, *Space Sci. Rev.*, this volume.

Author Index

LIST OF PARTICIPANTS

Jean Audouze, *Institut d'Astrophysique de Paris*, audouze@iap.fr
Timothy Beers, *Michigan State University*, beers@msupa.pa.msu.edu
Peter Bochsler, *University of Bern*, bochsler@phim.unibe.ch
Roger Cayrel, *Observatoire de Paris*, cayrel@obspm.fr
Corinne Charbonnel, *Lab. d'Astrophysique de Toulouse*, corinne@obs-mip.fr
Constantine Deliyannis, *Yale University*, con@astro.yale.edu
Tom Donahue, *University of Michigan*, tmdonahue@umich.edu
Douglas Duncan, *University of Chicago*, duncan@dei.uchicago.edu
Brian Fields, *University of Notre Dame*, bfields@cygnus.phys.nd.edu
Daniel Gautier, *Observatoire de Paris-Meudon*, daniel.gautier@obspm.fr
Johannes Geiss, *ISSI*, geiss@issi.unibe.ch
George Gloeckler, *University of Maryland*, gloeckler@umdsp.umd.edu
Craig Hogan, *Univ. of Washington*, hogan@centaurus.astro.washington.edu
Sergei Levshakov, *Ioffee Inst.*, lev@astro.ioffe.rssi.ru
Jeffrey Linsky, *University of Colorado*, jlinsky@jila.colorado.edu
Bernard Pagel, *Nordita*, pagel@nbivms.nbi.dk
Francesco Palla, *Oss. Astrofisico di Arcetri*, fpalla@arcetri.astro.it
Nikos Prantzos, *Institut d'Astrophysique de Paris*, prantzos@iap.fr
Francesca Primas, *University of Chicago*, primas@oddjob.uchicago.edu
Rafael Rebolo, *Inst. de Astrofisica de Canarias*, rrl@iac.es
Martin Rees, *University of Cambridge*, mjr@ast.cam.ac.uk
Hubert Reeves, *CEN de Saclay*, 100265.2126@compuserve.com
Robert Rood, *University of Virginia*, rtr@virginia.edu
Wallace Sargent, *Palomar Observatory*, wws@astro.caltech.edu
David Schramm, *University of Chicago*, dns@oddjob.uchicago.edu
Evan Skillman, *University of Minnesota*, skillman@astro.spa.umn.edu
François Spite, *Observatoire de Paris*, francois.spite@obspm.fr
Gary Steigman, *Ohio State University*, steigman@mps.ohio-state.edu
Gustav A. Tammann, *University of Basel*, tammann@ubaclu.unibas.ch
Elena Terlevich, *University of Cambridge*, et@ast.cam.ac.uk
Roberto Terlevich, *Royal Greenwich Observatory*, rjt@ast.cam.ac.uk
Friedrich Thielemann, *University of Basel*, fkt@quasar.physik.unibas.ch
Trinh Xuan Thuan, *University of Virginia*, txt@virginia.edu
Monica Tosi, *Osservatorio Astronomico di Bologna*, tosi@astbo3.bo.astro.it
Jim Truran, *University of Chicago*, truran@nova.uchicago.edu
David Tytler, *University of California*, dtytler@ucsd.edu
Sylvie Vauclair, *Observatoire Midi-Pyrenées*, svcr@obs-mip.fr
Alfred Vidal-Madjar, *Institut d'Astrophysique de Paris*, vidalmadjar@iap.fr
Rudolf von Steiger, *ISSI*, vsteiger@issi.unibe.ch
Terry Walker, *Ohio State University*, twalker@pacific.mps.ohio-state.edu
Tom Wilson, *MPI für Radioastronomie*, p073twi@mpifr-bonn.mpg.de

Space Science Series of ISSI

Kluwer Academic Publishers – Dordrecht / Boston / London

The manufacturer's authorised representative in the EU is Springer
Nature Customer Service Centre GmbH, Europaplatz 3, 69115 Heidelberg,
Germany. If you have any concerns regarding our products, please
contact ProductSafety@springernature.com

Printed and bound by CPI Group (UK) Ltd, Croydon, CR0 4YY
29/04/2026
02099472-0005